Species
The Evolution of the Idea
Second Edition

T0179088

Species and Systematics

The *Species and Systematics* series will investigate the theory and practice of systematics, phylogenetics, and taxonomy and explore their importance to biology in a series of comprehensive volumes aimed at students and researchers in biology and in the history and philosophy of biology. The book series will examine the role of biological diversity studies at all levels of organization and focus on the philosophical and theoretical underpinnings of research in biodiversity dynamics. The philosophical consequences of classification, integrative taxonomy, and future implications of rapidly expanding data and technologies will be among the themes explored by this series. Approaches to topics in *Species and Systematics* may include detailed studies of systematic methods, empirical studies of exemplar taxonomic groups, and historical treatises on central concepts in systematics.

Editor in Chief: Kipling Will (University of California, Berkeley)

Editorial Board
Sandra Carlson (University of California, Davis, USA)
Marcelo R. de Carvalho (University of Sao Paulo, Brazil)
Darren Curnoe (University of New South Wales, Australia)
Malte C. Ebach (University of New South Wales, Australia)
Christina Flann (Netherlands Centre for Biodiversity Naturalis, The Netherlands)
Anthony C. Gill (Smithsonian Institution, USA)
Mark S. Harvey (Western Australian Museum, Australia)
David R. Maddison (Oregon State University, Corvallis, USA)
Olivier Rieppel (The Field Museum, Chicago, USA)
Felix Sperling (Strickland Museum of Entomology, Edmonton, Alberta, Canada)
David M. Williams (The Natural History Museum, London, UK)
René Zaragüeta i Bagils (University of Paris 6, France)

Science Publisher:
Charles R. Crumly, CRC Press/Taylor & Francis

For more information visit:
www.crcpress.com/Species-and-Systematics/book-series/CRCSPEANDSYS

Species

The Evolution of the Idea
Second Edition

John S. Wilkins

CRC Press
Taylor & Francis Group
Boca Raton London New York

CRC Press is an imprint of the
Taylor & Francis Group, an **informa** business

CRC Press
Taylor & Francis Group
6000 Broken Sound Parkway NW, Suite 300
Boca Raton, FL 33487-2742

First issued in paperback 2020

ISBN 13: 978-0-367-65736-9 (pbk)
ISBN 13: 978-1-138-05574-2 (hbk)

Library of Congress Cataloging-in-Publication Data

Names: Wilkins, John S., 1955- author.
Title: Species : the evolution of the idea / John S. Wilkins.
Description: Second edition. | Boca Raton : Taylor & Francis, 2018. | Series: Species and systematics | Revised edition of: Species : a history of the idea. Berkeley : University of California Press, c2009. | Includes bibliographical references.
Identifiers: LCCN 2017041905 | ISBN 9781138055742 (hardback : alk. paper)
Subjects: LCSH: Species--History. | Species--Philosophy.
Classification: LCC QH83 .W527 2018 | DDC 577--dc23
LC record available at https://lccn.loc.gov/2017041905

**Visit the Taylor & Francis Web site at
http://www.taylorandfrancis.com**

**and the CRC Press Web site at
http://www.crcpress.com**

*In memory of David Lee Hull, 1935–2010,
from whom I learned much, and by whom I was
always encouraged, despite our disagreements.*

Contents

SECTION I   The Historical Development of "Species"

SECTION II   Modern Debates

SECTION III   Philosophical Discussions of the Species Concept

List of Figures

List of Tables

Series Preface

The *Species and Systematics* book series is a broad-ranging venue where authors can provide the scientific community with comprehensive treatments of the history and philosophy of fundamental concepts in systematic biology, phylogenetics, and the science of taxonomy. The series also intends to connect historical perspectives to new ideas and emerging technologies that have implications for the field. The series is committed to stimulating discussion among students and researchers in biology on controversial and clarifying ideas related to the future course we are charting in biodiversity research.

There are many approaches to the study of biological diversity and to embrace this, future volumes in *Species and Systematics* may include detailed development and comparisons of existing and novel methods in systematics and biogeography, empirical studies that provide new insight into old questions and raise new questions for biologists and philosophers of science, and historical treatises on central and reoccurring concepts that benefit from both a retrospective and a new perspective. Some volumes will address a single important concept in great depth, giving authors the freedom to present ideas with their own slant, while others will be edited collections of shorter papers intended to place alternative views in sharp contrast.

Kipling Will
Berkeley, CA

Preface to the First Edition

The history of research into the philosophy of language is full of *men* (who are rational and mortal animals), *bachelors* (who are unmarried adult males), and *tigers* (though it is not clear whether we should define them as feline animals or big cats with a yellow coat and black stripes).

Umberto Eco[1]

"What sort of insects do you rejoice in, where you come from?" the Gnat inquired.

"I don't *rejoice* in insects at all," Alice explained, "because I'm rather afraid of them—at least the large kinds. But I can tell you the names of some of them."

"Of course they answer to their names?" the Gnat remarked carelessly.

"I never knew them do it."

"What's the use of their having names," the Gnat said, "if they won't answer to them?"

"No use to *them*," said Alice; "but it's useful to the people that name them, I suppose. If not, why do things have names at all?"

"I can't say," the Gnat replied. "Further on, in the wood down there, they've got no names ..."

Lewis Carroll[2]

Why should we look at one concept in science, out of the context of the larger theories, practices and societies in which it occurs? Why trace "species?" This sort of question is raised by both philosophers and historians when histories of scientific ideas are written.

Philosophers tend to dislike history for several reasons. One is that they often address issues and ideas as if the opponent is sitting across the symposium table from them, no matter whether that opponent lived last week, last century or last millennium. Philosophers of science often treat history as a source of anecdotes to illustrate some more general point, such as the way the Copernican Revolution changed philosophical understanding, or how genes overcame vitalism. Famously or infamously, Imre Lakatos "rationally reconstructed" the history of scientific ideas in a footnote, because history is messy, and failed to clearly illustrate the philosophical point.

Historians tend to dislike intellectual histories, because such histories treat ideas as free-floating objects ("free-floating rationales" as Dennett calls them) independent of the individual psychologies and life histories, and of the social conditions in which they were raised and elaborated. Also, histories of ideas are too easy to do. All you need do is find some apparent resemblance between ideas at time a and time b,

---

[1] Eco 1999, 9.
[2] Carroll 1962, 225.

and you have a narrative. Historians, rightly, want to see actual historical influences, and the effects of social and cultural contexts, the differing *epistemes* at work.

Both professions can go too far. I think history comes in a number of scales, which following a practice in ecology, I will call *alpha history, beta history*, and *gamma history*. Alpha history is done by investigating archives, and looking at locales and artifacts. It is hard and local work, and will give the *data* of the larger scale histories. Beta history is done by covering a restricted period, or biography, or event. It relies on the alpha material, and synthesizes it into a narrative explanation of the subject. Gamma history, though, is out of fashion. Rather than being a "life and times" or "history of the period," it attempts to take alpha and beta historical work and synthesize a grand-scale narrative. And because a *really* grand scale narrative is almost impossible to do by one person, it pays to limit the subject to something manageable. This book is at the edge (some might uncharitably say, over the edge) of that limit. But if gamma history is not worth doing, why then is alpha and beta history?

Philosophy of science has become increasingly grounded in history. It is becoming the norm for philosophers of science to appeal closely to the historical development, failures as well as successes, of a given discipline or problem. Majorie Grene and Ian Hacking are perhaps the exemplars of this approach, although David Hull has also made a plea for actual examples in philosophy of biology.[3] And historians of science such as Polly Winsor and Jan Sapp have offered excellent case studies and narratives of all three kinds for philosophers to use. There is a shift towards this now, and that might justify a conceptual history at this time. However, there's another reason for writing this now, and that is that if philosophers don't do this, and historians don't, the scientists will, and have. A major target of this book is the scientist-developed *essentialism story* of the past 50 years. Polly Winsor and Ron Amundson, among others, have written critiques of the view that before Darwin, every biologist was held in thrall to Aristotle's essentialist biology, but there is no overall summary of this. Also, the essentialism story is used to justify or critique various species conceptions by the biologists themselves. History has a role in scientific debate.

Generally, scientists have a "rolling wall of fog" that trails behind them at various distances for different disciplines, above which only the peaks of mountains of the Greats can be seen. In medical biology, for instance, this wall is about five years behind the present. Little is cited before that, and those works that are, are cited by nearly everyone. So, there is a tendency for what Kuhn called "textbook history" to become the common property of all members of the discipline. However, taxonomy is an unusual discipline, in that the classical works are more widely cited and appealed to than in most other sciences. The ideas of an eighteenth-century Swede or French author can carry weight in a way that the genetics or physics of that time do not. Partly this is because a large element of taxonomy is conceptual: logical and metaphysical ideas, which change slowly, carry probative force. So asking "what is a species?" is to ask a philosophical and historical as well as a scientific question, and how the notion of species arrived at the present debate in part defines that debate.

Doing this kind of history is rather like trying to work out the past from a series of old photographs in a box in the attic you got from your grandparents. Faces appear

---

[3] Hull 1989.

in various guises, resemblances recur, and it is almost impossible to identify exactly who is whose child, friend or mere passers-by. Nevertheless, having that box of snapshots, one is the richer for it in understanding both the past and the present.

So, I seek absolution from each of these three professions—philosophy, history, and biology. I believe I show there is a basic error involved in the essentialism story that can be resolved by a conceptual history. Scientific history is at least partially conceptual, so I don't think it is illicit to write a conceptual history. But the conceptual history of an *idea*? That might be too much. Well, this is not exactly the history of an idea. It is a combined history of *various* ideas and words that have a subtle ambiguity in philosophy *and* biology. And it is my claim that this ambiguity has confused the present debate over species in both fields. I will summarize the argument here, so that it is clear what the issues this book addresses are.

The essentialism story is a view that has taken biologists and philosophers by storm. Primarily advocated, and largely developed—I hesitate to say invented—by Ernst Mayr, it is the view that there are basically two views of biological taxa in general, and species in particular. One is the view deriving from Plato and Aristotle, on which all members of a type were defined by their possession of a set of necessary and sufficient properties or traits, which were fixed, and between which there was no transformation. This is variously called *essentialism, typological* or *morphological thinking*, and *fixism*. The other is a view developed in full by Charles Darwin, in which taxa are populations of organisms with variable traits, which are polytypic (have many different types), and which can transform over time from one to another taxon, as the species that comprise them, or the populations that comprise a species, evolve. There are no necessary and sufficient traits. This is called *population thinking*.

I will argue that the essentialism story is false. It is based, as Polly Winsor has shown, on a misreading of the *logical* tradition in which *species* is a class that is differentiated out of a larger class, or *genus*, by a set of necessary properties all members bear, as being a claim about *biological* species, which are not so defined, but are instead diagnosed by morphological characters. There is a clear distinction between the *formal* definitions of logical species and the *material* characters and powers of the biological organisms of a biological species pretty well from the beginning of modern natural history, but arguably even from Aristotle onwards. This is the first claim.

A second related claim is that living species were always understood to include or require a *generative power* rather than morphological similarity or identity, which was always held to be a way of identifying them at best. I call this the *generative conception of species*, and it was held as much in the classical writings of Aristotle and Theophrastus, as by the moderns, and the present views of species are equally within that ancient tradition. There are plenty of cases of medieval, early modern, and recent pre-Darwinian authors employing some variant of this conception, allowing for deviation from types. Moreover, the use of diagnostic characters neither requires essences of a causal or material kind, nor implies that all diagnostic characters are borne by every member of the species. In the diagnostic sense, Darwin was as much a formal essentialist as Linnaeus, and modern taxonomists are still today. In short, generation, not definition, is what counts for living species.

My third claim is that the notions of fixity of species and essence and type are decoupled from each other. Types are not the same as essences, and they had a number of roles to play in classification. Fixity of species was invented by John Ray in the seventeenth century, and repeated by Linnaeus, and it was solely based on piety, not metaphysics. Where material essences were employed, for instance by Nehemiah Grew, they were considered to be causal essences for biological structures rather than of taxa, and this was also the view of the ideal morphologists of the early nineteenth century after Goethe. If essences play any role in taxonomy at all, it must be well after Darwin, and it is possible it never was really held before the essentialist story developed, or was invented, around 1958.

So much for the history. Philosophically, essentialism underwent a revival in the 1960s in the philosophy of language, in part because of a reaction to Popper's philosophical attack on *methodological* essentialism outside biology. This was, I believe, a reason why essentialism became a problem in philosophy of biology. Platonism, however, is a recurring theme in both biology and philosophy in the scientific period, and it is possible that philosophical interpretations of biology in the period were motivated in part by the increasing atomism and materialism of writers like Locke and Boyle, who were influential on many biologists, such as Buffon and Lamarck.

There is a distinction to be had between the naming of species and their underlying biological descriptions. Part of the modern debate rests on a confusion between nomenclature and the reference of names on the one hand, and the accounts given of the formation and maintenance of species. Species "nominalists," who I refer to as *species deniers* in the biological traditions, deny that the names refer to anything biologically real (which is not to argue against scientific realism—the reality referred to here is simply the status of species as objects in biology; whether they are real in a metaphysical sense is outside our scope here). Species *taxa* realists hold that the taxa are real objects in biology. Species *category* realists also believe that the rank of species is a real phenomenon, in that species have a unique and peculiar organization that other taxa, below and above that rank, do not have.

In summary, then, we have three claims that this book is intended to demonstrate: the logical and the natural species are distinct ideas that largely share only a term; there was a single species "concept" from antiquity to the arrival of genetics, the *generative conception*; and types are neither the same as essences nor something that changes much with Darwin.

The book also lists the modern "species concepts" (or rather, definitions of the word and concept *species*) in play. It will become clear that most of these are partial definitions, some of which are simply not operationally applicable. It is my belief that each biologist doing systematics will construct for themselves a conceptual sandwich out of these elements, one that suits the tastes required for studying the group of organisms they do. So the latter part of the book is a kind of conceptual delicatessen. Biologists, like anyone else engaged in an intellectual enterprise, tend towards monism of theoretical ideas: so *species* conceptions that work for them must work in all other cases. This is especially obvious in the "animal bias" of workers like G. G. Simpson and Theodosius Dobzhansky, who simply denied that things that were not species the way sexual animals were species, were in fact species at all. I hope to show here that the term has a wider application than in a few leading cases. The philosophical arguments, however, must wait for another book.

Preface to the Second Edition

He who ventures to write on the origin of species, ought to define what a species is, so ought he to do who describes species, no matter whether he considers his task finished when the description has been made, or whether he intends to make use of the described species to build up a more or less elaborate system. In other words: the systematist, as well as the evolutionist, ought to state clearly what he means by a species.

As a matter of fact, neither of them usually does.

Johannes Paulus Lotsy[4]

… this book was not actually a "safe" book for a historian to write. It flaunts the conventions of disciplinary specialization; it has a large chronological range and deals with a considerable variety of subject matters; and while it might not be strictly interdisciplinary, it takes in the history of science, religion, philosophy, theology, and more. This makes it vulnerable within a highly specialized profession …

Peter Harrison[5]

My motivation for revising and expanding the first edition of this book is due to several factors. One is that continued exploration of the concept of species has led me to modify some of my conclusions. The main historical novelty of this edition is that I have become convinced that the reason for the introduction of *species* in the first place was the attempts by theologians in the sixteenth century to determine how many kinds of animals were included on the Ark, which necessitated that there be a fundamental kind from which all other apparent species could be formed after the Deluge. There are several errors of fact that also needed to be corrected, particularly my misidentification of Claude Thomas Alexis Jordan as Karl Jordan.[6]

The second reason is that the philosophical issues were not adequately explained, as one reviewer noted, for non-philosophical readers, and this led me to consider that the ancient and modern philosophy of biology accounts of species and kinds in general needed to be covered. Given that this topic covers historical, philosophical, and scientific issues, it is important that readers from each of these specialties are given sufficient cues to be able to follow these issues.

Finally, modern debates on the nature of species, both the biological entities and the categorical class, are in part philosophical and in part scientific, ironically, often made out by scientists when they discuss *species*. As there have been several recent books since the first edition was published, readers are due an account of these debates, and so I have summarized them in a final chapter (coauthored with Brent Mishler), in which I take the liberty of giving my own views and conclusions.

---

[4] Lotsy 1916, 13.
[5] Harrison 2016, 743f
[6] Pointed out, a bit stridently, by Mallet 2010.

Arthur O. Lovejoy once referred to the inquiry of *philosophical semantics*.[7] It is this inquiry I am engaged upon here. Not merely the history of ideas, but the philosophies underlying them, are what philosophers of science, historians of science, and scientists themselves must rely upon to understand the evolution of science itself. Otherwise we live in an eternal present, and repeat the debates of the past.

I hope that these revisions will be welcomed.

BIBLIOGRAPHY

Carroll, Lewis. 1962. *Alice's Adventures in Wonderland and Through the Looking Glass.* Harmondsworth: Penguin. Original edition, 1865/1871.

Eco, Umberto. 1999. *Kant and the Platypus: Essays on Language and Cognition.* London: Vintage/Random House.

Harrison, Peter. 2016. The modern invention of "science-and-religion": What follows? *Zygon: Journal of Religion and Science* 51 (3):742–757.

Hull, David L. 1989. *The Metaphysics of Evolution.* Albany: State University of New York Press.

Lotsy, Johannes Paulus. 1916. *Evolution by Means of Hybridization.* The Hague: Martinus Nijhoff.

Lovejoy, Arthur O. 1936. *The Great Chain of Being: A Study of the History of an Idea.* Cambridge, MA: Harvard University Press. Reprint, 1964.

Mallet, James. 2010. Species: A history of the idea. John S. Wilkins. *Integrative and Comparative Biology* 50 (2):251–252.

---

[7] Lovejoy 1936, 14: "… a study of the sacred words or phrases of a period or movement, with a view to a clearing up of their various shades of meaning, and an examination of the way in which confused associations of ideas arising from these ambiguities have influenced the development of doctrines, or accelerated the insensible transformation of one fashion of thought into another, perhaps its very opposite. It is largely because of their ambiguities that mere words are capable of this independent action as force in history."

Acknowledgments

FIRST EDITION

This book has taken me the better part of a decade to write, and I have done so with the aid of many people. I am consistently surprised and pleased to find that even those individuals with whom I have great disagreements over species concepts turn out to be extremely nice, helpful and above all scholarly individuals. I never met Mayr, but those I know who have say the same thing about him. So, I am deeply indebted to the following for criticism, editing, material, suggestions and the occasional expression of outright incredulity:

Gareth J. Nelson and Neil Thomason, who were all that a doctoral candidate could desire of thesis advisors, and whose critical comments have helped make my thesis, from which this evolved, a much better work;

Quentin Wheeler, Malte Ebach, and Chuck Crumley for editorial advice and help.

In alphabetical order, Noelie Alito, Mike Dunford, Malte Ebach, Greg Edgcombe, Dan Faith, Michael Ghiselin, Paul Griffiths, Colin Groves, John Harshman, Jody Hey, Jon Hodge, David Hull, Jon Kaplan, Mike S.-Y. Lee, Murray Littlejohn, Brent Mishler, Larry Moran, Staffan Müller-Wille, Ian Musgrave, Gary Nelson, Mike Norén, Gordon McOuat, Massimo Pigliucci, Tom Scharle, Kim Sterelny, Neil Thomason, Charissa Varma, John Veron, Quentin Wheeler, David Williams, and Polly Winsor, who all provided information, criticism, advice, and assistance, some considerable;

Members of the audiences at the 2000 Australasian Association for the History, Philosophy and Social Studies of Science Association conference at the University of Sydney and the 2001 International Society for the History, Philosophy and Social Studies of Biology conference in Hamden, Connecticut, and the Systematics Forum at the Melbourne Museum, run by Robin Wilson; Michael Devitt also forced me to reassess the nature of essentialism at a later workshop, although without much success.

Polly Winsor, David Hull, Joel Cracraft, Norman Platnick and Ward Wheeler graciously granted time for an interview, and Joel has been of particular help with the phylogenetic conception. I must especially thank the late Herb Wagner, John Veron, Mike Dunford, Tom Scharle, and Scott Chase for technical information provided. Other acknowledgements are made in the notes. I sincerely apologize to anyone I have left unjustly unthanked.

I owe a debt to the Walter and Eliza Hall Institute of Medical Research in Melbourne, which was my employer during my PhD candidacy, and to the Biohumanities Project (Federation Fellowship FF0457917), under the direction of Paul Griffiths, at the University of Queensland, where the bulk of the work on the late Middle Ages and Renaissance, and some of the work on the eighteenth century was done, with the assistance of the library there. I must make special mention of the *Gallica* project at the Bibliothèque nationale de France (gallica.bnf.fr), the *Internet Archive* (www.archive.org), *Botanicus* (www.botanicus.org), and the *Complete Works of Charles Darwin* project (darwin-online.org.uk), overseen by John van Whye. Without these resources, many texts would have remained inaccessible to me in Australia.

Therefore, in the light of this gracious assistance from so many people, it follows that all misunderstandings, errors and incoherencies that remain, and there will be many, are my own fault.

SECOND EDITION

I must thank the University of California Press for permission to revert the rights so that I could publish in the series *Species and Systematics* again, which has been moved to CRC Press, and to Kim Robinson at UCP for helpful advice. As always, I am in the debt of my first and present editor Chuck Crumly.

Thanks to Gal Kober, Marc Ereshefsky, Quentin Wheeler, and Brent Mishler for pushing me on the chapter on phenomenal species. I must also acknowledge all the specialists with whom I have had long—often frustrating for us both—arguments on this topic leading to the first edition and the PhD upon which it was based. I must especially thank David Williams of the National History Museum, Malte Ebach, Gareth Nelson, and Brent Mishler for general encouragement, none of whom can be fairly thought to agree with me on all matters.

The following people have also provided information, assistance, or encouragement for this edition: Glenn Branch, Joachim Dagg, Donald Forsdyke, Laurent Penet, Staffan Müller-Wille, David Williams, Polly Winsor, and Frank Zachos. Also, I forgot to acknowledge David-Braddon Mitchell and John P. McCaskey in the first edition, for which I am duly ashamed. Again, I apologize to those left out this time.

I also must thank Richard Carter and Alexa Quinn for reading and commenting on the manuscript of this edition.

A blessing is bestowed upon the Biodiversity Heritage Library (www.biodiversity library.org) for making many of the crucial texts available in facsimile form.

Many thanks to Alexis Pollard also, who gave me space to work and encouragement to do so.

No thanks are offered to any funding agency. I did this on my own dime.

Author

John S. Wilkins earned his PhD at the University of Melbourne, Australia. He has researched and taught at the University of Queensland, the University of Sydney, the University of New South Wales, and the University of Melbourne. He is the author of *Species: A History of the Idea* (University of California Press, Berkeley, 2009, the first edition of this book), *Defining Species* (Peter Lang, New York, 2009), *The Nature of Classification* (Palgrave Macmillan, London, 2013, with Malte C. Ebach), and the editor of *Intelligent Design and Religion as a Natural Phenomenon* (Ashgate, Burlington, Vermont, 2010). He is the author or coauthor of 28 papers and 12 book chapters. Dr. Wilkins is currently an Honorary Fellow at the University of Melbourne School of Historical and Philosophical Studies, where he lectures.

Prologue

THE RECEIVED VIEW

Progress, far from consisting in change, depends on retentiveness. Those who
cannot remember the past are condemned to repeat it.

George Santayana[8]

First, then, we have the problem involved in the origin of species. As a preliminary to that, logic demands that we should define the term. It may be that
logic is wrong, and that it would be better to leave it undefined, accepting the
fact that all biologists have a pragmatic idea at the back of their heads. It may
even be that the word is undefinable. However, an attempt at a definition will
be of service in throwing light on the difficulties of the biological as well of the
logical problems involved.

Julian Huxley[9]

The problem of species is a long-standing one in natural history. Since the development of modern taxonomy and classification, naturalists, botanists, zoologists, and
all the other various terms for aspects of biology as we have known it or now know it
have tried to define clearly what it is they are taking about when they talk of "species."
There is a Received View of the species concept that is largely the result of work published by biologists themselves.[10] Polly Winsor[11] traces the origin of the Received View
history to Arthur Cain's[12] reliance on a 1916 "misinterpretation" of Aristotle[13] and his
subsequent writings on the history of taxonomy.[14] These influenced both Simpson
and Hull, and also Mayr's later development of the story.[15] Similar conclusions have
also been reached by Ron Amundson concerning typology[16]—transcendentalism and

---

[8]   Santayana 1917, 284.
[9]   Huxley 1942, 154.
[10] There is a tendency in the philosophy of science to refer to an older viewpoint or school of thought
      as the "Received View." The Semantic Conception of Theories presented by Suppe 1977, 1988 is
      contraposed to a Received View. Hull himself contrasts his evolutionary view to the Received View
      of Popper and Hanson [Hull 1988b, 484]. This tactic does run the risk of falling into the mistake of
      assuming a global hegemony of views against the "radical" nature of one's own, but until recently this
      view of species has indeed universally been received even by those who *want* to defend essentialism
      [such as Ruse 1987, 1998]. Amundson 2005 calls a similar tradition the Synthetic Historiography, in
      a broader context. Winsor 2003, independently also called something similar to this the "received
      view." Another more restricted term for Mayr's narrative is "The Essentialist Story" [Levit and
      Meister 2006].
[11] Winsor 2003.
[12] Cain 1958.
[13] Joseph 1916; cf. Winsor 2001. In fact, Joseph had not made the misinterpretation claimed, but had
      been read wrongly by Cain and Hull, as we shall see below.
[14] Cain 1959a, Cain 1959b.
[15] Cain was a student of Mayr's, according to Winsor.
[16] Amundson 1998; Amundson 2005.

idealism were less essentialistic than the Received View indicated, a view in part revised by Hull himself.[17] The Received View appears in fragmentary form throughout a number of publications over the past seventy years, and we shall look at three examples.

George Gaylord Simpson was perhaps the greatest of the paleontological evolutionary theorists of the twentieth century. His influential text *Principles of Animal Taxonomy*[18] included a chapter entitled "The development of modern taxonomy." According to Simpson, Scholastic logic, based on Aristotle, relied on the "essence" of a "species," which consists of its "genus" plus its "differentia," which was a logical relationship.[19] It was this that Linnaeus adopted,[20] and it has several serious faults: one is that "properties" and "accidents" (Simpson's scare quotation marks) are excluded from the "essence" *and therefore the definition* (my emphasis) of the species. Another is that this method produces a classification of *characters* not *organisms* (his emphasis). From Linnaeus to Darwin there was an increasing emphasis on empirical rather than a priori classification.

According to Simpson, there was a typological tradition stemming from Plato and coming via Aristotle, Neo-Platonic, scholastic, and Thomist philosophy into biological taxonomy. He writes:

> The basic concept of typology is this: every natural group of organisms, hence every natural taxon in classification, has an invariant, generalized or idealized pattern shared by all members of the group. The pattern of the lower taxon is superimposed upon that of the higher taxon to which it belongs, without essentially modifying the higher pattern. ... Numerous different terms have been given to these idealized patterns, often simply "type" but also "archetype," "Bauplan" or "structural plan," "Morphotypus" or "morphotype," "plan" and others.[21]

In contrast, modern taxonomy relies on common descent,[22] statistical properties of populations,[23] and biological relationships. Simpson took his historical information on this subject from a paper by A. J. Cain.[24] Cain, following the interpretation of Aristotle and the later logicians in H. W. B. Joseph's influential text,[25] had said that Linnaeus based his conception of species on Aristotelian definition of essences, on "*a priori* principles agreeable to the rules of logic."[26] He later reversed this position after retirement and time to properly read Linnaeus and Ray.[27] Nevertheless, his earlier papers[28] became the foundation for the Received View.

---

[17] Hull 1983.
[18] Simpson 1951.
[19] Simpson 1961, 36f.
[20] *Op. cit.* 38.
[21] Simpson 1961, 47.
[22] Simpson 1961, 52–54.
[23] Simpson 1961, 65.
[24] Cain 1958; cf. Winsor 2001.
[25] Joseph 1916.
[26] Page 147, quoted by Winsor 2001, 249.
[27] For example, in Cain 1993, Cain 1994, Cain 1995, Cain 1999.
[28] Cain 1958, Cain 1959b, Cain 1959a.

Ernst Mayr is someone whose ideas and narratives have been extremely influential, and therefore with whom we will deal later in some detail. He has often put forward a narrative history of species concepts, and his ideas are the most widely known and accepted. According to him, the species concept begins in biology with Linnaeus,[29] because before then, apart from Ray, nobody believed species were stable entities. Later Darwin held that species were fluid,[30] but in the intervening period there was a tendency of various writers like Cuvier, De Candolle, Godron,[31] and von Baer to treat them as real and definite entities, united, as von Baer says, by common descent from original stock. The realization that species exhibited a "supraindividualistic bond"[32] and that members of species reproduce only with each other came slowly. The older view was "typological" and it is "the simplest and most widely held species concept."[33] Typology, according to Mayr, as for Simpson, is due to the influence of Plato, and those who follow him are trying to define a species in terms of "typical" or "essential" attributes:

> Typological thinking finds it easy to reconcile the observed variability of the individuals of a species with the dogma of the constancy of species because the variability does not affect the essence of the eidos [the Greek term translated as "species"] which is absolute and constant. Since the eidos is an abstraction derived from human sense impressions, and a product of the human mind, according to this school, its members feel justified in regarding a species "a figment of the imagination," an idea.[34]

In contrast, he says, a species today is regarded as a gene pool rather than a class of objects.

Mayr presents this history in more detail in his *The Growth of Biological Thought*.[35] Here again, the distal source of essentialism is Plato, now via Aristotle, to John Ray and Linnaeus. Again, it is only with the increasing emphasis on empiricism and the development first of evolution and then of genetics that naturalists begin to realize that species are gene pools, reproductively isolated from each other. In the end, the biological species concept is triumphant.

A third history, for contrast, is that provided by David Hull, who became the leading philosopher of biology of his generation. He made it a point to focus on the actual history and biology of his subjects rather than on a "rational reconstruction" or textbook history, as was sometimes the practice of prior philosophical scholarship of science. In his seminal paper "The effect of essentialism on taxonomy – two thousand

---

[29] Mayr 1957, 2f.
[30] Mayr 1957, 4.
[31] Dominique-Alexandre Godron, a French botanist (1807–1880). According to Mayr [1982], his major interest was in the nature of introduced species and the effects of cross-breeding [Godron 1854]. A species fixist, in his *De l'espèce et des races dans les êtres organisés* [Godron 1859, 51] he wrote that interbreeding in the wild, without human intervention, was sufficient to mark species, and that races were only caused by such intervention.
[32] Mayr 1957, 8.
[33] Mayr 1957, 11.
[34] Mayr 1957, 12.
[35] Mayr 1982, 254–279.

years of stasis,"[36] he presented the Received View, in part relying on Joseph's book,[37] in part on Mayr and Simpson. The foundations of classification rest on Aristotle's notion of definition,[38] while a contrary tradition, beginning with Adanson, treated taxa as disjuncts of properties, which themselves became the essence of a species in classical Aristotelian form. The moderns—Simpson, Dobzhansky, Mayr—moved to a reproductive isolation conception, founded on evolution. In a much later work, he devoted a chapter[39] in which the Received View is expanded and in many ways revised: Plato is relegated to the background and Aristotelians offered a view of species as types from which deviations were possible. After Aristotle, we reach Linnaeus, then Buffon, Lamarck, Cuvier, Geoffroy, and then the ideal morphologists until we meet Darwin, followed by the New Systematics group—Dobzhansky, Mayr, Julian Huxley, and so forth.

In each of these, and others[40] the Received View runs roughly thus:

> Plato defined Form (*eidos*) as something that had an essence, and Aristotle set up a way of dividing genera (*genē*) into species (*eidē*) so that each species shared the essence of the genus, and each individual in the species shared the essence of the species. Linnaeus took this idea and made species into constant and essentialistic types. Darwin overcame this essentialism. Later naturalists, under the influence of genetics, discovered the biological species concept, in which species are found to be populations without essences, but with common ancestry. Population thinking replaces typological essentialism.

THE FAILURE OF THE RECEIVED VIEW

In between Aristotle (*d*322 BCE) and Linnaeus's first works on classification (*fl*1750 CE) there is a 2000-year gap. What happened in that intervening period? It was a period of active philosophy, as it includes the bulk of the classical period, not to mention the high period of Arab science and scholarship,[41] the medieval debates on logic

---

[36] Hull 1965a, Hull 1965b.

[37] Hull, personal communication. As Hull's ideas have been so influential, it should be noted that at about this time he also published a brief history [Hull 1967] in which he went into more detail, and which cites Joseph in note 3. In it, he notes that the observation that things do not always breed true was made by Aristotle and Theophrastus, and that therefore species do allow variation and divergence from the type. Even so, Hull still skips from Aristotle to Cesalpino, with only a short allusion to the Great Chain of Being.

[38] Hull 1965a, 318f.

[39] Chapter 3, "Up from Aristotle," Hull 1988b.

[40] For example, Wiley 1981, 70–72, to take one example at random.

[41] Which I do not cover here. Some secondary sources indicate that not much happened in the Islamic high period with respect to biology, but given my experience with commentators on the western sources, I tend to doubt it. Somebody with access to the material might like to investigate this period. One possibly important thinker is the ninth century writer, Al-Jāhiz (776–868), who compiled an extensive *Book of Animals* (Kitab al-Hayawan). Al-Asmaʻi (740–828) wrote a number of texts, mostly anthological, on animals and plants. The twelfth century author, Al-Marwazī (1056–1124), wrote *The Nature of Animals* [cf. McDonald 1988, Egerton 2012]. Islamic authors often attempt to make out that these medieval Islamic thinkers are directly plagiarized by European authors such as Darwin [e.g., Bayrakdar 1983]. However, as with Osborn's work [1894] this is based on perceived similarities rather than actual lineal influences.

and universals, and the Renaissance. Are we to think that there was no change or development in these ideas in that time? I undertook to investigate this gap, expecting no more than a continuous narrative supporting the Received View. The results were surprising. In most respects, the Received View is incorrect, seriously incomplete, or simply false. Effectively every philosophical issue raised about biological species in the modern period was raised in one form or another about philosophical, or logical, species during this interregnum. Moreover, the crucial mediate link between Aristotle and modern biology was not Linnaeus, or even Aristotle's own writings. It was, rather, the theologians of the sixteenth century who, following an increasing rationalism in the Christian intellectual tradition from the tenth century, attempted to give a naturalistic account of the capacity of Noah's Ark, and in so doing introduced the foundation for a basic *rank* of living kinds.[42] In effect, the species problem is a Christian invention, not an Aristotelian one.[43]

Further, the Received View tended to ignore the Great Chain of Being and the universals debates, which are of great importance in the way later writers such as Cusa deal with the notion of essence and definition. It is *not* the case that typology and essentialism were bound together, and in many ways typology as it was actually discussed is more in line with Mayr's "population thinking" than he ever admitted. Winsor[44] refers to the typological conception as the "method of exemplars," which we shall see is a much better characterization.[45]

Few truly novel *formal conceptual* elements of species concepts have arisen since the eighteenth century. Biologists and philosophers have been dealing from the same deck of conceptual cards ever since. Of course, many *technical* concepts are novel—for example, Templeton's genetic concept relies upon Mendelian genetics, and Wu's genic concept relies on molecular genetics, both of which are much later additions to the repertoire of biology.[46] It is also open to doubt that ideas that are merely formally similar are in fact "the same" ideas as later ones. But the similarities are themselves instructive; if an older debate dealing with formally similar notions is resolved into a few opposing viewpoints, we may learn from them what to expect in modern debates.

Despite popular expectation, Darwin and his successors did not add much to the species debate except to raise in sharp relief problems brought about by the introduction of the notion of speciation and the subsequent mutability of classifications. Nevertheless, Darwin acts as a focal point for what follows, and even the formalists had to address his conceptions. Since the "Modern Synthesis" of 1930–1942, the only truly novel *philosophical* ideas about species have been the Individuality Thesis and the refinement of notions of class inclusion and hierarchies, which themselves rest

---

[42] Allen 1949.

[43] In the first edition, I suggested the species concept was a neo-Platonist invention. Having further reviewed the material I now think that, while neo-Platonism was significant in the context of Bishop Wilkins and John Ray, it is not the case that it caused the species *concept*.

[44] Winsor 2003.

[45] See also Amundson [1998]. Challenges to the typological/essentialist account begin, so far as I can see, with Farber, Platnick and Nelson, and Panchen in the period from 1970 to 1985 (references cited below).

[46] Templeton 1989, Wu 2001a, Wu 2001b. I am indebted to Kim Sterelny for this point.

on prior work (although I keep finding people expressing the view that *logical* species are individuals; for example, Boëthius, Aquinas, and Cusa, and in the biological
tradition, even Mivart discusses a version of it, in which species are like organisms).
So far as I can tell, there are precursors or forerunners for everything else, at least
in the sense of resemblance, if not direct and demonstrated descent. Even clustering
concepts like the Homeostatic Property Cluster notion of Boyd have antecedents. It
is often the case that these ideas are continually reinvented, especially by biologists.
Equally, however, it is often the case that these modern alternatives are not directly
descended from the precursors, although there is evidence of indirect descent. For
example, Rutgers geneticist Jody Hey expressed the view[47] that species are merely
conventional objects used for communication, a reinvention in some respects of John
Locke's earlier conventionalism (as we will discuss). He depends extensively upon
W. V. O. Quine's philosophy of language in *Word and Object*.[48] Quine is, of course,
dealing with the same issues of the empiricist, linguistically directed, philosophical
tradition that Locke began, and knows his Locke very well, but Hey understandably
seems unaware of it. A geneticist can be excused for not knowing the detailed history
of an epistemological episode in philosophy, but this pattern is repeated even when
the originator of a view, for instance Hennig, *is* aware of his philosophical precursors
(including in this case Woodger and Gregg[49]). Later followers of that initial account
often do not realize that there ever was such a prior history.

For this reason, it is important to at least sketch some of the main pre-biology and
early biology historical accounts of species, and this of course means beginning in
the western tradition with Plato and Aristotle. I seek to be excused for this western
bias, despite much interesting literature in Persian and Asian cultures on classification of organisms, because it is from the western tradition and not the others that the
modern species problem derives, and on which it depends. For the same reason, I
will pass over the work of cultural anthropologists on classification in non-western
indigenous cultures[50] and the work done on the developmental and psychological
origins of taxonomic concepts.[51] Scientific concepts, whatever their etiology in our
biological substrate, are primarily subject to change due to the institutions and culture of science and cultural influences such as philosophy. We all share whatever
psychological and cognitive foundations there are of classification and categorical
notions, but the scientific debate has gone its own way, and so biological biases are
not determinants of such categories. Cultural evolution, including the evolution of
science,[52] has its own dynamics, perhaps biased by psychology and evolutionary
adaptations and heritage, but it is usually decoupled from them.[53]

---

[47] Hey 2001b, Hey 2001a.
[48] Quine 1960.
[49] Among others. Nicolai Hartmann appears to have been a philosophical influence on Hennig too
[Hennig 1966, 22]. See Varma 2013 for the details.
[50] See Bulmer 1967, Atran 1985, Atran 1990, Ellen 1993, Atran 1995, Atran 1998, Atran 1999.
[51] Millikan 1984, Sperber and Wilson 1986, Estes 1994, Sperber et al. 1995, Keil 1995, Griffiths 1997,
Eco 1999, Gil-White 2001, Hey 2001b. Despite its title, Ghiselin's paper "On psychologism in the
logic of taxonomic controversies" [Ghiselin 1966] does not address the psychological origins of taxa,
but the introduction into taxonomy of epistemological conventionalism.
[52] Hull 1988a, Hull 1988b.
[53] Campbell 1965, Toulmin 1972.

This is an essay in the history of ideas (although *conceptual history* is prefer-
able), and in particular the ideas that came before and might be demonstrated or
fairly thought to have contributed to the ideas in play in biological thought about
species and classification. "History of ideas" has become uncommon and somewhat
disparaged by professional historians, and this is understandable given the whiggish,
presentist bias much of it has exhibited. I am well aware of the problems faced by the
historian of ideas, as described by John Greene:

> Of all histories the history of ideas is the most difficult and elusive. Unlike things, ideas
> cannot be handled, weighed, and measured. They exert a powerful force in human his-
> tory, but a force difficult to estimate.[54]

But history of ideas is necessary if we are to understand why ideas are as they are
in internal terms; that is, in terms of the *content* of the ideas. There is also a need
for an externalist history of ideas, of the *context*, but this is much harder to do, and
in any event cannot be done without at least some awareness of the internal history.
Without defending this further, the history of ideas is at least intrinsically interesting
for those seeking to explain current scientific ideas, given that science, as an intellectual set of traditions, relies on the notions of the past as a starting point in the way
it develops them further. Knowing the past may also help scientists to avoid repeating
it unnecessarily.[55]

We shall not refer much to the usual labels and banners applied to the various
thinkers; terms such as *idealist, transcendentalist, empiricist,* and so on. This is
because it is my experience that such absolute distinctions ride roughshod over the
complexities and similarities of these thinkers. Calling somebody a "transcendentalist" implies they were interested only in Platonic ideas. In the case of a Cuvier
or an Agassiz, that is a hard claim to sustain—they attended closely to empirical
evidence, and for all their shortcomings, neither merely made inferences *sub specie
aeternitatis,* as Plato did. And in abandoning these hard distinctions I have found,
I believe, a similarity of conception I call *the generative conception of species* that
runs through most, if not all, pre-Darwinian thinkers and researchers on the topic
of *species,* and which is found even in post-Darwinians, and Darwin as well. As
Amundson noted, we encounter a "conceptual tangle"[56] when we attend to the actual
history of concepts.

As always in the history of ideas, much of the earlier material must be drawn in
terms of resemblance and succession. Because one author—say Porphyry—discusses
ideas that are similar to another's—say Aristotle—is not in itself reason to believe
either that there is a direct relationship of intellectual ancestry or even an indirect
relationship unless the one cites or refers to the other (as Porphyry does Aristotle).
Classical authors often failed to provide a proper modern scholarly apparatus of citations and imputed credit. Prior to the "biological problem" of species beginning with

---

[54] Greene 1963, 11.
[55] I am influenced by David Hull's evolutionary conception of science science [Hull 1983, Hull 1986,
Hull 1988a, Hull 1988b, Hull 1990], and have tried to present my own evolutionary account of science
and culture in this vein before [e.g., Wilkins 2002, Wilkins 2008].
[56] Amundson 1998, 159.

Bauhin, Cesalpino, and John Ray, this lineal descendency must therefore be taken as an approximation. With that caveat, let us proceed.

BIBLIOGRAPHY

Allen, Don Cameron. 1949. *The Legend of Noah. Renaissance Rationalism in Art, Science and Letters.* Urbana: University of Illinois.

Amundson, Ron. 1998. Typology reconsidered—Two doctrines on the history of evolutionary biology. *Biology and Philosophy* 13 (2):153–177.

—. 2005. *The Changing Role of the Embryo in Evolutionary Biology: Structure and Synthesis. Cambridge Studies in Philosophy and Biology.* New York: Cambridge University Press.

Atran, Scott. 1985. The early history of the species concept: An anthropological reading. In *Histoire du Concept D'Espece dans les Sciences de la Vie*, edited by Scott Atran, 1–36. Paris: Fondation Singer-Polignac.

—. 1990. *The Cognitive Foundations of Natural History.* New York: Cambridge University Press.

—. 1995. Causal constraints on categories and categorical constraints on biological reasoning across cultures. In *Causal Cognition: A Multidisciplinary Debate*, edited by Dan Sperber et al., 205–223. Oxford/New York: Clarendon Press/Oxford University Press.

—. 1998. Folk biology and the anthropology of science: Cognitive universals and the cultural particulars. *Behavioral and Brain Sciences* 21 (4):547–609.

—. 1999. The universal primacy of generic species in folkbiological taxonomy: Implications for human biological, cultural and scientific evolution. In *Species, New Interdisciplinary Essays*, edited by R. A. Wilson, 231–261. Cambridge, MA: Bradford/MIT Press.

Bayrakdar, Mehmet. 1983. Al-Jahiz and the rise of biological evolution. *Islamic Quarterly* 27 (3):307–315.

Bulmer, Ralph. 1967. Why is the cassowary not a bird? A problem among the Karam of the New Guinea highlands. *Journal of the Royal Anthropological Institute* 2 (1):5–25.

Cain, Arthur J. 1958. Logic and memory in Linnaeus's system of taxonomy. *Proceedings of the Linnean Society of London* 169:144–163.

—. 1959a. The post-Linnaean development of taxonomy. *Proceedings of the Linnean Society of London* 170:234–244.

—. 1959b. Taxonomic concepts. *Ibis* 101:302–318.

—. 1993. *Animal Species and Their Evolution.* Princeton, NJ: Princeton University Press.

—. 1994. Numerus, figura, proportio, situs: Linnaeus's definitory attributes. *Archives of Natural History* 21:17–36.

—. 1995. Linnaeus's natural and artificial arrangements of plants. *Botanical Journal of the Linnean Society* 117 (2):73–133.

—. 1999. John Ray on the species. *Archives of Natural History* 26 (2):223–238.

Campbell, Donald T. 1965. Variation and selective retention in socio-cultural evolution. In *Social Change in Developing Areas, a Reinterpretation of Evolutionary Theory*, edited by H. R. Barringer et al. Cambridge, MA: Schenkman.

Eco, Umberto. 1999. *Kant and the Platypus: Essays on Language and Cognition.* London: Vintage/Random House.

Egerton, Frank N. 2012. Medieval millennium. In *Roots of Ecology*, 17–32. Berkeley: University of California Press.

Ellen, Roy F. 1993. *The Cultural Relations of Classification: An Analysis of Nuaulu Animal Categories from Central Seram.* Cambridge, UK: Cambridge University Press.

Estes, William K. 1994. *Classification and Cognition.* New York/Oxford: Oxford University Press/Clarendon Press.

Ghiselin, Michael T. 1966. On psychologism in the logic of taxonomic controversies. *Systematic Zoology* 15:207–215.

Gil-White, Francisco. 2001. Are ethnic groups biological "species" to the human brain? Essentialism in our cognition of some social categories. *Current Anthropology: A World Journal of the Human Sciences* 42 (4):515–554.

Godron, Dominique-Alexandre. 1854. *Florula Juvenalis ou énumération des plantes étrangères qui croissent naturellement au Port Juvénal, près de Montpellier, précédée de considérations sur les migrations des végétaux*. 2nd ed. Nancy: Grimblot et Veuve Raybois.

—. 1859. *De l'espèce et des races dans les êtres organisés et spécialement de l'unité de l'espèce humaine*. Vol. 1. Paris: J.B. Baillière et Fils.

Greene, John C. 1963. *Darwin and the Modern World View: The Rockwell Lectures*, Rice University. New York: New American Library. Original edition, 1961.

Griffiths, Paul E. 1997. *What Emotions Really Are: The Problem of Psychological Categories*. Chicago: University of Chicago Press.

Hennig, Willi. 1966. *Phylogenetic Systematics*. Translated by D. Dwight Davis and Rainer Zangerl. Urbana: University of Illinois Press.

Hey, Jody. 2001a. *Genes, Concepts and Species: The Evolutionary and Cognitive Causes of the Species Problem*. New York: Oxford University Press.

—. 2001b. The mind of the species problem. *Trends in Ecology and Evolution* 16 (7):326–329.

Hull, David L. 1965a. The effect of essentialism on taxonomy—Two thousand years of stasis. *British Journal for the Philosophy of Science* 15 (60):314–326.

—. 1967. The metaphysics of evolution. *British Journal for the History of Science* 3 (12):309–337.

—. 1983. Darwin and the nature of science. In *Evolution from Molecules to Men*, edited by D. S. Bendall, 63–80. Cambridge, MA: Cambridge University Press.

—. 1986. Conceptual evolution and the eye of the octopus. *Studies in Logic and the Foundations of Mathematics* 114:643–665.

—. 1988a. A mechanism and its metaphysics: An evolutionary account of the social and conceptual development of science. *Biology and Philosophy* 3 (2):123–155.

—. 1988b. Science as a process: An evolutionary account of the social and conceptual development of science. Chicago: University of Chicago Press.

—. 1990. Conceptual Selection. *Philosophical Studies* 60 (1–2):77–87.

Hull, David Lee. 1965b. The effect of essentialism on taxonomy—Two thousand years of stasis (II). *British Journal for the Philosophy of Science* XVI (61):1–18.

Huxley, Julian. 1942. *Evolution: The Modern Synthesis*. London: Allen and Unwin.

Joseph, Horace William Brindley. 1916. *An Introduction to Logic*. 2nd ed. Oxford: Clarendon Press. Original edition, 1906.

Keil, Frank C. 1995. The growth of causal understandings of natural kinds. In *Causal Cognition: A Multidisciplinary Debate*, edited by Dan Sperber et al. Oxford/New York: Clarendon Press/Oxford University Press.

Levit, Georgy S. and Kay Meister. 2006. The history of essentialism vs. Ernst Mayr's "Essentialism Story": A case study of German idealistic morphology. *Theory in Biosciences* 124:281–307.

Mayr, Ernst. 1957. Species concepts and definitions. In *The Species Problem: A Symposium Presented at the Atlanta Meeting of the American Association for the Advancement of Science*, December 28–29, 1955, Publication No 50, edited by Ernst Mayr, 1–22. Washington, DC: American Association for the Advancement of Science.

—. 1982. *The Growth of Biological Thought: Diversity, Evolution, and Inheritance*. Cambridge, MA: Belknap Press of Harvard University Press.

McDonald, M. V. 1988. Animal-books as a genre in Arabic literature. *Bulletin (British Society for Middle Eastern Studies)* 15 (1/2):3–10.

Millikan, Ruth Garrett. 1984. *Language, Thought, and Other Biological Categories: New Foundations for Realism*. Cambridge, MA: MIT Press.

Osborn, Henry Fairfield. 1894. *From the Greeks to Darwin: An Outline of the Development of the Evolution Idea*. Columbia University Biological Series. I. New York: Macmillan.

Quine, Willard Van Orman. 1960. *Word and Object, Studies in Communication*. Cambridge, MA: Technology Press of the Massachusetts Institute of Technology.

Ruse, Michael. 1987. Biological species: Natural kinds, individuals, or what? *British Journal for the Philosophy of Science* 38 (2):225–242.

—. 1998. All my love is toward individuals. *Evolution* 52 (1):283–288.

Santayana, George. 1917. *The Life of Reason, or the Phases of Human Progress*. Vol. 1. New York: Scribner. Original edition, 1905.

Simpson, George Gaylord. 1951. The species concept. *Evolution* 5 (4):285–298.

—. 1961. *Principles of Animal Taxonomy*. New York: Columbia University Press.

Sperber, Dan et al., eds. 1995. *Causal Cognition: A Multidisciplinary Debate*. Oxford/New York: Clarendon Press/Oxford University Press.

Sperber, Dan and Deirdre Wilson. 1986. *Relevance: Communication and Cognition*. Oxford: Blackwell.

Suppe, Frederick. 1977. The search for philosophic understanding of scientific theories. In *The Structure of Scientific Theories*, edited by Frederick Suppe, 3–232. Urbana: University of Illinois Press.

—. 1988. *The Semantic Conception of Theories and Scientific Realism*. Urbana: University of Illinois Press.

Templeton, Alan R. 1989. The meaning of species and speciation: A genetic perspective. In *Speciation and Its Consequences*, edited by D. Otte and J.A. Endler, 3–27. Sunderland, MA: Sinauer.

Toulmin, Stephen. 1972. *Human Understanding*. Cambridge, UK: Cambridge University Press.

Varma, Charissa Sujata. 2013. Beyond set theory: The relationship between logic and taxonomy from the early 1930 to 1960. PhD dissertation, Institute for the History and Philosophy of Science and Technology, University of Toronto.

Wiley, Edward Orlando. 1981. *Phylogenetics: The Theory and Practice of Phylogenetic Systematics*. New York: Wiley.

Wilkins, John S. 2002. Darwinism as metaphor and analogy: Language as a selection process. *Selection: Molecules, Genes, Memes* 3 (1):57–74.

—. 2008. The adaptive landscape of science. *Biology and Philosophy* 23 (5):659–671.

Winsor, Mary Pickard. 2001. Cain on Linnaeus: The scientist-historian as unanalysed entity. *Studies in the History and Philosophy of the Biological and Biomedical Sciences* 32 (2):239–254.

—. 2003. Non-essentialist methods in pre-Darwinian taxonomy. *Biology and Philosophy* 18 (3):387–400.

Wu, Chung-I. 2001a. Genes and speciation. *Journal of Evolutionary Biology* 14 (6):889–891.

—. 2001b. The genic view of the process of speciation. *Journal of Evolutionary Biology* 14 (6):851–865.

Section I

The Historical Development of "Species"

1 The Classical Era: Science by Division

The history of the species concept can be divided into a pre-biological and a post-biological history, which is how the Received View has always treated it. But the two histories overlap substantially, and it is much better to consider instead the history of the species idea that applies to any objects of classification—the tradition of universal categorization and philosophical logic—and, independently, the particular history of the species idea that applies solely to biological organisms. Even though, for example, Linnaeus in his third volume of his *Systema Naturae*[1] famously applied the notion *species* to minerals as well as organisms, his biological usage included elements not included in the mineralogical case. We must therefore separate universal (logical or formal) categorical and biological taxonomic notions of *species*.[2]

We can thus distinguish between two broad kinds of taxonomy. *Universal categories* are what the philosophical tradition from Plato to Locke sought (but which continues through to the considerations of sensory impressions, or *qualia*, by the logical positivist and phenomenological philosophical schools) and in which species are any distinguishable or naturally distinguished categories with an essence or definition. Then there is the *biological taxonomy* that develops from this tradition as biology itself develops from the broader field known as "natural history."[3] These biological notions of species do not necessarily refer to reproductive communities, nor do they in the medical definitions of species of diseases of the period,[4] but we do need to recognize that "species" develops a uniquely biological flavor around the seventeenth century.[5]

These two conceptions form what might be regarded as part of or an entire research program, in Lakatos' sense. The universal taxonomy program that began with Plato culminates in the attempt to develop a classification of not only all natural objects, but all possible objects. It continues in the modern projects of metaphysics

---

[1] Linné 1788–1793 Vol. 3.

[2] This account owes its outline and many details to Nelson and Platnick 1981, and Panchen 1992.

[3] "History" is, in this older sense, the Greek word *historia*, which means an inquiry or investigation, which derives from the title, or rather the opening words, of Herodotus' *History*. Later it comes to mean "knowledge" or "learning."

[4] Cain 1999. The classification of diseases is called nosology, and it wrestles with many of the same issues that systematics does, especially in psychiatry. See Murphy 2006, Regier et al. 2009, Kendler and Parnas 2015.

[5] An excellent treatment of many of the themes discussed in this section, and a good aid to understanding the historical contexts, can be found in Mary Slaughter's wonderful book [Slaughter 1982]; a broader and more liberal treatment, but one that suffers from over-theorizing, is Michel Foucault's *The Order of Things* (*Les Mots et les Choses*) [Foucault 1970], particularly chapter 5. Much of the source material and a detailed discussion of the *logica combinatorialis* can be found in Rossi's treatment [Rossi 2000] relating mnemotechnical (*ars memoria*) traditions and the Universal Language Project.

and modal logics. The biological taxonomy project that developed out of it resulted in a program to understand the units of biology, to which Darwin contributed, and to which genetics later added. *Species* is a marker for both projects. Seen in that way, we might reasonably ask what the "core assumptions" of these projects are. In the universal case, it is classification (and science) by division. In the biological instance, it is, as I will argue, the marriage of reproduction or generation, with form, which I call the generative conception. This remains even today the basis of understanding *species*.

The general philosophical tradition of considering classification to be an indispensable aspect of science seems to end with Mill and Whewell,[6] and although there are later discussions on classification by Peirce[7] and Jevons,[8] the revivification of taxonomy in philosophy of science in recent decades appears, with one exception,[9] to be driven by biological systematics itself rather than through philosophy motivating biology. As a result, the philosophical foundations of taxonomy are critically incomplete and systematists often rely upon philosophers who disregard the matter almost entirely, like Popper.[10]

PLATO'S DIAIRESIS

As the Hellenic world began the process of expanding and consolidating, early Hellenic philosophers, known as the Milesians, had in the sixth century BCE begun to grapple with the problem of change—of generation and corruption—and what that meant for knowledge. This was essentially the core problem of Greek philosophy, for upon it rested the entire conception of the possibility of the knowledge of Nature (*phusis*).

Anaximander (c. 610–c. 546 BCE) argued that the world was composed of a single eternal substance, the *apeiron* (the unbounded), and that the world was only superficially changing. The Pythagoreans presumed that the foundation of the world was Number, and a table of contraries described by Aristotle in the *Metaphysics*— limited-unlimited, one-many, odd-even, light-dark, good-bad, right-left, straight-curved, male-female, square-oblong, rest-motion—accounted for the world.[11]

Later, Heraclitus (c. 535–c. 475 BCE) famously asked if you could step into the same river twice, as the river changed from step to step (although he is often misunderstood to answer in the negative; rather, he thought there was something eternal that preserved reference, noting that "Nature loves to hide," fragment 10). Various philosophers attempted to divide the world into its constituent elements and the eternal forms of reality. This led to the idea of classification, through the uncovering of the eternal *logos*, a term or class or order.

---

[6] Noted by David Hull in discussion. See Wilkins and Ebach 2013.
[7] See Atkins 2006 on the manner in which Peirce was influenced by Louis Agassiz in his classification of phenomena and eventually the sciences themselves. The direction of the influence from natural history to metaphysics and logic is the reverse of the Received View's account.
[8] Jevons 1878.
[9] Woodger 1937.
[10] Who only mentions it dismissively: Popper 1957, §27, Popper 1959, 65, as noted by Hull 1988, 252.
[11] 986a15, Brumbaugh 1981, 36.

TERMS AND TRADITIONS

The modern and medieval word "species" is a Latin translation of the classical Greek word *eidos*, sometimes translated as "idea" or "form." Other significant terms are also translations of Greek terms: *genus* from *genos*, *differentia* from *diaphora*. Liddell and Scott tell us that *eidos* means "form" and is derived from the root word "to see," and *genos* means "kind" and is derived from the root word "to be born of."

We still find these senses in the English words "specify," "special," "spectacle," and "generation," "gene," and "genesis." Diairesis or "division" is an interesting term that will play a major role in our story. Another Greek term that is later adopted by the Latins is *synagoge*, or in Latin, *"relatum." Relata* are those features by which things group together. *Essentia*, the Latin term used for "essence," is a neologism for a phrase used by Aristotle, the "what-it-is-to-be." The Latin term *substantia* is occasionally translated as "essence" in natural historical works, but this is often misleading.

Greek Word, Plural Form	Classical Meaning	Latin Translation	English Words
eidos, eidē εἶδός, εἶδή	That which is seen, form, shape, figure; a class, kind, or sort	*Species, forma*	Species, idea, kind, sort (Locke), form
genos, genē γένος, γενή	Race, stock, family; a generation	*Genus* (pl. *genera*)	Genus, kind
diaphora διαφορά	Difference, distinction, variance, disagreement	*Differentia* (pl. *differentiae*)	Difference
diairesis διαίρεσις	A dividing, division	*Divisio*	Division
synagogē συναγωγή	A bringing together; a conclusion	Relatum (pl. *relata*)	Relation, affinity

See Liddell and Scott (1888). However, merely because words derive their etymology from older terms, or translate words in other languages, it does not immediately follow that they are the same terms with the same intension or extension, or that one author is influenced by another. By *eidos* and *species*, for example, different authors have meant *forms, kinds, sorts, species* in the logical sense of non-predicables (Ross 1949, 57), *biological species, classes, individuals,* and *collections,* both arbitrary and artificial, or natural and objective. In particular, *species* has meant both varieties of ideas and sense impressions, *species intelligibilis* (Spruit 1994–1995), and also the material form of the elements of the sacraments in the Roman Catholic tradition. This must be borne in mind or it will cause confusion when we consider the views of different authors. This is an instance of a more general problem, called "incommensurability" after Kuhn's thesis that terms in scientific theories can have different referents in the shift from one theory to another; Sankey (1998) has called this particular problem "taxonomic incommensurability." We should not make too much of this, but the terms used in classification shift in subtle and major ways that sometimes obscure the views each author is presenting. We are primarily concerned with the tradition outside biology that *has* impacted on the biological notions and usage.

The English plural of "species" singular is "species." The word "specie," often misused as the singular of "species," actually refers to small coinage. However, in Latin there is a singular (*species*) and plural form (*speciei*), which one translator signified (Porphyry 1975) by italicizing the ending thus: spec*ies*. This is clumsy for the purposes of this book, so here I will follow the rule of the biological writers of the past century and refer to "the" species or italicize the entire word for the concept, and leave the term unqualified for number except by context.

For the period from the Greeks to the beginnings of natural history in the sixteenth and seventeenth centuries, it may help to follow Locke's suggestion, to get a feel for the meanings and avoid anachronistic interpretations, by replacing *genos* and *genus* with "kind" in English, and *eidos* and *species* with "sort."

Thus the informal usage of, say, Theophrastus in talking indifferently about *genē* and *eidē*, can be seen as the same sort of informal usage an English speaker might make by stylistically mixing "kind" and "sort" in a discussion to avoid repetition. It is imperative to remember that these were not technical terms of biology until the modern period, in particular after Linnaeus.

Finally, it helps to remember that the adjectival forms of *species* and *genus* in English are "special" and "general." The general includes the special.

Classification at this time was uncritically applied to all things, whether artificial or natural (a distinction early Greeks would not have fully accepted anyway), conceptual, semantic, or empirical. The problem of how to properly classify things, including living things, is first recorded to be dealt with specifically by Plato in the *Sophist*.[12] Plato, regarded by many later thinkers, particularly during the Renaissance, as *The* Philosopher, founded the philosophical school known as the Academy in Athens. He proposed a method of binary division of contraries until the object being classified was reached; this was known as the *diairesis* (division, or as some call it, dichotomy), in a similar fashion to the Pythagoreans.[13] For example, he somewhat whimsically defined fly-fishing as a model for all such classification. He has the Stranger of the dialogue ask leading questions, such as whether the fisher has skill (*techne* = art) or not, defining art into two kinds, agriculture and tending of mortal creatures on the one hand, and art of imitation on the other, and later introduces a

---

[12] 219a–221a.
[13] Some believe that Plato was a Pythagorean who broke the mystery boundaries of the religion [Kahn 2001], and that some of Plato's work is a guarded expression of Pythagorean mysteries regarding the ratio, or *logos* of number.

fairly arbitrary distinction about acquisitive art, resulting in the following "final" definition of angling:

> STRANGER: Then now you and I have come to an understanding not only about the name of the angler's art, but about the definition of the thing itself. One half of all art was acquisitive—half of the acquisitive art was conquest or taking by force, half of this was hunting, and half of hunting was hunting animals, half of this was hunting water animals—of this again, the under half was fishing, half of fishing was striking; a part of striking was fishing with a barb, and one half of this again, being the kind which strikes with a hook and draws the fish from below upwards, is the art which we have been seeking, and which from the nature of the operation is denoted angling or drawing up (*aspalieutike, anaspasthai*).[14]

As Oldroyd[15] summarizes it, for Plato "angling is a coercive, acquisitive art, carried out in secret, in which live animals living in water are hunted during the day by blows that strike upwards from below!" (Figure 1.1). In addition to division, Plato also classified by grouping (*synagoge*), so that he divided things and grouped them according to their differences and similarities.[16] Plato's classification style here is clearly arbitrary. In order to force the division into dichotomies, he (through his "Stranger") selects the "right" connections for the next differentia, but nothing is obvious about these steps, and it is clear that he knows ahead of time what he wants to deliver. In short, this is question-begging, a party trick not unlike his "showing" that a slave boy "remembered" the proof of Pythagoras' Theorem in the *Meno*.[17]

Plato does not distinguish between classification of natural things and artificial things. For him, or rather his protagonist Socrates, issues of justice and social order are on a par with, and in fact transcend, issues of the natural world. Plato shows little deep interest in any aspect of the terrestrial world,[18] although the heavens, being as close to the eternal as possible, are recommended as philosophical objects of study.[19]

Clearly, there is considerable detail to Plato's views of Form, or Idea (*eidos*) that must be passed over here, but this is a well-understood field of history of philosophy, and as it is entirely outside the scope of the present treatment, I am forced to avoid it.[20] For Plato the *ideai* were metaphysical or ontological realities that were neither changeable nor in the transitory world, and which were preconditions for knowledge. This is the commencement point for the later tradition of forms as the basis for classification we find in Aristotle—for instance the famous "carve nature at its joints" passage in the *Phaedrus*:

---

14 Hamilton and Cairns 1961, 218e–221c, 963, Jowett's translation.
15 Oldroyd 1986, 42.
16 Pellegrin 1986.
17 82b–85b.
18 Kitts 1987.
19 *Timaeus* 27d–34b.
20 A good technical introduction to the received opinion on Plato's theory of Forms is provided by Windelband 1900, §35. For a more updated treatment see the introduction to Matthews 1972, or chapter 1 of Oldroyd 1986.

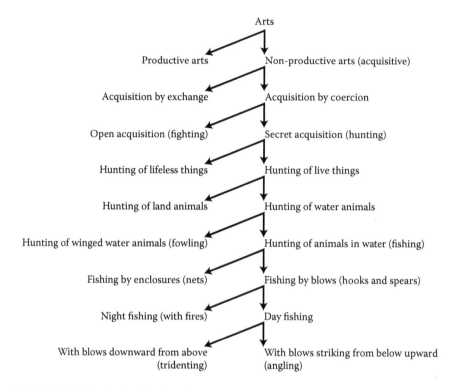

Arts

Productive arts   Non-productive arts (acquisitive)

Acquisition by exchange   Acquisition by coercion

Open acquisition (fighting)   Secret acquisition (hunting)

Hunting of lifeless things   Hunting of live things

Hunting of land animals   Hunting of water animals

Hunting of winged water animals (fowling)   Hunting of animals in water (fishing)

Fishing by enclosures (nets)   Fishing by blows (hooks and spears)

Night fishing (with fires)   Day fishing

With blows downward from above   With blows striking from below upward
(tridenting)   (angling)

FIGURE 1.1   Plato's classification of angling.

> Soc. The second principle is that of division into species according to the natural formation, where the joint is, not breaking any part as a bad carver might. [Jowett's translation][21]

Socrates then says he is a great lover of division and generalization, and follows anyone who can see the "One and the Many" in nature. Note, however, that Socrates includes human society in the term "nature" ("brought forth," πέφυκεν) here.

Nelson and Platnick[22] quote a passage from Plato's *The Statesman*[23] in which Plato presents an argument against incomplete groups that is similar in many ways to Aristotle's argument against privative groups given below, but this point is not mediated to later writers on logic. As Hull observes,[24] Plato's direct influence on biology is late, not until the seventeenth century by way of the Neo-Platonists. Indirectly, though, is another matter.

---

[21] 265e. Nichols translates it: "to cut apart by forms, according to where the joints have naturally grown, and not to endeavor to shatter any part, in the manner of a bad butcher" [Plato 1998].

[22] Nelson and Platnick 1981, 67f.

[23] 262d–263a.

[24] Hull 1967, 312.

ARISTOTLE: DIVISION, AND THE GENUS AND THE SPECIES

If, as Whitehead said, western thought is a series of footnotes to Plato, then biological thought is a series of footnotes to Plato's one-time pupil Aristotle of Stagira (384–322 BCE), a fact also noted by Darwin shortly before his death, when, in a letter to William Ogle thanking him for a copy of his translation of Aristotle's *Parts of Animals*, he wrote:

> From quotations which I had seen I had a high notion of Aristotle's merits, but I had not the most remote notion what a wonderful man he was. Linnaeus and Cuvier have been my two gods, though in very different ways, but they were mere school-boys to Aristotle. I never realized before reading your book to what an enormous summation of labor we owe even our common knowledge.[25]

Aristotle traveled widely throughout the Hellenistic world, tutoring the young Alexander before he became Great. He therefore lived at a time of increased travel and trade, as well as during the flowering of Greek thought and science. As such, he was able to access a great deal more information than had, for instance, the Pre-Socratics. As Hull notes,[26] we may quibble over whether Aristotle was a scientist, but that he behaved like one is not at issue. A large part of Aristotle's poor reputation results from the rhetoric of the Renaissance humanists, who sought to downplay the worth of the leading contemporary source of Scholastic philosophy and theology.

Aristotle wrote several works of a biological nature, the most prominent for our purposes being *On the Parts of Animals*, *The History of Animals*, and *On the Generation of Animals*, which came in the later medieval period, after being translated into Latin, to be known as the *Liber Animalium*.

ARISTOTLE ON CLASSIFICATION

In his formal works he employed the logical notions *genos* (genus), *eidos* (species), and *diaphora* (differentia).[27] These were not special biological notions; they were part of his project of a wider classificatory logic, outlined in the *Metaphysics*, the *Categories*, and the *Posterior Analytics*. However, Aristotle was not just an abstract philosopher, as he and his school undertook a number of dissections to establish the facts about many animals, although many of these texts seem not to have survived.[28] It does seem that Aristotle was not undertaking anything we would consider taxonomy, though, preferring a more teleological or finalist approach to classification of living things, and generally classifying when he did under habitat (land dwelling, water dwelling, and air dwelling) rather than morphology.[29]

---

[25] February 22, 1882 [Darwin 1888, 427, Vol. 2].
[26] Hull 1988, 75–77.
[27] Pellegrin 1986.
[28] Lennox 2001, chapter 5.
[29] Ably reviewed in detail, with an attempt to reconcile the biological practice with the views expressed in the more technical works, by Charles 2002, chapters 8 and 9.

He was quick to point out a problem with the simple Platonic method of dichotomous classification, although he did not reject the idea of division as such.[30] Many of the categories used in a Platonic diairetic classification were what Aristotle called "privative" categories—defined in terms of what they were *not*, rather than what they were. He proposed instead a method of the decomposition of broader categories into parts on the basis of how the parts differed, but he did not require that each division had to be a dichotomy, as Plato and the Academicians had. There could be many parts in each category. In the *Posterior Analytics*, he says, using the term *infimae species* for the most specific division of a topic:

> The authors of a hand-book on a subject that is a generic whole should divide the genus into its first *infimae species*—number e.g., into triad and dyad—and then endeavour to seize their definitions by the method we have described .... After that, having established what the category is to which the subaltern genus belongs—quantity or quality, for instance—he should examine the properties 'peculiar' to the species, working through the proximate common differentiae. He should proceed thus because the attributes of the genera compounded of the *infimae species* will be clearly given by the definitions of the species; since the basic element of them all [note: *sc.* genera and species] is the definition, i.e. the simple *infimae species*, and the attributes inhere essentially in the simple *infimae species*, in genera only in virtue of these.[31]

This method came to be known in the Middle Ages as *per genus et differentiam*[32]— by the general type and the particular difference. Something that was differentiable within a *genus* was known to the western tradition as a *species*. In scholastic philosophy, *species* represented a range of things we would now call propositions, sense impressions, and so forth, and this usage persisted through Leibniz to the philosophical discussions of Kant, Mill, and Russell. However, initially a species was merely something that could be differentiated out of a more general concept or term.

He extends his discussion in the *Metaphysics* by asking what it is that makes *man* (the logical species) a unity instead of a "plurality" such as *animal* and *two-footed*.[33] He argues that the differentiae of a genus can lead to its including species which are polar opposites in their specific differences, but, with respect to the genus itself, there is no differentiation. This makes sense only if each genus is divided further in terms other than the predicates that define the genus. Further, he rejects Plato's dichotomous approach, saying "... it makes in general no difference whether the specification is by many or few differentia, neither does it whether that specification is by a few or by just two...." Therefore, he asks whether the genus exists "over

---

[30] Pellegrin 1986.

[31] 96b15–24 [McKeon 1941], using G. R. G. Mure's older translation for emphasis of the terms *genus* and *species*. A more recent translation, Barnes 1984, is less clear, using terms such as "atomic," "primitive," and "simple" in preference to "infimae species." As an interpretation of the intent of Aristotle himself, the Barnes edition is probably better (although I have my doubts—it imports modern logical notions and so obscures Aristotle's preoccupations; but then, the older translations accreted much Scholastic and later connotations), but Aristotle has been mediated to the modern (post-medieval) tradition via the sorts of interpretation embedded in the Mure translation.

[32] "Through genus and a difference" [Blackburn 2008]. More fully, as *per genus proximum et differentiam specificam*—"through nearest class and specific difference" [Joseph 1916, 112].

[33] Book Z, chapter 12 (1037b–1038a).

and above the specific forms constitutive of it," and answers that it does not matter, because the definition is "the account derived just from the differentiae." In the end, we reach the "form and the substance," the last differentia, unless we use accidental features, in which case we will find that we have an incorrect division as evidenced by the differentiae being "equinumerous with the cuts." In short, a species is the form and the substance of the immediately preceding genus, when we reach the last differentia.

It is sometimes held that for Aristotle all classification was in terms of absolute definition and essence, and often this is true. But he did allow for an excess or deficiency of some organs or properties of organisms in *The Parts of Animals*,[34] and in the *Rhetoric*, one of the main topics or lines of argument was "the more and the less."[35] In the eighth book of the *Metaphysics*, he even says

> … it is important to understand the kind of differentiation, given that they are the principles of the beings of things. Some things, that is, are marked out by being more or, conversely, less F, by being dense, say, or rare and so forth, which are all instances of the surfeit/deficiency differentiation.[36]

Similarly, in the *History of Animals* he discusses differences being more or less the same property in respect of the genus, and "in short, in the way of excess or defect," "for the more and the less may be represented as excess and defect."[37] Species of birds and fish, for instance, may not properly instantiate their genus. The more and the less, however, refer to aspects of the *eidos* that can vary over a range, and which can be important for the organism's life.[38] The range is precise, and forms part of the differentia of that species.

In *The Parts of Animals*, Aristotle discusses why privative terms are not proper to classification.[39] He says that you cannot properly further divide a privative term, and that precludes it from being a generic (that is, general) term. Worse, a privative classification can include the same group under contradictory terms. Classifications, according to Aristotle, must say something direct and clear. Dividing the world into things that are, and are not, describable by some predicate, is at best only a partial classification, and the taxa that result are not good divisions of the broader genus. Some things that fall under a privative term must be species, but there is no genus out of which those species can be differentiated, and so the privative "genus" is illusory. Plato's mistake was, he thought, to assume that classifications must be made

---

[34] For example, 646b20.

[35] 1358a21.

[36] Book Iota, 1042b, Lawson-Tancred translation [Aristotle 1998]. In other translations, such as W. D. Ross' [McKeon 1941], the phrase is "the kinds of differentiae … characterized by the more and the less." See Franklin 1986 for a full discussion of Aristotle on variation within species.

[37] 486a–486b. D'Arcy Thompson translation [Barnes 1984].

[38] Lennox 2001, 178.

[39] Book I, chapter 3, 642b–643a.

dichotomously.[40] Aristotle was saying, as we would now describe it, that classification must always be cast in terms of proper sets and subsets. Partial inclusion is not
legitimate in a good classification, in effect because it does not make proper sets.

It is traditionally held that Aristotle was inconsistent in the way he used *genos*
and *eidos* between the logical and the biological writings[41] but Pellegrin, Balme, and
Lennox have shown otherwise.[42] In part, the problem arises because the common
view rests mainly on the post-medieval concepts that, we shall see, are derived out
of the later neo-Platonic revision of Aristotelian logic. Aristotle is only inconsistent
if understood to use the technical term *eidos* in the same way in the logical works as
he does in the context of biology, and he does not.

Phillip Sloan observed that the first edition of this book failed to distinguish
between three senses of *eidos*:

> the use of *eidos* as predicate of substance, as we encounter it in *Categories*; *eidos* as
> form that is the universal grasped in perception and referred to in language; and *eidos*
> as the individuated form-in-individual that in a living being is identified with *psuche*
> as the principle of life and vital action.[43]

He is correct, and I thank him for the elucidation, but I note that this strengthens
rather than diminishes my interpretation. The so-called "Aristotelian essentialism"
interpretation of biological taxonomy transports the *Categories* sense into the natural historical sense.

Pellegrin says that Aristotle was not aiming to produce a biological taxonomy in
History of Animals. Instead, he was producing general classifications, and animals
happened to be one domain in which he applied that method. What Aristotle treats as
genera and species do not answer directly to the modern, post-Linnaean, conceptions
of species, although this has sometimes been the default interpretation. We have seen
that for him a species is a group that is formed by differentiating a prior group formed
by a generic concept. Genera have essential predicates (or definitions), and so do species. Infimae species happen to be indivisible except into individual things, that is all.
In this respect, biological species are no different from any other kind. Pellegrin says,

> Aristotle thus conveys by the term *genos* the transmissible type that in our eyes char-
> acterises the species, and by *eidos* the model that is actually transmitted in generation.
> It would be necessary for these two terms to converge and become superimposed for
> the modern concept of a species to be born. For Aristotle, the [biological] species did
> not yet exist.[44]

---

[40] Balme says that in *Parts of Animals* 1.2–4 [Balme 1987, 19]:

> ... Aristotle concludes that diairesis can grasp the form if it is not used dichotomously as Plato used
> it but by applying all the relevant differentiae to the genus simultaneously; after that he explains
> the ways in which animal features should be compared so as to set up differentiae—by analogy
> between kinds (genē), by the more-or-less as between forms comprised within a kind (eidē).

[41] For example, Mandelbaum 1957.

[42] See Wiener 2015 for an argument that Aristotle is in fact doing taxonomy in his biological writings.

[43] Sloan 2013. See also the footnote to Peck's translation of *Generation of Animals* 731b30–33 [Aristotle
1942, 130–131n].

[44] Pellegrin 1986, 110.

Pellegrin argues that Aristotle's disagreement with Plato is not that classification by division is wrong, but that one should not proceed by dichotomous division into groups that are defined by a differentia and its contrary. He notes,

> Although Aristotle condemns dichotomy as used in the Academy, and does so in all the relevant texts, he does not reject division. For Aristotle, he says, the [infimae] species is a group, and is merely the least divisible group, or, in other words, the least inclusive classification. However, in the later logical tradition from which Ray and Linnaeus and others borrowed their systematic ranks, the smallest *group* category was the genus.[45] *Homo* was a species for them because all present men (women being included) were descendants of a single pair, Adam and Eve. Linnaeus' binomial nomenclature of the genus name and species epithet, as in *Homo sapiens*, was intended, like a personal name, to give the group (the surname, as it were) and a uniquely referring name of that individual parental pair (and their descendants). This distinction must be borne in mind to prevent eisegesis.[46]

Aristotle used the term "universal" (*katholou*, καθόλου[47]) to mean any term (predicate) that covered several things. In *On Interpretation*, he says:

> Some things are universal, others individual. By the term "universal" I mean that which is of such a nature as to be predicated of many subjects, by "individual" that which is not thus predicated. Thus "man" is a universal, "Callias" an individual.[48]

A universal need not therefore be something that is literally universally true, but a general term that is predicated of many things. It is important to bear this in mind when we discuss the medieval "universals debate" and the question of the individuality of species. A universal is any term that covers two or more subjects.

Aristotle's biological works and those of his student Theophrastus also strongly influenced the later development of biology, and especially early botany, but one particular doctrine was most influential—the doctrine of the souls of living things in *De Anima*. Soul (*psyche*) here means something like "motivating force"—plants have only a "nutritive soul,"[49] animals also have a "sensitive soul" capable of sensation,[50] and hence they must have an "appetitive soul," as do all organisms capable of sensation, because they must have some desire.[51] Some animals have in addition a "locomotory soul"[52] and one of those, Man, alone also has the power of rational thought, or a "rational soul."[53] Soul is the source of movement and growth, and it is the final

---

[45] A point made by Hugo De Vries in 1904 [published in de Vries 1912], intriguingly, although one should dispute his detailed claims about the genus being natural and not the species in Linnaeus' works.

[46] Pellegrin 1986, 48.

[47] The term is a portmanteau of κατά (according to) and ὅλος (whole). It could be translated here as "on the whole."

[48] 17a–17b, Edghill translation [Barnes 1984]. See also *Prior Analytics*, 24a.

[49] 413a21–35, 414a30.

[50] 413b4–9, 414b3.

[51] 413b21–24, 414b1–15.

[52] 413b3, 414b17, 415a7.

[53] 415a7–12.

cause of those faculties,[54] that for which things are generated. This forms the foundation for the later Great Chain of Being tradition, and for the initial classifications and explanations of Cesalpino and Bauhin.[55]

Aristotle has thus developed what we might think of as "science by definition." If one can define the proper attributes of a thing, and what divides it from other universals, then one has understood and explained it. All those who attempt this method of science prior to the seventeenth century are in effect practicing Aristotle's method. As a result, the terms *genos* (*genus*) and *eidos* (*species*) come to have several senses based on the notion of differentiating out the special classes from the general. This has given some historians of biology, particularly those writing from a modern scientific perspective, considerable trouble.

Two issues arise. One is whether Aristotle thought that species were immutable, and the other is whether Aristotle was an essentialist with respect to species. Lennox holds that Aristotle did not depend on species being eternal in the biological sense, that they were not fixed.[56] They could come into being and pass away. But the *kinds* (*genē*) do not come into being and pass away.[57] They are formed in virtue of having the differentia that distinguish them from their superordinate genus. To this extent, then, he is an essentialist and a fixist of kinds. But I do not think it is correct to saddle him with being a species (biological species) fixist. In a telling passage in *Generation of Animals* II.1, he notes that while individuals cannot be eternal, as they are subject to generation and corruption, he admits that if anything can be eternal on earth, it is the *eidos* of men and animals.

> These are the causes on account of which generation of animals takes place, because since the nature of a class [*genous*] of this sort is unable to be eternal, that which comes into being is eternal in the manner that is open to it. Now it is impossible for it to be so *numerically*, since the "being" of things is to be found in the particular, and if it really were so, then it would be eternal; it is, however, open to it to be so *specifically*.[58]

Note that he admits only the *possibility* rather than the necessity. Later, we will consider the rise of the "Aristotelian essentialism" story, but a key aspect of the ascription of taxic (that is, biological) essentialism to him rests on his distinction between the proper and accidental properties of things.

THE TRADITION OF THE TOPICS

In his *Categories*, sometimes called the *Topics*, Aristotle reduced the number of things that might be said of any thing to ten "topics" (τόποι, *topoi*, "places"):[59]

---

[54] 415b9–27.
[55] Sachs 1890.
[56] Lennox 1987, 2001.
[57] Lennox 2001, 154.
[58] Aristotle 1942, 731b730–733@131. Note that the term translated as *specifically* is *ginomenon* (as produced, or as born), which is related to *genos*, not *eidei* (see footnote b of that page).
[59] 1b25–2a4.

Of things said without any combination, each signifies either substance or quantity or qualification or a relative or where or when or being-in-a-position or having or doing or being affected.[60]

The topics, or as they are called in the later logics, *heads of predicables*, are important in the later history of the logical notion of species and genus, particularly in the Middle Ages. For Aristotle, they are (as the Middle Ages philosophers called them in Latin) *praedicamenta* (those which are said). In short, the topics are the things that can be said of some thing or things. When speaking of a thing, the topics are:

1. Substance (οὐσία, *ousia*, sometimes mistranslated as essence)—that which makes something a particular, and not general, thing.
2. Quantity (ποσόν, *poson*)—how much of the thing, in space or measure.
3. Quality (ποιόν, *poion*)—of what kind or quality the thing is.
4. Relation (πρός τι, *pros ti*, toward something)—the relation in which the thing stands to something else.
5. Place (ποῦ, *pou*, where)—the thing's position in its environment.
6. Time (πότε, *pote*, when)—the time of the thing.
7. Being-in-a-position (κεῖσθαι, *keisthai*, to lie)—the orientation of the thing.
8. Having or state (ἔχειν, *echein*, to have or be)—the properties of the thing.
9. Action (ποιεῖν, *poiein*, to make or do)—the changes the thing causes.
10. Being affected (πάσχειν, *paschein*, to suffer or undergo)—the changes made to the thing.

Aristotle ties the topics into the logic of classification. He says:

Whenever one thing is predicated of another as of a subject, all things said of what is predicated will be said of the subject also. For example, man is predicated of the individual man, and animal of man; so animal will be predicated of the individual man also—for the individual man is both a man and an animal. The differentiae of genera which are different and not subordinate one to the other are themselves different in kind. For example, animal and knowledge: footed, winged, aquatic, two-footed, are differentiae of animal, but none of these is a differentia of knowledge; one sort of knowledge does not differ from another by being two-footed. However, there is nothing to prevent genera subordinate one to the other from having the same differentiae. For the higher are predicated of the genera below them, so that all differentiae of the predicated genus will be differentiae of the subject also.[61]

Aristotle defines four "predicables" (that which is predicated of things): definition (*horos*), property (*idion*), genus, and accident (*symbebekos*).[62] Species (*eidos*) is not, in Aristotle's list, a predicable, because it is only true of individuals although it was added to the "heads of predicables" in the fourteenth-century debates on logic.

---

[60] Ackrill translation [Aristotle 1995].
[61] 1b10–1b24.
[62] 101b16–25.

It is important, therefore, to realize that Aristotle is concerned with the application of logical categories to words, and not to natural kinds in the later sense.

ARISTOTLE'S NATURAL HISTORY OF SPECIES

In *History of Animals*, Aristotle discusses what will produce variety in animal life—the main cause is locality, including climate, and hybridization.[63] In Africa ("Libya"[64]),

> animals of diverse species meet, on account of the rainless climate, at watering-places, and there pair together; and such pairs will breed if they be nearly of the same size and have periods of gestation of the same length.[65]

Zirkle says that Aristotle insisted that such hybrids were fertile[66]—but it is unclear from the context exactly how different these diverse species were supposed to be. Moreover, he typically does not use the formal technical terms *eidos* or *genos* when speaking of living kinds (except when he uses them as examples in formal contexts like the topics).

Aristotle goes on to say,

> Elsewhere also offspring are born to heterogeneous pairs; thus in Cyrene the wolf and the bitch will couple and breed; and the Laconian hound is a cross between the fox and the dog.

He then reports that "they" say that the Indian dog is a bitch-tiger third generation hybrid. Even if we allow that for Aristotle "species" here does not mean a strict biological species, and that travelers' tales are misleading him, he is not so obviously a mutabilist, either. However, he certainly restricts hybridization here to similar-sized animals.

The role of hybridization in the generation of new kinds of animals and plants is a continuing theme, especially in the Christian tradition, as we shall see.

Lennox notes that there is a great gap in practical biology between Aristotle and Theophrastus, and Albertus Magnus' *De Animalibus* in the twelfth century CE.[67] However, there continued to be philosophical treatments of the idea of division in living things throughout the remainder of the classical period. Aristotle's student Theophrastus offered one such.

THEOPHRASTUS, AND NATURAL KINDS

Theophrastus (370–c. 285 BCE) in his botanical work *Enquiry into Plants* (*Peri phutōn historias*) applied Aristotle's notions of biology and classification to the

---

[63] Book VIII, chapter 28 (605b22–607a7).
[64] In classical literature, Libya was a synecdoche for Africa west of Egypt.
[65] Barnes 1984.
[66] Zirkle 1959, 640.
[67] Lennox 2001, chapter 5.

wealth of new specimens being returned to the Greek Empire of Alexander from the new conquests. He is the first known author to classify plants in an overall system, and may thus be thought of as the first botanical systematist. His translator, Sir Arthur Hort (one of the translators of the first critical edition of the Greek New Testament), notes that Theophrastus' botanical work is guided by a

> constant implied question ... "what is its *difference*?", "What is its essential nature?", viz. "What are the characteristic features in virtue of which a plant may be distinguished from other plants, and which make up its own 'nature' or essential character?"[68]

In short, Theophrastus is applying Aristotle's notion of classifying into differentia. He may not have studied all the plants himself. Some descriptions, such as of the cotton plant, banyan tree, cinnamon bush, and so forth, are taken from reports by some of Alexander's followers who were trained observers. Still, what is most significant about Theophrastus in this context is that his method of classification was an attempt to discover the underlying essence of the kinds of plants based on evidence. His differentiations are almost entirely based upon anatomical features of the organisms, although he also allows the habitat (locality) to be employed; for instance, between water dwellers and land dwellers. He says,

> Now it appears that by a "part," [*meros*] seeing that it is something which belongs to the plant's characteristic nature, we mean something which is permanent either absolutely or when it has appeared (like those parts of animals which remain for a time undeveloped) permanent...[69]

But this is unsatisfactory, because some crucial diagnostic characters, like flowers, inflorescences, leaves, fruit, and so forth, and the shape of the shoot itself, are all temporary or seasonal. He concludes that the characteristics, or "parts," should be chosen from those that are most directly concerned with reproduction.[70] After setting out the parts he will use (branch, twig, root) he says, "Now, since our study becomes more illuminating if we distinguish different kinds (*eidē*), it is well to follow this plan where it is possible. The first and most important classes, those which comprise all or nearly all plants, are tree, shrub, under-shrub, herb."[71] Elsewhere he says

> Now let us speak of the wild kinds. Of these there are several classes (*eidē*) which we must distinguish (*diairein*) by the characteristics of each sub-division as well as by those of each class (*genos*) taken as a whole (*tois holois eidesi*).[72]

In the case of the dwarf-palm, he distinguishes it as a different *genos* from the ordinary palm.[73]

---

[68] Theophrastus 1916, xviii.
[69] I.1.ii.
[70] I.2.2.
[71] I.3.1.
[72] VI.1.2.
[73] II.6.10.

Hort notes in a footnote that Theophrastus

> uses *eidos* and *genos* almost indiscriminately. Here *tōn holōn genōn* means the same
> as *tois holois eidesi*; and below *genōn* and *eidōn* both refer to the smaller divisions
> called *merē* above.

Theophrastus is attempting, in a non-systematic and perhaps less careful manner,
to apply Aristotle's philosophy of classification to botany, but he uses these terms in
a more vernacular sense of "form" and "race" or "stock." In any case, his theory of
classification, if it can be called that, is morphological and seeks in this the "nature"
of the plants. *Species* as such has no special meaning for him here. To back up this
interpretation, note that in his *Metaphysics*,[74] he offers the same view as Aristotle:

> Knowing, then, [does] not [occur] without some difference.[75] For both, if [things] are
> other than each other, there is some difference, and, in the case of universals where the
> [things that fall] under the universals[76] are more than one, these too differ of necessity,
> be the universals genera or species.[77]
>
> In general, though, to perceive simultaneously the identical in many is [the task] of
> knowledge,[78] whether, in fact, it is said [of them] in common and universally or in some
> unique way with regard to each, as, for example, in numbers and lines, and animals
> and plants; complete is the one [consisting] of both. There are, however, some [knowl-
> edges] whose end is the universal (for therein is their cause), while of others it is the
> particular, those in respect of which division [can proceed] down to the individuals, as
> in the case of [things] done and [things] made: for this is how their actualized state is.[79]

Consequently, marking differences is how science is done.

Zirkle says that Theophrastus spent the entirety of Book II showing how species
could change from one to another.[80] Given the informal way in which Theophrastus
talks about species, we might consider this doubtful. In fact, what he talks about
there is *generation* (including spontaneous generation), "but growth from seed or
root would seem most natural,"[81] and especially vegetative propagation. Zirkle
appears to be referring to the "degeneration" of some cultivars into wild forms.[82] The
oak sometimes "deteriorates from seed,"[83] for example, so that the child is unlike the
parent. Sometimes wild trees like pomegranates, figs, and olives can spontaneously
change into a domestic form, and vice versa, and so on.[84] It is a very long stretch to

---

[74] Gutas 2010.
[75] Greek: *diaphoras*.
[76] Greek: *katholou*.
[77] 8b16–20.
[78] Greek: *epistēmē* (ἐπιστήμη), which is sometimes translated as "science" but which means more exactly "skill" or "knowledge."
[79] 8b24–9a4. "Actualized state" translates *energeia* (ἐνέργεια), "activity" or "operation."
[80] Zirkle 1959, 639.f
[81] II.1.1; see also Book I of *De causis plantarum*, 1.2–3, 5.1–2, but Theophrastus notes that this is some-times from "unnoticed seeds" in 5.3 and 5.4.
[82] For example, II.2.5.
[83] II.2.6, cf. *De causis plantarum*, I.9.1.
[84] II.3.1.

call these changes of species. In modern terms, these would either be genetic expression of latent varieties or somatic mutation and development.

One case that is ambiguous is the change from one-seeded wheat and rice-wheat in the third generation when they are bruised (in seed form?) before they are sown, but then he says "[t]hese changes appear to be due to change of soil and cultivation," and seasonal variation.[85] He does say, as Zirkle noted, that "the water-snake changes into a viper, if the marshes dry up" but it is unclear whether he thinks this is a species change in our sense. Instead, "when a change of the required character occurs in the climatic conditions, a spontaneous change in the way of growth ensues."[86] This is hardly a claim of mutability of species, either. The remainder of the book consists of a discussion of transplanting, grafting, watering, and planting. He does note that fruit trees are male and female.[87] Another case is the widely held view that gall wasps come from the seed of the wild fig, which causes their fruit to drop,[88] and he also follows Aristotle on the spontaneous generation of animals when the earth is warmed and "qualitatively altered" by the sun.[89] These are all that I can find that match Zirkle's claim. It will pay to be skeptical of the oft-repeated claim that species were *always* held to be changeable before the seventeenth century.

EPICUREANISM AND THE GENERATIVE CONCEPTION

The Aristotelian and Platonic traditions were not the only ones in the classical period that dealt with species. The atomists, and in particular the Epicurean tradition, also had an account of why forms are as they are. Epicurus' (341–270 BCE) own writings are largely lost, especially his *On Nature*. However, we have an account of Epicurean doctrines in the work *On the Nature of Things* by Lucretius (99 BCE–c. 55 BCE), a Roman disciple of Epicurus. Lucretius tied specific natures of things to the ways in which they came to be:

> If things could be created out of nothing, any kind of things could be produced from any source. In the first place, men could spring from the sea, squamous fish from the ground, and birds could be hatched from the sky; cattle and other farm animals, and every kind of wild beast, would bear young of unpredictable species, and would make their home in cultivated and barren parts without discrimination. Moreover, the same fruits would not invariably grow on the same trees, but would change: any tree could bear any fruit. Seeing that there would be no elements with the capacity to generate each kind of thing, how could creatures constantly have a fixed mother? But, as it is, because all are formed from fixed seeds, each is born and issues out into the shores of light only from a source where the right ultimate particles exist. And this explains why all things cannot be produced from all things: any given thing possesses a distinct creative capacity.[90]

---

[85] II.3.4.

[86] II.4.4.

[87] II.6.6.

[88] II.8.1–2.

[89] *De causis plantarum* I.5.5.

[90] Book I. 155–191 [Lucretius 1969, 38].

It is commonly understood that Lucretius gives a more or less faithful exposition in Latin of Epicurus' ideas expressed in Greek some two centuries earlier, in the period shortly after Aristotle, in the fourth century BCE.[91] This being so, we can suppose that something resembling the biological species concept existed by the fourth century BCE. The Epicurean view of species (which is not restricted to biological species—like Aristotle's, it applies to elemental forms of all things, but is here illustrated in terms of living things) relies upon the potential nature of the composite parts of things, a kind of *generative conception of species*. He goes on to say that things grow at the right season and are able to live because only then are the right "ultimate particles" (i.e., atoms) available to promote growth. Otherwise, everything could happen, such as children and trees maturing in an instant:

> But it is evident that none of these things happens, since in every case growth is a gradual process, as one would expect, from a fixed seed, and, as things grow, they preserve their specific character; so you may be sure that each thing increases its bulk and derives its sustenance from its own special substance.[92]

We shall see this generative notion of species being struck upon repeatedly in the history of the concept both before and after the term *species* attains a technical sense in biology (for example in both Cusa before it and Buffon after). It is of interest that Epicurus appears to reject Aristotle's acceptance of spontaneous generation,[93] given the role spontaneous generation plays in the later history of living species.

Lucretius further expounds the nature of species in the generalized sense of classification of all things in terms of the natures of the atoms that comprise those things, in I. 584–598. Again, he appeals to the natures *in potentia* of the constituents as determining the limits of a species.

> Furthermore, since in the case of each species,[94] a fixed limit of growth and the tenure of life has been established, and since the powers of each have been defined by solemn decree in accordance with the ordinances of nature, and since, so far from any species being susceptible of variation, each is so constant that from generation to generation all the variegated birds display on their bodies the distinctive markings of their kind, it is evident that their bodies must consist of unchanging substance. For, if the primary elements of things [i.e., atoms] could be overpowered and changed by any means, it would be impossible to determine what can arise and what cannot, and again by law each thing has its scope restricted and its deeply implanted boundary-stone; and it would be equally impossible for the generations within each species to conform so consistently to the nature, habits, mode of life, and movements of their parents.[95]

> ... every species that you see breathing the breath of life has been protected and preserved from the beginning of its existence either by cunning or by courage or by speed.[96]

---

[91] Sedley 2007.
[92] *Loc. cit.*
[93] Lennox 2001, chapter 10.
[94] Note that the Latin does not use the word *species*. Instead the word *saecla* (generations, races) is used by Lucretius.
[95] Lucretius 1969, 49.
[96] Lucretius 1969, 191.

Intriguingly, and [in]famously, Lucretius[97] and the Epicureans have an "evo-lutionary myth" of the origins of living species, and in it they suppose that these generative natures were not fixed in the initial period of life. The mixtures of the elemental particles were random, and so all kinds of organisms and monsters were born. Eventually, only those that could propagate remained in existence, and the others died out. It is sometimes held that Lucretius and the Epicureans therefore held a natural selection view of adaptation,[98] but in fact they suppose that the species are as they originally were formed by chance, and are thereafter kept to the limits of their generative potential by the elimination of the unfit. This is not selection as Darwin and Wallace proposed it—there is no variation except in the different but unchanging natures of the characters of the particles that by chance form the species themselves, not within the species.

The Epicureans therefore differed from Aristotle, who held that species were forms that are imposed upon the substance of things, instead holding that species are forms generated by the natures of their substances. For Aristotle, material sub-stance is malleable. For Epicureans, it is deterministic of the nature of the things it comprises. Of course, Aristotle, too, held that the four elements he proposed in the *Physics* contribute through the "material cause" to the nature of the objects, but he also allowed for formal, efficient, and final causes. Epicurus and his disciples seem not to allow for any determination of natures other than by the material atoms.

This explains a comment made in Boëthius' *Second Commentary* on Porphyry's *Introduction* (*Isagoge*) to Aristotle's *Categories* some four centuries later, and which is crucial to the transmission of the species problem to the medieval universals debate, and thence to the modern era:

> It is clear … that this happened to him [Epicurus], and to others, because they thought, through inexperience in logical argument, that everything they comprehended in rea-soning occurred also in things themselves. This is surely a great error; for in reasoning it is not as in numbers. For in numbers whatever has come out in computing the dig-its correctly, must without doubt also eventuate in the things themselves, so that if by calculation there should happen to be a hundred, there must also be a hundred things subject to that number. But this does not hold equally in argumentation; nor, in fact, is everything which the evolution of words may have discovered held fixed in nature too.[99]

Boëthius is complaining that the Epicureans and the "others" (other atomists, that is) have presumed that because they have been able to construct a coherent account, that what is said must be true of the things being spoken of. Aristotelians, and the neo-Platonists that followed, held that the physics of Aristotle was based upon observable features of the world, while Epicurus' atoms are mere speculations, and hence so are the things that depend upon them for their natures, such as species. This is somewhat ironic, given the clearly theoretical nature of the quintessence in Aristotle's cosmology, and even more so given the merely logical role that *genus* and *species* play in Aristotle's categorical logic. Aristotelian essence, in effect, represents

---

[97] 5. 837–877.
[98] The classical locus being Osborn 1894.
[99] Second commentary on Porphyry's *Isagoge*, Book I, section 2 [McKeon 1929, 73].

exactly the kind of reification to which Boëthius, who is defending Aristotle, objects. Although Aristotle's essence concept was not then considered to be an inappropriately abstracted notion when Boëthius wrote,[100] the issue had been raised by Porphyry himself as to whether species exist merely as abstract mental objects. The Epicureans were not well regarded during the period of Christian domination.[101] The tradition was held to be an impious and immoral philosophy, by Christian as well as by Jewish thinkers of the post-classical era. The direct influence of Epicurean generative conceptions is thus likely to be sparse before the Enlightenment.

THE HERMETIC TRADITION: SPECIES COME FROM LIKE

Even less immediately influential were the Hermetic writers, although they became significant during the Renaissance. Around 100CE, a text was written under the pseudonym of Hermes Trismegistus, today known as *Asclepius* I.[102] This tradition in part is the direct descendent of Plato and the Academy, with a strong veneer of mysticism and *gnosis* (sacred knowledge). In this text, nature is the matter that nourishes the forms (*species*) that are imposed on it by God,[103] and in the manner of Plato's souls in *The Republic*, different kinds (*species*) are realized according to the source. The god-kind produces gods, the daemon-kind produces daemons,[104] and the man-kind produces men.[105] Unlike Plato, however, the author allows that individuals might partake in many kinds "... though all individuals exactly resemble the type of their kind, yet individuals of each kind intermingle [*miscentur*] with all other kinds." A later scribe has interposed the comment that organic bodies receive their kinds by the fiat of the gods, and individual things receive form by the ministration of daemons.

In *Asclepius* II (c. 150–270 CE) by a different author, matter is considered "ungenerated" but to have a "generative power" and to be creative,[106] reiterated also in Book II (c. 270 CE). The Hermetic tradition is likely influenced strongly by the Stoics. The subsequent influence of the Hermetics was strong in the alchemical and mystery religions, and through them to the gnostic traditions of European thought, including the Kabbalah.[107] It is therefore a minor source for the later Great Chain of Being.

THE LATE CLASSICAL TRADITION OF NATURAL HISTORY

The classical period of biology divides mainly into two—the Peripatetic period deriving from Aristotle and Theophrastus, and the practical or encyclopedic period instituted by Pliny the Elder and Dioscorides, in which botanical and zoological

---

[100] Which Whitehead 1938 later called the *Fallacy of Misplaced Concreteness*.
[101] Clark et al. 2007. Nor the Jewish tradition: A heretic in the *Mishnah* (*Sanh.* 10:1) is called *apikoros*, which derives from Epicurus' name, presumably from the popular misconception that Epicurus denied the gods existed, or perhaps from Josephus' mention in the *Antiquities*, x.11.7 [Deutsch 1901–1906].
[102] Scott 1924.
[103] 3c.
[104] *Daemon* (Greek δαίμων) = spirits, not merely demons.
[105] 4.
[106] 15.
[107] Lelli 2007.

information was assembled for practical use. The latter tradition developed into the herbalists of the Middle Ages, in which medical information was the goal and rationale.[108] A later third tradition was the so-called etymological tradition that began with the *Physiologus* (c. 350)[109] and ended with Isidore of Spain's *Etymologiae* (*d*636 CE).[110]

Pliny the Elder (23–79 CE), who famously perished while observing from a boat the eruption of Vesuvius that buried Pompeii, wrote an encyclopedic account of all animals, birds, trees, and so on known to Roman science, called the *Historia naturalis* (*Natural history*). It was the standard reference work for some 1500 years, influencing the later bestiary tradition[111] and being the foundation for almost all medieval botanical works. In it, he refers to kinds of organisms almost always as "genera;" for instance, he speaks of two "kinds" (*genera*) of camel, the Bactrian and the Arabian,[112] or of kinds of lions,[113] or six kinds of eagles.[114] There is no emphasis placed on reproduction in Pliny, and most of the descriptions are morphological and behavioral. Likewise, the herbalist tradition that began with Dioscorides' (c. 40–90) *De materia medica* assumes that there are kinds and sorts, but makes no clear distinction between them. Even so, modern classifications can be fairly clearly mapped onto the species mentioned.[115] Pliny based much of his material on Theophrastus and Dioscorides.[116]

THE NEO-PLATONISTS: SPECIES AS A PREDICABLE

Aristotelian categories strongly influenced the neo-Platonists, who in turn influenced the medieval scholastics from whom Linnaeus drew his ranking categories. A clear example is the fourth or fifth century writer Martianus (or Felix) Capella. Martianus, who was by tradition a farmer in fifth-century CE Africa, but more probably was a wealthy landowner, did not explicitly deal with the classification of living things, and effectively repeated the abstract position of Aristotle's chapter 13 of the *Categories*. His text, whimsically entitled *The Marriage of Philology and Mercury*, was used as a major textbook of the medieval educational program that came to be known as the Quadrivium and the Trivium, for over a thousand years, surely a record for a purpose-written instructional textbook (excluding, perhaps, Euclid). Martianus wrote of a genus being a collection of forms under one name, and species are "man, horse, lion." He wrote, "we also call species forms" which have a "name and definition. ... The term and definition of genus are thus determined."[117]

In the neo-Platonic interpretation of Aristotle, mediated to medieval Christianity by Martianus, and Porphyry in the *Isagoge*, via Boëthius in the *Commentaries*,

---

[108] Stannard 1968, reprinted in Stannard et al. 1999.

[109] Diekstra 1985.

[110] Isidore of Seville 2005. Cf. Wood 2013.

[111] Pliny the Elder 1906, 1940–63.

[112] II.xxx. The Perseus edition of the Latin text of Mayhoff numbers chapters differently from the English translation of Bostock. The Latin numbering is used here.

[113] VIII.xx.

[114] X.iii.

[115] Dioscorides 1959, 663–679.

[116] Nordenskiöld 1929, 191.

[117] From the section "On rhetoric" in Book I, translation by H. E. Wedeck, from Runes 1962, 211–212.

a species was a member of a broader group—a genus—that was formed by a predicate. There was no necessity for any object to be a member of a single genus, and a species might be, with respect to some other predicate a genus in its own right. In short, species were predicate-relative individuals. They were whatever was differentiable out of the genus. This gave rise to Porphyry's dichotomous notion of classification.

Porphyry of Tyre (c. 234–c. 305 CE), a student of Plotinus (c. 204/5–270 CE), syncretized Plato's dichotomous method with Aristotle's logical division of predicables. Aristotle's conception of the *infimae species* was primarily a matter of logical analysis. Porphyry combined this and Plato's method of classification in the *Sophist* to produce what became known as *Porphyry's comb* or *tree* (*Arbor Porphyriana*) in the later Middle Ages (Figure 1.2). Porphyry treated species slightly differently than Aristotle. In place of the four predicables of Aristotle, Porphyry had five: *genus*, *difference*, *species*, *property*, and *accident*, replacing *definition* with *species*.[118]

Boëthius reported that Porphyry had raised the issue of whether species and genera exist only in the mind:

> As for genera and species, [Porphyry] says, I shall decline for the present to say (1) whether they subsist or are posited in bare [acts of] understanding only, (2) whether, if they subsist, they are corporeal or incorporeal, and (3) whether [they are] separated from sensibles or posited in sensibles and agree with them. For that is a most noble matter, and requires a longer investigation.[119]

This began what we now know as the "Universals" debate[120] and led, fairly directly, to the position that came to be known as nominalism.

---

[118] Cf. Barnes' commentary §0 in Porphyry 2003, 26–32, also see Joseph 1916, chapter IV. As noted, Joseph's book has been implicated in the adoption by Cain and Hull of the notion that pre-evolutionary species are timeless and static entities defined by their essences. It should be noted that Joseph is presenting a formal account of the pre-set theoretic logics from Aristotle until his day, and he does, in several places (pp 53n, 92–96), note the differences between Aristotle and Porphyry, but it would not have been obvious to anyone not familiar with the technical aspects of the medieval commentaries on Porphyry. Joseph's book is actually a very good late example of the treatment of logic in the Aristotelian tradition, surviving into the post-Darwinian era but aware of it (pp 473–475). Hull 1967, 310–313 cites a passage from the first edition of 1905 in which Joseph notes that the evolution of species is not thought of by biologists in the logical sense, but I cannot find it in either edition. However, Joseph does say this in the first edition:

> But now that the theory of organic evolution has reduced the distinction between varietal and specific difference to one of degree, the task of settling what is the essence of a species becomes theoretically impossible. It is possible to describe a type; but there will be hundreds of characteristics typical of every species [page 82].

[119] The passage is translated from *Anicii Manlii Severini Boethii In Isagogen Porphyrii commenta*, editio 2a, lib. I, ca. 10–11, Samuel Brandt, ed., p. 159 line 3–p. 167 line 20. Translation by Paul Spade, unpublished, used with permission. Barnes' direct translation of Porphyry [Porphyry 2003, 3, lines 10–15] reads:

> For example, about genera and species—whether they subsist, whether they actually depend on bare thoughts alone, whether if they actually subsist they are bodies or incorporeal and whether they are separable or are in perceptible items and subsist about them—these matters I shall decline to discuss, such a subject being very deep and demanding another and a larger investigation.

[120] Klima 2013. For a full discussion, see Aaron 1952.

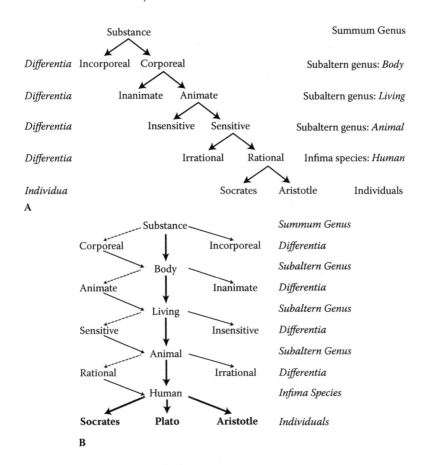

FIGURE 1.2 The Tree of Porphyry, compared with a cladogram. Version **A** has been adapted from Oldroyd[121] to make the logical isomorphism with a cladogram clearer—the traditional form of Porphyry's tree is more like the one Oldroyd presents (**B**).   *(Continued)*

Why did Porphyry even raise this question? It seems to come out of nowhere. Aristotle had no doubt that the *eidos* was real—it was the form of the thing that existed as a material object. Plato had no such doubt either—to him *only* Ideas were real.[122] I suspect that Porphyry was responding to the debates over atomism that bridge the period of Aristotle and the Epicureans, to his day, around 500 years' duration. Plotinus had addressed the arguments of the Gnostics in his *Enneads*, which Porphyry had edited. It is also possible that the topic had indeed been raised by

---

[121] Oldroyd 1983, 29.

[122] And interestingly, until late in the seventeenth century, "Realism" denoted a realism about *ideas*; what we would consider Idealism today [Blackmore 1979, Aaron 1952].

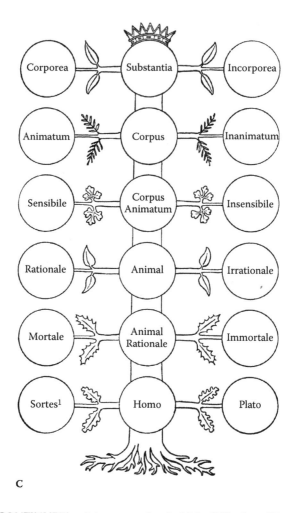

C

FIGURE 1.2 (CONTINUED)   A late example of which, **C**,[123] where "Socrates" has ironically been corrupted as "Sortes." See also the discussion by Barnes,[124] who locates the earliest such tree in the late Middle Ages, not in Porphyry's own text. The terms used are those of the late medieval scholastic tradition.

---

[123]   Baldwin 1901, 714, v2.
[124]   Porphyry 2003, 109f.

the Roman Epicureans, who discussed the nature of sensation extensively, and that the Aristotelians in the person of Porphyry are attempting to defend the essentialist account against atomism.[125]

AUGUSTINE: THE MUTABLE IN GOD'S DESIGN

Augustine of Hippo's (354–430 CE) *Civitas Dei* (*The city of God*) was finished in about 426 CE, and espouses a Christianized neo-Platonism of sorts, which is not surprising, as Augustine lived and worked in the Roman part of Africa from which neo-Platonism sprang. Although not directly interested in matters of natural history, he did assert that

> all forms of mutable things, whereby they are what they are (of what nature soever they be) have their origin from none but Him that is true and unchangeable. Consequently, neither the body of this universe, the figures, qualities, motions, and elements, nor the bodies in them from heaven to earth, either vegetative as trees, or sensitive also as beasts, or reasonable also as men, nor those that need no nutriment but subsist by themselves as the angels, can have being but from Him who has only simple being.[126]

Forms (species) are thus maintained by the direct action of God, rather than any internal or innate quality. Augustine's focus is, as the title suggests, on heaven, and bodies of organisms are of interest only so far as they are relevant to resurrection. This lack of interest in the natural world extends until the late Middle Ages, as we shall see. This chapter was influential upon Peter Lombard (c. 1100–1160/4 CE) in the much-discussed *Four Books of Sentences*[127] where the same points are made.

One text of Augustine's that has been interpreted to mean he held to an "evolutionary" view is in his commentary on the meaning of Genesis:

> Where, then, were they [*plants, when they were created*]? Were they in the earth in the "reasons" or causes from which they would spring, as all things already exist in their seeds before they evolve [*develop—JSW*] in one form or another and grow into their proper kinds in the course of time? ... it appears [*from Scripture—JSW*] ... that the seeds sprang from the crops and trees, and that the crops and trees themselves came forth not from seeds but from the earth.[128]

---

[125] Barnes agrees [Porphyry 2003, 312–317], with more warrant than I have, that the traditional claim of Stoic influence has no basis in fact, and conjectures (pp. 356–358) that there are many "Epicurean touches" in Porphyry. Preus 2002 notes that Plotinus, Porphyry's teacher, had made some passing comments on form (*eidos*) in the *Enneads* (V.9.6) which gives an Epicurean-style generative account of species, in which *logoi* are the generative powers "in the seed" and of every part of an organism. He says

> Some call this power in the seeds 'nature,' which was driven thence from those prior to it, as light from fire, and it turns and enforms the matter, not relying on the help of those much-discussed mechanisms (levers), but by imparting the *logoi*. [Preus' translation, 46]

[126] Augustine 1962, in Book VIII, chapter 6.
[127] Book I, Disc. III, chapter I, McKeon 1929, 190f.
[128] *De Genesi Ad Litteram*, (*The literal meaning of Genesis*) c. 390 CE, Book V, chapter 4 [Augustine 1982, 151f].

However, this is best understood that God created secondary powers that spontaneously generate plants at the right time. This passage does not license an interpretation of Augustine as propounding either fixism or mutablism of species.

BIBLIOGRAPHY

Aaron, Richard Ithamar. 1952. *The Theory of Universals*. Cambridge, UK: Cambridge University Press.

Aristotle. 1942. *Generation of Animals*. Translated by A. L. Peck, *Loeb Classical Library*. London/Cambridge, MA: W. Heinemann/Harvard University Press.

—. 1995. *The Complete Works of Aristotle: The Revised Oxford Translation*, edited by Jonathan Barnes. Bollingen Series; 71:2. Princeton, NJ: Princeton University Press.

—. 1998. *The Metaphysics*. Translated by Hugh Lawson-Tancred. London: Penguin.

Atkins, Richard Kenneth. 2006. Restructuring the sciences: Peirce's categories and his classifications of the sciences. *Transactions of the Charles S. Peirce Society: A Quarterly Journal in American Philosophy* (4):483–500.

Augustine, Saint, Bishop of Hippo. 1962. *The City of God*. Translated by John Healey and R. V. G. Tasker. London: J M Dent.

—. 1982. *The Literal Meaning of Genesis*. Translated by John Hammond Taylor, *Ancient Christian Writers*; no. 41–42. New York: Newman Press.

Baldwin, James Mark, ed. 1901. *Dictionary of Philosophy and Psychology Including Many of the Principal Conceptions of Ethics, Logic, Aesthetics, Philosophy of Religion, Mental Pathology, Anthropology, Biology, Neurology, Physiology, Economics, Political and Social Philosophy, Philology, Physical Science, and Education and Giving a Terminology in English, French, German, and Italian*. 3 vols. New York/London: Macmillan.

Balme, D. M. 1987. The place of biology in Aristotle's philosophy. In *Philosophical Issues in Aristotle's Biology*, edited by Allan Gotthelf and James G. Lennox, 9–20. Cambridge, UK: Cambridge University Press.

Barnes, J., ed. 1984. *The Complete Works of Aristotle*. 2 vols. Vol. 1. Princeton, NJ: Princeton University Press.

Blackburn, Simon. 2008. "per genus et differentiam." In *The Oxford Dictionary of Philosophy*. Oxford: Oxford University Press.

Blackmore, John. 1979. On the inverted use of the terms 'realism' and 'idealism' among scientists and historians of science. *British Journal for the History of Science* 30:125–134.

Brumbaugh, Robert S. 1981. *The Philosophers of Greece*. Albany: State University of New York.

Cain, Arthur J. 1999. Thomas Sydenham, John Ray, and some contemporaries on species. *Archives of Natural History* 24 (1):55–83.

Charles, David. 2002. *Aristotle on Meaning and Essence*. Oxford: Oxford University Press.

Clark, Brett et al. 2007. The critique of intelligent design: Epicurus, Marx, Darwin, and Freud and the materialist defense of science. *Theory and Society* 36 (6):515–546.

Darwin, Francis, ed. 1888. *The Life and Letters of Charles Darwin: Including an Autobiographical Chapter*. 3 vols. London: John Murray.

de Vries, Hugo. 1912. *Species and Varieties: Their Origin by Mutation. Lectures Delivered at the University of California*. 3rd ed. Chicago: Open Court. Original edition, 1904.

Deutsch, Gotthard. 1901–1906. "APIKOROS (pl. APIKORSIM)." In *Jewish Encyclopedia*, edited by Cyrus Adler et al., 665–666. New York: Funk & Wagnalls.

Diekstra, F. N. M. 1985. The Physiologus, the bestiaries and medieval animal lore. *Neophilologus* 69 (1):142–155.

Dioscorides, Pedanius, of Anazarbos. 1959. *The Greek Herbal of Dioscorides: Illustrated by a Byzantine, A.D. 512*. Translated by John Goodyer and Robert William Theodore Gunther. New York: Hafner.

Foucault, Michel. 1970. *The Order of Things: An Archaeology of the Human Sciences*. London: Routledge Classics. Reprint, 2002.

Franklin, James. 1986. Aristotle on species variation. *Philosophy* 61:245–252.

Gutas, Dimitri. 2010. *Theophrastus on First Principles (known as his Metaphysics)*. Boston, Netherlands: Brill.

Hamilton, Edith, and Huntington Cairns, eds. 1961. *The collected dialogues of Plato: Including the Letters, Bollinger Series; no. 71*. Princeton: Princeton University Press.

Hull, David L. 1967. The metaphysics of evolution. *British Journal for the History of Science* 3 (12):309–337.

—. 1988. *Science as a Process: An Evolutionary Account of the Social and Conceptual Development of Science*. Chicago, IL: University of Chicago Press.

Isidore of Seville, Saint. 2005. *Etymologiae: The Etymologies of Isidore of Seville*. Translated by Stephen A. Barney. New York: Cambridge University Press.

Jevons, William Stanley. 1878. *The Principles of Science: A Treatise on Logic and Scientific Method*. 2nd ed. London: Macmillan. Original edition, 1873.

Joseph, Horace William Brindley. 1916. *An Introduction to Logic*. 2nd ed. Oxford: Clarendon Press. Original edition, 1906.

Kahn, Charles H. 2001. *Pythagoras and the Pythagoreans: A Brief History*. Indianapolis, IN: Hackett.

Kendler, K. S., and J. Parnas. 2015. *Philosophical Issues in Psychiatry: Explanation, Phenomenology, and Nosology*. Baltimore, MD: Johns Hopkins University Press.

Kitts, David B. 1987. Plato on kinds of animals. *Biology and Philosophy* 2 (3):315–328.

Klima, Gyula. 2013. The medieval problem of universals. In *The Stanford Encyclopedia of Philosophy*, edited by Edward N. Zalta. https://plato.stanford.edu/entries/universals -medieval/

Lelli, Fabrizio. 2007. Hermes among the Jews: Hermetica as Hebraica from antiquity to the renaissance. *Magic, Ritual & Witchcraft* 2 (2):111–135.

Lennox, James G. 1987. Kinds, forms of kinds, and the more and the less in Aristotle's biology. In *Philosophical Issues in Aristotle's Biology*, edited by Allan Gotthelf and James G. Lennox, 339–359. Cambridge, UK: Cambridge University Press.

Lennox, James G. 2001. *Aristotle's Philosophy of Biology: Studies in the Origins of Life Science*. Cambridge, UK/New York: Cambridge University Press.

Liddell, Henry George, and Robert Scott. 1888. *An Intermediate Greek-English Lexicon, Founded upon the Seventh Edition of Liddell and Scott's Greek-English Lexicon*. Oxford: Clarendon Press.

Linné, Carl von. 1788–1793. *Systema naturae per regna tria naturae, secundum classes, ordines, genera, species, cum characteribus, differentiis, synonymis, locis*. Edited by Jo. Frid. Gmelin. Editio decima tertia, aucta, reformata. ed. 3 vols. Lipsiae: Impensis Georg. Emanuel Beer.

Lucretius. 1969. *On the Nature of Things (De rerum natura)*. Translated by Martin Ferguson Smith. London: Sphere Books.

Mandelbaum, Maurice. 1957. The scientific background of evolutionary theory in biology. *Journal of the History of Ideas* 18 (3):342–361.

Matthews, Gwynneth. 1972. *Plato's Epistemology and Related Logical Problems*, edited by Mary Warnock, *Selections from Philosophers*. London: Faber & Faber.

McKeon, Richard. 1929. *Selections from Medieval Philosophers*. 2 vols. Vol. 1. New York: Charles Scribner's Sons.

—, ed. 1941. *The Basic Works of Aristotle*. New York: Random House.

Murphy, Dominic. 2006. *Psychiatry in the Scientific Image*. Cambridge, MA: MIT Press.

Nelson, Gareth J., and Norman I. Platnick. 1981. *Systematics and Biogeography: Cladistics and Vicariance.* New York: Columbia University Press.

Nordenskiöld, Erik. 1929. *The History of Biology: A Survey.* Translated by Leonard Bucknall Eyre. London: Kegan Paul, Trench, Trubner and Co.

Oldroyd, David R. 1983. *Darwinian Impacts: An Introduction to the Darwinian Revolution.* 2nd rev. ed. Kensington, NSW: University of New South Wales Press.

—. 1986. *The Arch of Knowledge: An Introductory Study of the History of the Philosophy and Methodology of Science.* Kensington, NSW: New South Wales University Press.

Osborn, Henry Fairfield. 1894. *From the Greeks to Darwin: An Outline of the Development of the Evolution Idea.* Columbia University Biological Series. I. New York: Macmillan.

Panchen, Alec L. 1992. *Classification, Evolution, and the Nature of Biology.* Cambridge, UK/New York: Cambridge University Press.

Pellegrin, Pierre. 1986. *Aristotle's Classification of Animals: Biology and the Conceptual Unity of the Aristotelian Corpus.* Translated by Anthony Preus. Rev. ed. Berkeley, CA: University of California Press.

Plato. 1998. *Phaedrus.* Translated by James H. Nichols. Ithaca/London: Cornell University Press.

Pliny the Elder. 1906. *Naturalis Historia.* Edited by Karl Friedrich Theodor Mayhoff. Lipsiae: Teubner.

—. 1940–63. *Natural History.* Translated by H. Rackham et al. Vol. 1–10. London: Heinemann.

Popper, Karl R. 1957. *The Open Society and its Enemies.* 3rd ed. London: Routledge and Kegan Paul.

—. 1959. *The Logic of Scientific Discovery.* Translated by Karl Popper et al. London: Hutchinson.

Porphyry. 1975. *Isagoge.* Translated by Edward W. Warren. Toronto: Pontifical Institute of Mediaeval Studies.

—. 2003. *Porphyry's Introduction, Clarendon Later Ancient Philosophers.* Oxford/New York: Oxford University Press.

Preus, A. 2002. Plotinus and biology. In *Neoplatonism and Nature: Studies in Plotinus' Enneads,* edited by Michael F. Wagner, 43–55. Albany, NY: State University of New York Press.

Regier, Darrel A. et al. 2009. The conceptual development of DSM-V. *American Journal of Psychiatry* 166 (6):645–650.

Ross, David. 1949. *Aristotle.* 5th ed. London: Methuen/University Paperbacks.

Rossi, Paolo. 2000. *Logic and the Art of Memory: The Quest for a Universal Language.* London: Athlone.

Runes, Dagobert D, ed. 1962. *Classics in Logic: Readings in Epistemology, Theory of Knowledge and Dialectics.* New York: Philosophical Library.

Sachs, Julius V. 1890. *History of Botany (1530–1860).* Translated by Henry E. F. Garnsey and Isaac B. Balfour. Oxford: Clarendon Press. Original edition, 1875.

Sankey, Howard. 1998. Taxonomic incommensurability. *International Studies in the Philosophy of Science* 12 (1):7–16.

Scott, Walter, ed. 1924. *Hermetica: The Ancient Greek and Latin Writings which Contain Religious or Philosophic Teachings Ascribed to Hermes Trimegistus.* Boulder, CO: Hermes House.

Sedley, David N. 2007. *Creationism and its Critics in Antiquity. Sather Classical Lectures.* Berkeley, CA: University of California Press.

Slaughter, Mary M. 1982. *Universal Languages and Scientific Taxonomy in the Seventeenth Century.* Cambridge, UK/New York: Cambridge University Press.

Sloan, Phillip R. 2013. The species problem and history. *Studies in History and Philosophy of Science Part C: Studies in History and Philosophy of Biological and Biomedical Sciences* 44 (2):237–241.

Spruit, Leen. 1994–1995. *Species Intelligibilis: From Perception to Knowledge.* 2 vols. Leiden/New York: Brill.

Stannard, Jerry. 1968. Medieval reception of classical plant names. In *Actes du XIIe Congrès International d'Histoire des Sciences, 21–31 Août 1968, Paris.* Paris.

Stannard, Jerry et al. 1999. *Herbs and Herbalism in the Middle Ages and Renaissance.* Aldershot/Brookfield VT: Ashgate Variorum.

Theophrastus. 1916. *Enquiry into Plants, and Minor Works on Odours and Weather Signs.* Translated by Arthur Hort, *Loeb Classical Library.* London: Heinemann.

Whitehead, Alfred North. 1938. *Science and the Modern World.* Pelican ed. Harmondsworth: Penguin.

Wiener, Chad. 2015. Dividing nature by the joints. *Apeiron* 48 (3):285–326.

Wilkins, John S., and Malte C. Ebach. 2013. *The Nature of Classification: Kinds and Relationships in the Natural Sciences.* Edited by Steven French, *New Directions in the Philosophy of Science.* London: Palgrave Macmillan.

Windelband, Wilhelm. 1900. *History of Ancient Philosophy.* Translated by Herbert Ernest Cushman. New York: Scribner. Reprint, Dover 1956.

Wood, Jamie. 2013. Isidore of Seville. In *The Encyclopedia of Ancient History.* Chichester, UK: Wiley.

Woodger, Joseph Henry. 1937. *The Axiomatic Method in Biology.* Cambridge, UK: Cambridge University Press.

Zirkle, Conway. 1959. Species before Darwin. *Proceedings of the American Philosophical Society* 103 (5):636–644.

2 The Medieval Bridge

BOËTHIUS: THE NATURE OF THE SPECIES IN LOGIC

In his *On division* (*De divisione*, c. 505 CE), Boëthius (480–524/6 CE) discussed the nature of classification by division. In an extended introduction to the topic, he sets out influentially the basis for the "classificatory logic" of the next 1500 years, and as he wrote in Latin, this was almost as influential in the development of western thought as Augustine's works were. Nearly all his examples are based on animal/ human and similar biological cases, but it should not be thought this is a biological concept, any more than Aristotle's.

He distinguishes division of genus into species from whole into part, and utterances into "proper signification." His version of the *diairesis* is not binary, however. "Every division of a genus into its species has to be made into two or more parts, but there cannot be infinitely many or fewer than two parts."[1] The matter of accidents and essences applies here to utterances (that is, of the meanings of propositions) not species, and he notes that while a genus is the whole of which a species is a part, more universal utterances such as equivocation are not wholes in nature.[2]

ISIDORE OF SEVILLE: METAMORPHOSES

Isidore (c. 560–636), was one of the first of the encyclopediasts after Pliny. He wrote an extensive summary of the state of knowledge in the seventh century, basing his accounts on prior writings and their supposed etymologies (hence the title *Etymologiae*), and discussed human and animal species briefly in the midst of discussions about centaurs and other mythological beasts. In a short section entitled "Metamorphoses (*De transformatis*)," he writes

> many creatures naturally undergo mutation and, when they decay, are transformed into different species—for instance bees, out of the rotted flesh of calves, or beetles from horses, locusts from mules, scorpions from crabs.[3]

However, he does not follow this up with anything more general, and it merely follows in the classical tradition of monstrosities.

---

[1] 877C [Kretzmann and Stump 1988, 14].
[2] 878D–879A.
[3] Isidore of Seville 2005, 246 [XI.iv.3].

UNIVERSALS VERSUS NOMINALISM: SPECIES
ARE IN THE UNDERSTANDING

There was little advancement on biological species concepts in the medieval period, which is usually delimited from 500 to 1500 CE. This is not because the medievals were unimportant—very far from it. The revival of the universals debate in the late eleventh century by Roscelin (of Compiègne, c. 1050–c. 1125), and Peter Abelard (1079–1142) was critical in bringing the notion of genera and species to the forefront of western thinking.[4] Once there, the idea was taken up by the nascent biological sciences in the seventeenth century, but so far as I can tell, nominalism did not directly influence biological practice, although it did influence John Locke in the seventeenth century in ways that then affected natural historians. To understand the significance of this, we must understand the philosophical arguments.

Universals were terms or classes that covered many individual objects ("individuals") but which remained one; in other words, a universal was something that existed across many things. How it did that was the subject of a long debate from Porphyry onward. As we have seen, Porphyry and Boëthius presented the problem—whether universals exist in the mind or in reality—but did not resolve it. Over the next few centuries, three positions were set up:

1. Realists
   a. Moderate Realists
   b. Strong Realists (sometimes called "Extreme" or "Robust" Realists[5])
2. Nominalists[6]

Realists held that what united all the individuals under a universal term existed in reality. In other words, a universal was a feature of the world. *Moderate Realists* held that universals existed in the individual objects; roughly the straight Aristotelian metaphysical position. *Strong Realists* held that universals could exist (and did) even if nothing existed that fell underneath them. For example, a Moderate Realist might say that Goodness existed in every good person, but if there were no good people (excluding God for the purposes of argument) there was no Goodness in the world. A Strong Realist would say that Goodness existed whether or not there were any good beings. This is pretty well Plato's view (under the heading The One and the Many). This split basically set the options open until the fourteenth century, when nominalism was formulated.

Nominalists, in the early fourteenth century, held that universals are just concepts in the mind (or in words, which went by the phrase *flatus vocis* or "breath of the voice"; i.e., a mere sound), and that all that is real are the named individual objects—this man, that horse, that rock. Consequently, nominalists denied the existence outside the mind of logical categories. Realists rejected this and held that knowledge (*scientia*) required the existence of universals outside the mind.[7] The nominalists

---

[4] Leff 1958. An excellent overview of the universals debate is given by Aaron 1952.
[5] Brower 2016, 733.
[6] Dutton 2006.
[7] Brown 1999.

held that universals existed solely in the understanding, or at least that is the nominalist position as later understood. Probably Roscelin did not assert this, but merely rejected Realism, and Ockham's version might be better called Conceptualism rather than an outright denial of universals, as he did not deny that universals did not exist *as concepts.*

The conception of species in this debate centered largely on the objects of knowledge rather than on anything particularly biological. However, as evidenced in the ninth century in John Scotus Eriugena's (or Duns Scotus; 815–c. 877) *De divisione naturae* (*The division of nature*), biological examples are used to illustrate the discussions.[8] These issues come to the fore again in late twentieth century philosophy of biology under the banner of species-as-individuals.

Peter Abelard (or Abelaird; 1079–1142), perhaps the best of the philosophers before the rediscovery and dissemination of Greek texts via the Arab translation in the twelfth and thirteenth centuries, was influenced by the nominalist claim, and had studied under Roscelin. He convinced his lecturer William of Champeaux to modify his Realist position, for example. In his *Gloss on Porphyry*, he asserted that universals are names only, akin to proper nouns but applicable to many things rather than one thing, based on a common likeness. The universal noun does not refer to a *thing* that is universal, for only individuals exist. His views led to his being declared heretical by later writers, including Bernard of Clairvaux.

For Abelard, *species* is a purely logical notion, and he says

> … considering the nature of species in man, I find at once from the nature of the species the argument for proving animal.[9]

That is, if you understand what sort a man is, then you know that the sort includes the species Animal, just as Aristotle's *De anima* asserted, and this is a logical truth.

William of Ockham (or Occam; c. 1300–1349) is perhaps the most enduringly influential of the medieval nominalists. For him, logical species are just a way to recollect similar individuals already encountered, and a general term cannot be abstracted from a *single* individual, but only a number of individuals encountered, as he says in *The Seven Quodlibita*.[10] Hence to assert that something coming from a distance is an animal, one must already have that concept by recollecting prior individual animals. This demonstrates that the nominalist is attempting a "bottom-up" form of classification, based on observed cases. Moreover, species, logical or otherwise, are things which have the same "power,"[11] but concepts of them are of the "second intention" (that is, are categorized by the mind):

> … that concept is called a second intention which signifies precisely intentions naturally significative, or which sort are *genus, species, difference,* and others of this sort [i.e., heads of predicables] for as the concept of man is predicated of all men …, so too one common concept, which is the second intention, is predicated of first intentions

---

8 Book IV [McKeon 1929, 107–141].
9 McKeon 1930, 211.
10 Quod. 1, ques. 13 [McKeon 1930, 365].
11 Quod. V, ques. 2.

[a conception of a thing formed by the first or direct application of the mind to the individual object; an idea or image] which stand for things, as in saying, *man is a species*, *ass is a species*, *whiteness is a species*, *animal is a genus*, *body is a genus*; in the manner in which *name* is predicated of different names ... and this second intention thus signifies first intentions naturally, and can stand for them in a proposition, just as the first intention signifies external things naturally.[12]

For Ockham, the general term is the name of a number of concepts formed from what might later be called "sense impressions." To say something is a universal is to say it is the name of a number of individual names, each of which "stands for" the "external" (to the mind) individual. The second intention term is thus not a universal in itself, but neither is it an individual; it's just a name.

Another thing that occurred in the period was the advancement of species to a predicable. In the later Middle Ages, Porphyry's tetrad—*genus, difference, species, property*, and *accident*—was much debated[13] and it was not exactly followed by all, as there was debate over whether species should be considered "predicate-types," because, as Green-Pedersen says, the species "can only be predicated about individuals, and there can be no science about individuals," a view held consistently through the Middle Ages.[14] In the fourteenth, and increasingly in the fifteenth, century, though, the suggestion is made in the commentaries that the species is an addition (*annexum*) to the genus. This, in effect, would make *species* into a class concept like *genus*, which it was not before. Abelard, as early as the twelfth century, seems to accept the Porphyrian/Boëthian taxonomy of classes as including species.

THE HERBALS AND THE BESTIARIES: MEANING AND MORAL SPECIES

In the later Middle Ages, particularly in the two centuries leading up to the modern scientific era, bestiaries and herbals developed, acting as precursors to biological classification.[15] Herbals were repositories of useful pharmacological knowledge, arranged alphabetically most of the time, and derived from the tradition of Pliny the Elder and Dioscorides.

The herbal traditions developed in an attempt to map the classical names of plant species onto known colloquial names in order to ensure that medical preparations were reliable.[16] The earliest known herbal is by Crataeus, physician to Mithridates VI, king of Pontus (111–64 BCE), but only fragments of it survive in Dioscorides' work. Herbals underwent a decline in popularity in the later Middle Ages but had a resurgence with the development of printing in the sixteenth century.

Bestiaries were usually a combination of moral tales, in which animals represented the virtues or vices, and exotic geography books.[17] Typically, they relied upon

---

[12] Quod IV ques. 19.
[13] Green-Pedersen 1984, 118–121.
[14] Green-Pedersen 1984, 120.
[15] Arber 1938.
[16] Stannard et al. 1999, chapters 1 and 2.
[17] McCulloch 1962.

the *Physiologus*, which remained widely read until the late Middle Ages, and had been translated into Latin in the fourth century, into French in the thirteenth century, and Middle English shortly before that. The work was a deliberately theological Christian work, in which each animal illustrated vices or virtues.

Another source was Isidore of Seville's (560–636) *Etymologiae* (c. 630),[18] which purported to give the linguistic origins of the names of beasts and all other matters, to make a theological point. As Wirtjes, the translator of the Middle English edition of the *Bestiary*, says, bestiaries were intended to show Nature as a second Book of God alongside the Bible, to show the Christian how to live morally.[19] As such, they do not express much of a view about the nature of biological species. A medieval bestiary, known in fact as *The* Bestiary, simply refers to "kindes,"[20] and Hugh of Fouilloy's (1111?–1172?; probably not the author anyway) *The Aviary* (*Aviarum*),[21] written between 1132 and 1152, refers to "species" of hawks: "There are two forms [*species*] of the hawk, namely, the tame and the wild." Clearly there is no generative or biological conception in play here. Species are just sorts of things. In an anonymous contemporary work, *The Book of Beasts*,[22] that was itself based on the ever-present *Physiologus* and Pliny via the work of Solinus (*fl* third century),[23] the author several times *does* exhibit the notion of "species" as being a kind of beast that breeds true, such as stags, birds in general, and shellfish.[24] However, he also includes hyenas as a kind of animal, even though it appears he thought of the animal as a hybrid.[25] In the century following the early twelfth century, colloquial bestiaries became common and popular.

There are two major, and a number of minor, exceptions to the moral form of bestiaries in this period. Minor exceptions include Thomas van Campitré (Thomas Cantimpratensis; 1210?–1293) in his compilation entitled *Liber de natura rerum*[26] and Konrad von Megenberg (1309?–1374) in a colloquial text *Puch der natur* (*Book of Nature*).[27] The two major exceptions are the emperor Frederick II and his acquaintance, Albertus Magnus.

FREDERICK II, THE HERETIC FALCONER

Frederick II of Hohenstaufen (1194–1250) was, to say the least, an interesting man. Excommunicated twice, although he was the Holy Roman Emperor (and King of Sicily and Jerusalem), for challenging the supremacy of the Pope over secular power, he nevertheless deviated from the Norman tradition and installed a court of cultural

---

[18] Isidore of Seville 2005.
[19] Wirtjes 1991, lxviii–lxxix.
[20] White 1954.
[21] Fouilloy 1992.
[22] Clark 2006.
[23] Solinus 1895.
[24] Clark 2006, 39, 107, 214.
[25] Walter Raleigh, in his 1614 *History of the World* [Raleigh 1614], denies that the hyena is a true species [White 1954, 31n], because it was a hybrid form of cat and dog, and so would not have been on Noah's Ark.
[26] Cantimpratensis 1973.
[27] *Puch* is an archaic spelling of *Buch*. The critical edition is von Megenberg 1861; cf. Stresemann 1975, 8.

and scientific sophistication.[28] He was greatly taken by the recent translations of Aristotle and Averroës (Ibn Rushd, 1126–98) by Michael Scot (1175–1235) of various natural history texts, including the *Liber animalium* (*De animalibus historia, De partibus animalium, De generatione animalium*) of Aristotle. He wrote *De arte venandi cum avibus* (*The art of hunting with birds*), published around 1248, and is reputed to have replied to one of the letters of the Mongol Khans[29] demanding that he submit to the Khan's power, that he would gladly resign his throne if we were allowed to be the Khan's falconer.[30] Records of the period from 1239–1240 in his court indicate that falconry took a close second place for him only to the affairs of government. Inspired by Aristotle's *De animalibus*, his passion was falconry, recently adopted by the European aristocracy from the Arabs, but he was disappointed in Aristotle's lack of accuracy. He notes in the Preface to *De arte*:

> *Inter alia*, we discovered by hard-won experience that the deductions of Aristotle, whom we followed when they appealed to our reason, were not entirely to be relied upon, more particularly in his descriptions of the characters of certain birds.
>   There is another reason why we do not follow implicitly the Prince of Philosophers: he was ignorant of the practice of falconry—an art which to us has ever been a pleasing occupation, and with the details of which we are well acquainted. In his work "Liber Animalium" we find many quotations from other authors whose statements he did not verify and who, in their turn, were not speaking from experience. Entire conviction of the truth never follows mere hearsay.[31]

This statement is revolutionary, and marks the beginning of the end of the Middle Ages mindset toward natural history, albeit for the purposes of a sport. He also says "Our purpose is to present the facts as we find them." Unfortunately, due to the author's heterodoxy, this view was not directly widely influential, apart from informing Albertus Magnus. Still, it does represent an attention to the actual facts of observation of organisms, which probably meant that such things as breeding compatibility between types were beginning to take priority over form, or arbitrary distinctions such as the one in [Pseudo-]Hugh's *Aviarum*.

In Book I of *De arte venandi*, Frederick adopts Aristotle's division of birds in *On Animals* into waterfowl, "whose organs are so fashioned that they may remain for indefinite periods immersed in water," land birds, and "neutral" birds which may live in either habitat (chapter 2), because it suits the usage of falconry experts, and then notes that "they may also be divided into various genera and these again into a number of species."[32]

He notes in chapter 3 that birds "may be classified in still another manner—as raptorial and nonraptorial species." In the Bologna *ms*, he notes that the raptorial birds "are the eagles, hawks, owls, falcons and other similar genera." But in the Vatican Codex, the paragraph reads thus:

---

[28] But see Abulafia 1988, 252ff for a dissenting account of how "Norman" Frederick's court really was, defending its sophistication and culture.
[29] Batu Khan.
[30] Abulafia 1988, 267.
[31] Wood and Fyfe 1943, 3f.
[32] Wood and Fyfe 1943, 7.

It was the habit of Aristotle and the philosophers to classify objects into positive and negative groups and to begin their discussions with the positive. Since it is our purpose to give special attention to raptorials, we shall first consider the nonrapacious (or negative) varieties; afterward we shall consider at length raptorial birds.

Since this is in fact opposed to Aristotle's method of classification, inasmuch as it allows for privative groups, this paragraph is perhaps more based upon the medieval understanding of logic from Boëthius than directly upon Aristotle's own works, and may be considered spurious. Possibly the distinction had not become popularly recognized (among educated non-scholars).

Later in that chapter he promises a treatise on the "genera into which raptores [sic] are divided, and the species in each genus," which was never produced. But he does note that "the same genera and species are given different names by diverse authors. Sometimes the same bird may have a variety of synonyms; and the same name applied to diverse birds so dissimilar that one cannot establish the true identity of a species simply by its name." He therefore notes, in complete anticipation of later taxonomic problems, that

a description of the essential characters of individual birds [i.e., of a species] is more difficult to furnish, whether they resemble or are different from another in the shape of the limbs, the movements they make, the way they feed, the care of their young, their mode of flight, and their style of defense. Let it, however, be remembered that, in general, their bodily conditions and their other peculiarities are due to definite causes.[33]

This is an amazing statement for the thirteenth century. Not only does he allow that species must vary in their traits, but that they are *caused* to vary. He goes on to say that different localities will have different genera and species, or a location may be the only habitat of species not found elsewhere. He even says that a genus might be found in many localities but with "a different color, or varying in other respects." It is not entirely exaggeration when Stresemann says of him that no direct observer among ornithologists until Konrad Lorenz in 1933 surpassed him in "variety of experience and acuteness of interpretation."[34]

The way he refers to species is so clearly in line with modern usage that he might be considered to have been the first to give a truly biological account. This is in part due to the fact that he had practical concerns—breeding birds. When discussing bird reproduction, he notes

Nature in her endeavor to preserve the race by the continuous multiplication of individuals has decreed that every species of the animal kingdom, whether it progresses by the use of wings or walks on the ground, shall take pleasure in sexual union so that they may seek instinctively to bring about such enjoyment.[35]

---

[33] Wood and Fyfe 1943, 10.
[34] Stresemann 1975, 11.
[35] Chapter 13-E [Wood and Fyfe 1943, 49].

Hence also, members of species know "instinctively" how to build nests of a "special design," and raise their children.

His empirical bent is further exhibited when he discusses the old myth that the barnacle goose arises by spontaneous generation from dead wood, a view still held in the sixteenth century by Julius Scaliger.[36] Frederick says

> There is also a small species known as the barnacle goose, arrayed in motley plumage... of whose nesting haunts we have no certain knowledge. There is, however, a curious popular tradition that they spring from dead trees. It is said that in the far north old ships are to be found in whose rotting hulls a worm is born that develops into the barnacle goose. This goose hangs from the dead wood by its beak until it is old and strong enough to fly. We have made prolonged research into the origin and truth of this legend and even sent special envoys to the North with orders to bring back specimens of those mythical timbers for our inspection. When we examined them we did observe shell-like formations clinging to the rotten wood, but these bore no resemblance to any avian body. We therefore doubt the truth of this legend in the absence of corroborating evidence. In our opinion this superstition arose from the fact that barnacle geese breed in such remote latitudes that men, in ignorance of their real nesting place, invented this explanation.[37]

As it happens, *Branta leucopsis* breeds in the northern parts of Europe, in hilltops, cliffs, slopes, and islands with nearby coasts or rivers with plenty of grass and other vegetation to graze. It spends winters in saltmarshes, lowland fields near coasts, and offshore islands with suitable grassland.[38]

So, Frederick is inclined to think that species are caused by generation from parents via sex, rather than accept the view of spontaneous generation for some species, a view held well into the eighteenth century.[39] Further, he notes that "productive Nature" formed organs for each species that are benevolent for one species but malevolent for another, and that it

> must be held, then, that for each species, and each individual of the species, Nature has provided and made, of convenient, suitable, material, organs adapted to individual requirements. By means of these organs the individual has perfected the functions needful for himself. It follows, also, that each individual, in accordance with the particular form of his organs and the characteristics inherent in them, seeks to perform by means of each organ whatever task is most suitable to the form of that organ.[40]

All that is missing here is a claim that the most fit will become the most widespread form, and we would have an anticipation of Darwin. However, he regards identification of species to be a matter of recognition of plumage, and at one point disagrees with those who reject a particular variety of peregrine based on its plumage

[36] From Hacking 1983, 70. Cf. also page 37.
[37] Wood and Fyfe 1943, 51f.
[38] Cramp 1980.
[39] Farley 1977.
[40] Chapter 23-I [Wood and Fyfe 1943, 57].

as being a true peregrine, showing that identification marks were insufficient to act as the *definiens* or essence of a species.[41]

ALBERTUS MAGNUS ON BEASTS AND PLANTS

Albert of Lauingen, later known as "The Great," is better known as Albertus Magnus (1193–1280). He was a Dominican friar who, among other things, taught Thomas Aquinas and was among the first to comment upon the works of Aristotle that were freshly coming out of the Arab tradition. Previously, much logical discussion in the European intellectual world was founded upon Peter of Lombard's *Sentences*, and the books of Boëthius. Now, works like Aristotle's *De motibus animalium*, which Albert found in Italy, and *Historia animalium* became commonplace (see above).

Albert also had a youthful interest in falconry, which was the proximate source of the love of nature evidenced in several medieval authors, including, as we have seen, Frederick II.[42] Unlike Frederick, though, he was and remained an orthodox member of the Church, later being given the title "Doctor Universalis," and his ideas were taught in subsequent educational institutions. He was directly familiar with Frederick's work on falconry. Albert may also have learned personally from Frederick's falconers, to whom the emperor had given him access.

His major work on natural history is *De animalibus*, Books 22–26 of which cover beasts and man. He also produced a long text, based more upon his own observations than *De animalibus* had been, on plants, *De vegetabilibus*. In both, he relied a lot on Pliny, but occasionally he corrects some of the more mythological accounts of Pliny. Nevertheless, the Chimera, the Manticore, and other mythological beasts appear in *De animalibus* along with some acute and accurate observations, recounted in James Scanlon's introduction to the cited edition and translation. Occasionally, he gave several separate descriptions for the same animal (e.g., *Alces alces*, the European elk, being named as "Alches," "Aloy," and "Equicervus").

In *De animalibus*, Albert describes, species by species, various animals known, after a short tract on man, "the most perfect of all animals," reflecting the consensus of the period.[43] In this tract, he notes that human reproduction is due to sexual intercourse "by which the potentialities of the two sexes are inextricably mingled."[44] He cites *De coitu* of Constantinus Africanus (c. 1010–1087), a Dominican monk who translated Arabic medical works, who, he said, held that

> the Creator unmistakenly wanted the animal kingdom to endure as a stable entity, never to die out. To this end He ordained the class of animals to be continually renewed by the coupling of the sexes and reproduction of the species so that none would be lost.[45]

---

[41] Wood and Fyfe 1943, 122.
[42] Stresemann 1975, chapter 1, Albertus Magnus 1987, 4, 19.
[43] Books 22 to 26.
[44] Book 22 [Albertus Magnus 1987, 59].
[45] A modern translation of *De coitu* is Boëthius... Delany 1969.

In Tract 2, he begins with the Quadrupeds, since these are the class of animals from which the most significant domesticated animals come. The internal order is alphabetically arranged in Latin, beginning with "Alches" (Elk). His classifications also include whether the animals are "wet" or "dry" in the Galenic scheme. However, he allows for a higher genus than the biological species level itself, for example where when discussing the "Cefusa" mentioned in Solinus' *Collecta rerum memorabilium* (one of his major sources, see above), possibly a gorilla, he assigns it to "the genus of simians."[46] This is not a precursor to Linnaean binomials, though. Each species is briefly described, except for the oddities (the elephant, for instance), common animals (the hedgehog, the wolf), and domesticated animals. The entry on horses runs for 41 manuscript pages, and includes details on care, diseases, and so forth. In cases of hybrids, such as the Mule (Mulus[47]), Albert notes that it is sterile because it is "produced from male and females sperms that are quite dissimilar in nature," although he still regards it as a species. He correctly notes that female mules may occasionally give birth to viable young, but ascribes the cause to the heat of the country countering the internal cold of the animal. He says of apes (Symia) that the animal "admits of many species." He includes a monkey as an ape, not making a distinction between tailed and non-tailed simians.

In Book 23, he describes bird species, which although they are animals, are different from the "general sense" of that term.[48] Again, the coverage he gives varies according to their oddity, commonplace nature, or domestication, with Falcons and Hawks given extensive coverage, following Frederick. He notes that different species of falcon have different-colored feet and plumage.[49] He describes ten species of "noble falcons" and three "inferior falcons," and four species of hybrids,[50] noting that while only four crossbred species are known, there are probably more, and also in other groups of species such as goshawks, sparrow hawks, and eagles.[51] Oddly, he includes bats (*Verspertilio*) among the birds, presumably because they fly, although he notes that they look like winged mice. That this is a purely arbitrary Aristotelian system based on overall habitat is confirmed by the inclusion of crocodiles, hippopotamuses, seals, dolphins, and whales among the fishes in the book on aquatic animals.[52] It is clear that for Albert, *species* is a kind term that relates shared properties, but which is maintained by sexual reproduction. We may take this as the best of late medieval opinion.

Albert also addressed the Barnacle Goose question, noting that he and his friends had bred one with a domestic goose, and that the spontaneous generation account is "altogether absurd as I and many of my friends have seen them pair and lay eggs and hatch chicks."[53] Raven documents many of his correct observations, and some of his repeating falsehoods, and Stannard shows that of all the plant species Albert lists,

---

[46] Albertus Magnus 1987, 93.
[47] Albertus Magnus 1987, 160.
[48] Albertus Magnus 1987, 188.
[49] Albertus Magnus 1987, 224.
[50] Albertus Magnus 1987, chapter 16.
[51] Albertus Magnus 1987, 247.
[52] Book XXIV.
[53] Raven 1953, 67f, quoting Book 23, 19.

nearly all can be identified with a modern species.[54] Nevertheless, Albert did list in Book VII 54–64, five ways plant species could change their species: by improvement or deterioration of seed, by being cut down and shoots being of another species, when cuttings grow as vines rather than oaks ("in Alvernia"), when a tree grows rotten and springs forth a different plant, and by grafting. Amundson considers this a kind of species mutability,[55] but I suspect it is due more to the fact that in cultivation, "species" is not the same thing as a *biological* species, but rather a synonym for "kind." As Stannard notes

> In medieval nomenclature, *genus, species, varietas,* and *forma* were used interchangeably. Sometimes Albert uses *genus* to denote what we would accept as a *genus* ... but in other places his *genus* is practicably equivalent to our *species* ...[56]

The persistence of the pre-scientific tradition of the herbals, and of spontaneous generation, has something to do with the fact that as a heretic, Frederick's work was not a safe source for later workers—indeed later writers accused him of atheism and heresy on several counts. But since Albert's work *was* used, and indeed he was canonized, it becomes hard to account for this purely in those terms. Rather, it seems that what stopped Frederick, Albert, and others such as Roger Bacon from influencing natural history and inaugurating an early empiricism was the fact that natural history was not seen as an end in itself until much later, but rather as an adjunct to theology and homiletics. That said, it is worth noting that a scientific attitude was not beyond the reach of a medieval educated man like Frederick, and that acuity of observation, when driven by immediate interests for which accurate information was required, was something that could be attained in a way that is quite modern. The shift to factual science can thus be seen to be driven by practical matters—in this case, falconry. However, there was a decline in empirical work on animals and plants in the fourteenth century and for a century after.[57]

ST. THOMAS: [LOGICAL] SPECIES AS INDIVIDUALS

Thomas Aquinas' (1225–1274) discussion in the *Summa* (Book I, Q. 86)[58] does not materially advance the matter and can be treated for our purposes here as straight transmission of the prior Scholastic logic. Aquinas treats it as a question of knowledge, and the "species" he considers is that of *species intelligibilus* (answer 1):

> Our intellect cannot know the singular in material things directly and primarily. The reason of this is that the principle of singularity in material things is individual matter, whereas our intellect, as I have said above, understands by abstracting the intelligible species from such matter. Now what is abstracted from individual matter is the universal. Hence our intellect knows directly the universal only.

---

[54] Stannard 1979.
[55] Amundson 2005, 36.
[56] Stannard 1980, 366n.
[57] Most likely due to the Black Death. A similar issue arises with medieval logic.
[58] Aquinas 1947.

When he does discuss the more general sense of species, he repeats the standard view, although he makes an interesting observation about infimae species being individuals:

These [*species infimae* or *specialissimae*] are called individuals, in so far as they are not further divisible formally. Individuals however are called particulars in so far as they are not further divisible neither materially nor formally.[59]

So there are two senses of "individual" in play here, says Thomas. One is simply that it is atomic—not further formally divisible. But there are also individuals that are not *materially* divisible, and they are particulars. For Thomas, the infimae species is a formal, but not a material, individual. Of course, the question whether species are individuals and what kind of individual they might be is a vexed modern issue, as we shall see.

It is not clear whether Aquinas was a Realist or a Conceptualist on universals. It may be that he held both views at different times or on different issues.[60]

Aquinas allows that living species can be generated by spontaneous generation from putrefaction.[61] However, he says

Species, also, that are new, if any such appear, existed beforehand in various active powers; so that animals, and perhaps even new species of animals, are produced by putrefaction by the power which the stars and elements received at the beginning. Again, animals of new kinds arise occasionally from the connection of individuals belonging to different species, as the mule is the offspring of an ass and a mare; but even these existed previously in their causes, in the works of the six days. ... Hence it is written (Eccles. 1:10), "Nothing under the sun is new, for it hath already gone before, in the ages that were before us."[62]

In other words, species only express what potential they have already been given by God at creation, as had Augustine. This hardly licenses Zirkle's claim[63] that Aquinas held a mutabilist view. Also see his *De principiis naturae*, in which he restates this view of potentials expressed in generation.

BIBLIOGRAPHY

Aaron, R. I. 1952. *The Theory of Universals*. Cambridge, UK: Cambridge University Press.
Abulafia, David. 1988. *Frederick II: A Medieval Emperor*. London/New York/Ringwood, Vic.: Allen Lane/Penguin.
Albertus Magnus. 1987. *Man and the Beasts (De animalibus, Books 22–26)*. Binghamton, NY: Medieval & Renaissance Texts & Studies, Center for Medieval and Early Renaissance Studies.

---

[59] *In lib. X Met.* Lect 10, 21–23. Quoted in McKeon 1930, 498.
[60] See Brower 2016 for a discussion.
[61] *Summa* I.73.1. Obj 3.
[62] *Loc. cit.* Reply to Objection 3.
[63] Zirkle 1959, 640.

Amundson, Ron. 2005. *The Changing Role of The Embryo in Evolutionary Biology: Structure and Synthesis. Cambridge Studies in Philosophy and Biology*. New York: Cambridge University Press.

Aquinas, Thomas. 1947. *Summa theologica*. First complete American ed. New York: Benziger.

Arber, Agnes. 1938. *Herbals: Their Origin and Evolution. A Chapter in the History of Botany 1470–1670*. 2nd ed. Cambridge, UK: Cambridge University Press. Reprint, 1970, Darien, CT: Hafner.

Brower, Jeffrey. 2016. Aquinas on the problem of universals. *Philosophy and Phenomenological Research* 92 (3):715–735.

Brown, Stephen F. 1999. Medieval Christian philosophy: Realism versus nominalism. In *The Columbia History of Western Philosophy*, edited by Richard H. Popkin, 271–278. New York: Columbia University Press.

Cantimpratensis, Thomas. 1973. *Liber de natura rerum*. Berlin: de Gruyter.

Clark, Willene B. 2006. *A Medieval Book of Beasts: The Second-Family Bestiary—Commentary, Art, Text and Translation*. Woodbridge: Boydell.

Cramp, Stanley. 1980. *Handbook of the Birds of Europe, the Middle East and North Africa: the Birds of the Western Palearctic*. Vol. 2, *Hawks to Bustards*. Oxford: Oxford University Press.

Delany, Paul. 1969. Constantinus Africanus' "De Coitu": A translation. *The Chaucer Review* 4 (1):55–65.

Dutton, Blake D. 2006. Universals (Medieval). In *Continuum Encyclopedia of British Philosophy*, edited by A.C. Grayling et al., 3240–3243. Oxford: Bloomsbury.

Farley, John. 1977. *The Spontaneous Generation Controversy from Descartes to Oparin*. Baltimore, MD: Johns Hopkins University Press.

Fouilloy, Hugh of. 1992. *The Medieval Book of Birds: Hugh of Fouilloy's Aviarium*. Translated by Willene B. Clark. Binghamton, NY: Medieval & Renaissance Texts & Studies.

Green-Pedersen, Niels Jørgen. 1984. *The Tradition of the Topics in the Middle Ages*. München: Philosophia Verlag.

Hacking, Ian. 1983. *Representing and Intervening: Introductory Topics in the Philosophy of Natural Science*. Cambridge, UK: Cambridge University Press.

Isidore of Seville, Saint. 2005. *Etymologiae: The Etymologies of Isidore of Seville*. Translated by Stephen A. Barney. New York: Cambridge University Press.

Kretzmann, Norman, and Eleonore Stump. 1988. *Logic and the Philosophy of Language*. Vol. 1, *Cambridge Translations of Medieval Philosophical Texts*. Cambridge/New York/Melbourne: Cambridge University Press.

Leff, Gordon. 1958. *Medieval Thought from Saint Augustine to Ockham*. Harmondsworth, UK: Penguin.

McCulloch, Florence. 1962. *Mediaeval Latin and French Bestiaries*. Rev. ed. University of North Carolina Studies in the Romance Languages and Literatures; no. 33. Chapel Hill, NC: University of North Carolina Press.

McKeon, Richard. 1929. *Selections from Medieval Philosophers*. 2 vols. Vol. 1. New York: Charles Scribner's Sons.

—. 1930. *Selections from Medieval Philosophers*. 2 vols. Vol. 2. New York: Charles Scribner's Sons.

Raleigh, Walter. 1614. *The History of the World*. 6 vols. Vol. 1. London: William Stansby for Walter Burre.

Raven, Charles Earle. 1953. *Natural Religion and Christian Theology: Science and Religion*. Vol. I. Cambridge, UK: Cambridge University Press.

Solinus, C. Julius. 1895. *Collectanea Rerum Memorabilium*. 2nd ed. Weidmann: Berolinum.

Stannard, Jerry. 1979. Identification of the plants described by Albertus Magnus' *De vegetabilibus* lib. VI. *Res Publica Litterarum* 2:281–318.

—. 1980. Albertus Magnus and medieval herbalism. In *Albertus Magnus and the Sciences: Commemorative Essays*, edited by James A. Weisheipl, 355–377. Toronto: Pontifical Institute of Mediaeval Studies.

Stannard, Jerry et al. 1999. *Herbs and Herbalism in the Middle Ages and Renaissance.* Aldershot/Brookfield, VT: Ashgate Variorum.

Stresemann, Erwin. 1975. *Ornithology from Aristotle to the Present.* Translated by Hans J. Epstein and Cathleen Epstein. Cambridge, MA: Harvard University Press.

von Megenberg, Konrad. 1861. *Das Buch der Natur von Konrad von Megenberg: Die Erste Naturgeschichte in Deutscher Sprache.* Stuttgart: Karl Aue. Original edition, 1475–1481.

White, Terence Hanbury. 1954. *The Book of Beasts: Being a Translation from a Latin Bestiary of the Twelfth Century.* New York: Putnam.

Wirtjes, Hanneke. 1991. *The Middle English Physiologus.* Oxford/New York/Melbourne: Published for the Early English Text Society by the Oxford University Press.

Wood, Casey A., and Florence Marjorie Fyfe, eds. 1943. *The Art of Falconry, being the De arte venandi cum avibus of Frederick II of Hohenstaufen.* Stanford University CA, London: Stanford University Press/H. Milford.

Zirkle, Conway. 1959. Species before Darwin. *Proceedings of the American Philosophical Society* 103 (5):636–644.

3 Species and the Birth of Modern Science

NICHOLAS OF CUSA: CONTRACTED SPECIES

Nicholas of Cusa (1401–1464) in his *On learned ignorance* represents a bridge between the medievals, who were rediscovering Aristotle but using the categories bequeathed to them by the neo-Platonists, and the Renaissance era.[1] Cusa was an eclectic, and his comments on categories show an influence from Pythagorean as well as neo-Platonic sources, and he was an early proponent of the universal character or language that we consider below.[2] He held, for example, that ten is the supreme number, and that all unity is found in it.[3] He thus justifies the Aristotelian ten *topoi* (topics), and says,

> the universe is contracted in each particular through three grades. Therefore, the universe is, as it were, all of the ten categories [*generalissima*], then the genera, and then the species. And so, these are universal according to their respective degrees; they exist with degrees and prior, by a certain order of nature, to the thing which actually contracts them. And since the universe is contracted, it is not found except as unfolded in genera; and genera are found only in species.[4]

He proposes that the universe is not discretely differentiated, but that species exist, though they vary by degrees. Cusa is naturally aware of the nominalist debate, and takes a pretty standard Aristotelian view of the matter:

> But individuals exist actually; in them all things exist contractedly. Through these considerations we see that universals exist actually only in a contracted manner. And in this way the Peripatetics [Aristotelians] speak the truth [when they say that] universals do not actually exist independently of things. For only what is particular exists actually. In the particular, universals are contractedly the particular.[5]

In effect, he is saying that species are particulars (individuals), ignoring the fourteenth-century discussions. All general things, such as universals, and indeed the entire universe, only actually exist "in a contracted way,"[6] as particulars, although things of the same species share in a specific nature:

---

[1] Hopkins 1981.
[2] Rossi 2000.
[3] Hopkins 1981, Book II, chapter 6, §123.
[4] Hopkins 1981, §124.
[5] *Loc. cit.*
[6] Hopkins 1981, §125.

For example, dogs and the other animals of the same species are united by virtue of the common specific nature which is in them. This nature would be contracted in them even if Plato's intellect had not, from a comparison of likenesses, formed for itself a species. Therefore, with respect to its own operation, understanding follows being and living; for [merely] through its own operation understanding can bestow neither being nor living nor understanding.[7]

Species are real, independent of the mind, and living species are caused by their "common specific nature." Cusa thus answers Porphyry's question: the understanding gathers species and genera together through comparison, so that these universals are likenesses of nature. Genera and species exist both in the mind and in nature.

Therefore, in understanding, it unfolds, by resembling signs and characters, a certain resembling world, which is contracted in it.[8]

Later, in Book III, Cusa defines the universe itself as existing "contractedly in plurality," unlike God, who is "the Oneness of the Maximum" existing "absolutely in itself":

Now, the many things in which the universe is actually contracted cannot at all agree in supreme equality; for then they would cease being many. Therefore, it is necessary that all things differ from one another—either (1) in genus, species, and number or (2) in species and number or (3) in number—so that each thing exists in its own number, weight, and measure. Hence, all things are distinguished from one another by degrees, so that no thing coincides with another. Accordingly, no contracted thing can participate precisely in the degree of contraction of another thing, so that, necessarily, any given thing is comparatively greater or lesser than any other given thing. Therefore, all contracted things exist between a maximum and a minimum, so that there can be posited a greater and a lesser degree of contraction than [that of] any given thing.[9]

No individual member of a species, since it would be a contracted thing, can therefore exhibit or instantiate all the features of the species, and so there is in the actual organisms of a species (or members of any non-biological species) variation in the degree to which they participate in the specific essence. According to Cusa, then, there is variation both within and between taxa. Here we see the beginnings of the "type" concept—while the type itself is definable in terms of some necessary and sufficient conditions, "members" of the type can diverge from it or not instantiate it fully. The only limit "of species, of genera, or of the universe" is "the Center, the Circumference, and the Union of all things."

We see also the underlying assumption of the Great Chain of Being. Cusa holds that each genus has a species that is "highest," which is coincidental with the lowest

---

[7] Hopkins 1981, §126.
[8] This brings to mind Donne's much later use of the phrase, "the world's contracted thus" in his poem "The Sunne Rising" (c. 1605). Donne was a one for lamenting the loss of the older medieval categories of thought, famously complaining in his "Anatomy of the World" (1611) that all coherence was gone with the loss of Aristotelian physics and Ptolemaic astronomy [Kuhn 1959, 194].
[9] Hopkins 1981, §182.

species of the superordinate genus, "so that there is one continuous and perfect universe." Since the "maximum union" is God, species of lower and higher genera are not united in something that cannot vary, but in a "third species" in which individuals differ by degrees. He relates this to the *scala naturae* explicitly, in "the books of the philosophers" (that is, in Aristotle and his successors). He says,

> Therefore, no species descends to the point that it is the minimum species of some genus, for before it reaches the minimum it is changed into another species; and a similar thing holds true of the [would-be] maximum species, which is changed into another species before it becomes a maximum species. ... Accordingly, it is evident that species are like a number series which progresses sequentially and which, necessarily, is finite, so that there is order, harmony, and proportion in diversity....[10]

The gradualism of the Great Chain is evident, as also is the influence of the neo-Platonic view of classification. Of particular note is that Cusa's examples are biological ones. In fact, Cusa apparently suggested (but did not do) experimental work on plants to uncover their natures.[11] More importantly, he allows for a formal and gradual change from one species of a higher level to a species of the next level below. This is not in any sense evolutionary, but it lays some groundwork for an evolutionary account to develop later.

MARSILIO FICINO: THE PRIMUM OF THE GENUS

Marsilio Ficino (1433–1499) was responsible for a number of neo-Platonic texts being translated and published under Cosimo de Medici. Among other texts, he oversaw texts by Porphyry, Proclus, Plotinus, and Dionysius the Areopagite; all source texts for the neo-Platonic philosophy.[12]

In his discussion of the genus–species distinction, he is at pains to view the logical progression of Aristotle as being a description of the actual[13] progression of things, and God, of course, is the source of all things. In his *Five Questions Concerning the Mind*, Ficino writes:

> The motion of each of all the natural species proceeds according to a certain principle. Different species are moved in different ways, and each species always preserves the same course in its motion so that it proceeds from this place to that place and, in turn, recedes from the latter to the former, in a certain most harmonious manner.[14]

Here we have again the generative notion of species that we saw in Lucretius. However, unlike the materialist account of Lucretius, Ficino's view is based upon the "principle" of the species and of the *primum* of a genus (the ontological principle *primum in aliquo genere*). Each genus has what later came to be thought of as a *type species*, a first and highest example of the kind, according to Ficino, the *primum*,

---

[10] Hopkins 1981, §§185–187.
[11] Morton 1981, 98f.
[12] Cassirer et al. 1948, 185.
[13] But not temporal. Ficino is dealing with atemporal, formal, progression.
[14] Cassirer, Kristeller, and Randall 1948, 194.

which is the species that is purely of the genus, with no defining essences other than those of the genus.[15] God is, of course, the *primum* of the genus Being, and from him all being flows, in the standard neo-Platonic way. Things do what they do because they share the limits of their species and are constrained by the end of that genus. The *primum* does it best of all.

With elements, plants, and "brutes," Ficino gives the Aristotelian accounts—the elements have heaviness and so they fall. Plants and animals have a nutritional and generative power (compare Aristotle's nutritive faculty, *De Anima* 413a20–33, and the discussion of reproduction in *De generatione animalium*). Species are perpetuated because they have an end, or rather, because to be a species is to have an end. Ficino presents this as a prelude to discussions of the mind that do not concern us here.

THE GREAT CHAIN OF BEING

That the philosophical view deriving from the neo-Platonists had reached popular culture by the eighteenth century is evidenced in Alexander Pope's *Essay on Man* (1733), section VII:

> Vast chain of being! Which from God began;
> Natures ethereal, human, angel, man,
> Beast, bird, fish, insect, who no eye can see,
> No glass can reach; from infinite to thee;
> From thee to nothing.—On superior powers
> Were we to press, inferior might on ours;
> Or in the full creation leave a void,
> Where, one step broken, the great scale's destroy'd:
> From Nature's chain whatever link you like,
> Tenth, or ten thousandth, breaks the chain alike.

This view is known as the *Great Chain of Being*, and it has a history that arises from Aristotelian concepts, through the neo-Platonists, into the Middle Ages and the revival of Aristotle in the fourteenth through to the sixteenth centuries.[16] The predicables that defined organic species were an ascending scale of increasingly "perfect" features: from being, to growth, to animation, to rationality. Raymond Lull (1232–1316) is perhaps the exemplar of this view.[17] In his view, the chain of being was a series of steps in a staircase to heaven (Figure 3.1A).[18] Bovillius (Charles de Bouvelles, 1479–1553) illustrated the moral dimension in a 1510 woodcut (Figure 3.1B).[19]

---

[15] Cassirer, Kristeller, and Randall 1948, 189.
[16] Lovejoy 1936, Kuntz and Kuntz 1988.
[17] Rossi 2000, Ragan 2009.
[18] Ragan says that the woodcut was added to the 1512 print of *Liber de ascensu et decensu intellectus* (written in 1304) [Llull 1512]. Gontier 2011 assigns the woodcut to Alonso de Proaza.
[19] Bovelles 1510, 119, Ferrari 2011. Figures from Archive.org.

A

FIGURE 3.1 Representations of the Great Chain. **A.** Raymond Lull's chain of being conception as a stairway to heaven. The steps are: stone, flame, plant, animal, man, heaven, angel, God. The image is not in the original manuscript versions of the 1304 edition of *Liber de ascensu et decensu intellectus*, and was added in the 1512 printed version by Alonso de Proaza. *(Continued)*

B

FIGURE 3.1 (CONTINUED) Representations of the Great Chain. **B.** Following Lull, Bovillus (Charles de Bovelles) illustrated a slightly simpler, and more morally directed, scale: as one adds predicates (being, vitality, sensibility, rationality) one ascends the scale from minerals, vegetables, sensible animals, and rational beings. Humans can morally decline down that scale too, as they lose their intellectual function (virtue), their sensibility (luxury), their life (appetite), and motion, leaving only bare existence. The properties here answer to Aristotle's "souls."

The Great Chain view consisted of several related theses held in varying ways by its adherents. One of these, named by Lovejoy *the principle of plenitude*, has it that the world is as full of all the things as it could be, or, as Lovejoy himself stated it,

> ... the universe is a *plenum formarum* in which the range of conceivable diversity of *kinds* of living things is exhaustively exemplified, but also any other deductions from the assumption that no genuine potentiality of being can remain unfulfilled, that the extent and abundance of the creation must be as great as the possibility of existence and commensurate with the productive capacity of a 'perfect' and inexhaustible Source, and that the world is better, the more things it contains.[20]

---

[20] Lovejoy 1936, 52.

In short, everything that can be, is, and the world is made to be everything it can be. This is the source Leibniz's doctrine of the *lex completio* that Voltaire so wickedly caricatured in his *Candide* as the teachings of Dr Pangloss. It is found in Plato's writings, but not in Aristotle, who famously wrote in the *Metaphysics*, "it is not necessary that everything that is possible should exist in actuality," and "it is possible for that which has a potency not to realize it."[21]

However, the second plank of the Great Chain is the *law of continuity* (Leibniz calls it the *lex continui*)—that all qualities must be continuous, not discrete. While Aristotle did not make all things linear, arranged in a single ascending series, he did require that there be no sudden "jumps," from which the medieval claim *natura non facit saltus* (nature does not make leaps) came, and which we shall meet again. Aristotle's version did not itself insist that one would classify a single living being in one and only one series, nor that an organism that is graded as superior in one respect must be superior in all, but that became the general impression later. Lovejoy says

> It will be seen that there was an essential opposition between two aspects of Aristotle's influence on subsequent thought, and especially upon the logical method not merely of science but of everyday reasoning. ... He is oftenest regarded, I suppose, as the great representative of a logic which rests upon the assumption of the possibility of clear divisions and rigorous classification. Speaking of what he terms Aristotle's "doctrine of fixed genera and indivisible species," Mr. W. D. Ross has remarked that this was a conclusion to which he was led mainly by his "close absorption in observed facts." Not only in biological species but in geometrical forms—... he had evidence of rigid classification in the nature of things. But this is only half of the story about Aristotle; and it is questionable whether it is the more important half. For it is equally true that he first suggested the limitations and dangers of classification, and the non-conformity of nature to those sharp divisions which are so indispensable for language and so convenient for our ordinary mental operations. ...
>
> From the Platonic principle of plenitude the principle of continuity could be directly deduced. If there is between two given natural species a theoretically possible intermediate type, that type must be realized—and so on *ad indefinitum*; otherwise there would be gaps in the universe, the creation would not be as "full" as it might be, and this would imply the inadmissible consequence that its Source or Author was not "good," in the sense which that adjective has in the *Timaeus*.[22]

From Aristotle's notion of an ontological scale, the higher beings were less potential and more determinate (God, the *ens perfectissimum*, could not be otherwise than he is), "all individual things may be graded according to the degree to which they are infected with [mere] potentiality."[23] It is perhaps arguable if this idea really does exist in Aristotle's writings, or that he intended it; but whether or not he did, this is the idea that was formulated by the Neo-Platonists, and various passages in his works led to a serial classification of organisms.[24]

---

[21] *Metaphysics* II, 1003a 2, and XI, 1071b 13, quoted in Lovejoy 1936, 55.

[22] Lovejoy 1936, 57f.

[23] Ross, 1949, 178, after a discussion of the Metaphysics; quoted by Lovejoy 1936, 59.

[24] Lovejoy 1936, 61–66 on Plotinus' construction of the chain. Singer 1950, 40–41 on the Neo-Platonic view.

His philosophy influenced many biologists; especially in this context Bonnet, as we shall discuss in the next chapter. Later, in the late eighteenth century, at first in Buffon, then his pupil Lamarck and Erasmus Darwin in England, but also in Pierre Maupertuis,[25] Jean-Baptiste Robinet,[26] Denis Diderot,[27] Paul Henri Dietrich d'Holbach,[28] and Kant, the Great Chain became temporalized. The ladder became a pathway.

The *lex completio* and the *lex continui*, taken together, do not allow for temporal change as, for example, Leibniz could not allow that there could be more monads later than earlier.[29] Others, most famously Voltaire, rejected the gradation scale for the same reason. Things did change, and they weren't always for the best. One or the other had to go. As we shall see with Lamarck, the *lex completio* was the first of the two major planks of the Great Chain to be rejected in biology. The Great Chain does imply that there is no real distinction between species, as all intermediate gaps are filled. This influenced Linnaeus in his view of natural classification.[30]

PETER RAMUS AND THE LOGIC OF WHOLES AND PARTS

Peter Ramus (1515–1572) was something like a protostructuralist, apparently not very consistent or even clear in his logical writings.[31] However, he plays a role here for several reasons. One is that he mediates the anti-Aristotelian views of the Renaissance humanists like Lorenzo Valla to the post-Reformation age, and in so doing introduced to intellectual thought both a distinction between artificial and natural logic, but also the tree of Porphyry, which is also occasionally referred to as the Tree of Ramus (Figure 3.2). A shameless self-promoter, he was censured by the religious authorities of Catholic France, and was eventually killed in the St. Bartholomew's Day Massacre for having converted to Calvinism.

Ramus was greatly dependent upon the terms in which medieval logic was discussed, but he had an idiosyncratic view of the genus-species distinction. He treated general terms as based on clusters or groups of simples, and called these genera. He concluded that species were therefore the same thing as individuals or particulars, and said that men and women were different species, for which he was condemned and in which he was followed by various late-sixteenth-century authors. He based this apparently on the view that words had a single denotation, and that the denotation was of material accidents:

A species is the thing itself concerning which the genus answers [in replying to the question, What is it?]: thus the genus man answers concerning Plato, the genus dialectician concerning this particular dialectician.[32]

[25] Terrall 2002.
[26] Robinet 1768, Robinet et al. 1777–1783.
[27] Diderot 1754.
[28] Holbach and Mirabaud 1770.
[29] Lovejoy 1936, 256–258.
[30] Stevens 1994.
[31] Ong 1958. Ong's book is singularly bad tempered, but erudite. The author, a Jesuit and a neo-Thomist, is attacking the vogue Ramus' anti-Aristotelian views were having at the time.
[32] *Training in Dialectic*, 1543, fol. 14. Quoted in Ong 1958, 204.

P. RAMI DIALECTICA.

TABVLA GENERALIS.

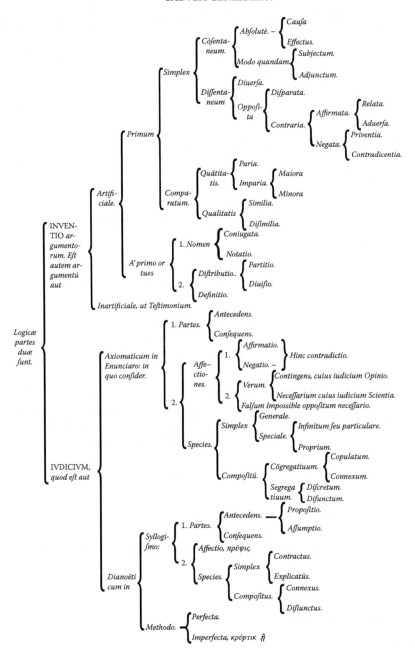

FIGURE 3.2   The Tree of Ramus. Ramus' conception of the logical topics was presented as a tree diagram or table by his editor and disciple Johann Thomas Freige in 1576.[33] Notice that species here is a judgment that is axiomatic in the expression of a term, and has subcategories.

Ramus seems to think that the general term (*genus*) is an answer to a given question of the constitution (*species*) of a particular thing. He has no notion of descending genera and species, nor that one genus (e.g., *dialectician*) might be a subaltern species in another (*man*). Either he misunderstood the scholastic tradition, which is Ong's interpretation, or he was redefining the logical enterprise. Ramus' ideas were influential in setting the education curriculum for the Trivium for much of the following century, and influenced the Port Royal Logic.[34] However, he seems to have had little direct effect upon biological classification as it developed, apart from the fact that many of the early modern naturalists would have been taught his views.

NOAH'S ARK AND THE CREATION OF THE SPECIES RANK

At this point in our narrative, we still have neither a uniquely *biological* notion of species nor of a fixed rank in logic. As species concepts evolved, at some point both of these must arise. The origin of *species* the rank, is, ironically, in the changing interpretation of the book of *Genesis*, especially the creation story of chapter 1, verses 21, 24 to 26 (Table 3.1), and the Flood narrative of chapter 6, verse 9 to chapter 9, verse 17 (Table 3.2).

The shift from rationalism to empiricism in theology is driven by the Reformation insistence on the plain meaning of the scriptures. However, the shift *to* rationalism in theology occurred much earlier, in the twelfth century.[35] Then, the issue was between reason and revelation. In the post-Reformation age, it is between observational evidence and revelation. Allen notes that Abelard was the first "to completely surrender to the charms of reason," but that it was the influence of Avicenna, in the Islamic tradition (then recently translated into Latin) who started the issue of faith *versus* reason in the West.[36]

In the period leading up to the modern era, the interpretation of the Bible was a mixture of literal readings, allegory, mystical meanings, analogy, and metaphor. In the twelfth century, for example, Hugh of St. Victor (d. 1141), wrote *De arca Noe morali*,[37] in which he allegorized the Ark as the Church, stating that

> … this spiritual building … is Noah's Ark. … this Ark denotes the Church, and the Church is the body of Christ …[38]

However, by the fifteenth century, in the wake of the Reformation, biblical texts were more literally interpreted and treated as factual, historical accounts.[39] Spanish bishop Alonso [or Alfonso] de Madrigal, known as Tostado (c. 1400–1455), wrote, in his commentary on *Genesis*:

---

[33] Freedman 1993.

[34] Ramus 1756, Hotson 2007, 39.

[35] Allen 1949, chapter 1.

[36] Allen 2003, 6–7.

[37] Hugh of Saint Victor 1962.

[38] *Op cit*. I.7.

[39] For a compelling account of this shift and the contribution of Protestant theology to empirical natural philosophy, see Harrison 2015.

TABLE 3.1
The Creation Account, with Kind Terms Highlighted (Genesis 1:21, 24–26)

[21] Creavitque Deus cete grandia, et omnem animam viventem atque motabilem, quam produxerant aquae **in species suas** et omne volatile **secundum genus suum**. Et vidit Deus quod esset bonum. *[24]* Dixit quoque Deus: Producat terra animam viventem **in genere suo**, jumenta et reptilia et bestias terrae **secundum species suas**. Factumque est ita. *[25]* Et fecit Deus bestias terrae **juxta species suas**, et jumenta et omne reptile terrae **in genere suo**. Et vidit Deus quod esset bonum. [*Vulgate* 1682]	*[21]* God created the large sea creatures, and every living creature that moves, with which the waters swarmed, **after their kind**, and every winged bird **after its kind**. God saw that it was good. *[24]* God said, "Let the earth bring forth living creatures **after their kind**, cattle, creeping things, and animals of the earth **after their kind**," and it was so. *[25]* God made the animals of the earth **after their kind**, and the cattle **after their kind**, and everything that creeps on the ground **after its kind**. God saw that it was good. [*NIV*]

TABLE 3.2
The Animals on the Ark (Genesis 6: 17–20)

[17] Ecce ego adducam aquas diluvii super terram, ut interficiam omnem carnem, in qua spiritus vitae est subter caelum: Universa quae in terra sunt, consumentur. *[18]* Ponamque foedus meum tecum: et ingredieris arcam tu, et filii tui, uxor tua, et uxores filiorum tuorum tecum. *[19]* Et ex cunctis animantibus universae carnis bina induces in arcam, ut vivant tecum: masculini sexus et feminini. *[20]* De volucribus **juxta genus suum**, et de jumentis **in genere suo**, et ex omni reptili terrae **secundum genus suum**: bina de omnibus ingredientur tecum, ut possint vivere. [*Vulgate* 1682]	*[17]* I am going to bring floodwaters on the earth to destroy all life under the heavens, every creature that has the breath of life in it. Everything on earth will perish. *[18]* But I will establish my covenant with you, and you will enter the ark—you and your sons and your wife and your sons' wives with you. *[19]* You are to bring into the ark two of all living creatures, male and female, to keep them alive with you. *[20]* Two **of every kind** of bird, **of every kind** of animal and **of every kind** of creature that moves along the ground will come to you to be kept alive. [NIV]

And because Noah took care of the animals and gave them food which was kept in the apothcea on the second level, there was a stairway from the habitation of Noah to this place so that he could descend and take up food. So he gave them food, walking between the apes, dragons, unicorns, and elephants, who thanks to God did not harm him but waited for him to give them nourishment at the proper time. The divine pleasure saw to it that there was a great peace among these animals; the lion did not hurt the unicorn or the dragon the elephants, or the falcon the dove. There was also a vent in the habitation of the tame animals and another in that of the wild animals through which dung was conveyed to the sentina. Noah and his sons collected the dung and cast it by mean of an orifice into the sentina so that the animals would not rot in their own offal. One could also believe that the odour of the dung was miraculously carried off so that the air was not corrupted and men and animals were not slain by the pest. So the men in the ark laboured daily and had no great time for leisure.[40]

---

[40] Translated in Cohn 1999, 38f.

Thus began the tradition of treating the scriptural texts realistically rather than allegorically, in the Christian tradition. The contrast with Hugh of St. Victor's exposition three centuries earlier is striking. Arguably this is due to the emphasis in Renaissance humanism of taking texts in their historical context and authorial intent.[41]

Jean Borrell, also known (after a local hawk[42]) as Johannes Buteo (French Catholic, 1490–1560/1572?), attempted in his tract *De arca Noe* to work out the size of the Ark, the food required, and the number of kinds of animals it could contain; he gave 93 such beasts, but not the bird kinds (Table 3.3).[43] The term he mostly uses is *genera* but he does use the term *species* as well.

Buteo knew there were many living animals that were not included on the Ark, but as his comment on mice (generated from corruption, what came later to be known as "spontaneous generation"[44]) shows, he expected some kinds to generate out of the mud. Moreover, he is unsure if other kinds should be included, such as mules:

> There are those, however, who think that mice were not brought in the ark nor anything of that family since they are born of corruption, as they also say of mules because they come from another kind [*ex alio genere procreatur*].[45]

This began a tradition of calculating the logistics of the Ark, which became, over the next century, a pressing problem, since more and more species were being discovered.

The task was taken up by Walter Raleigh in his 1614 *History of the World*. Raleigh argued that the Ark was of sufficient capacity to include the beasts (but not fishes).

> … it is manifest, and undoubtedly true, that many of the Species, which now seeme differing and of severall kindes, were not then *in rerum natura*. For those beasts which are of mixt natures, either they were not in that age, or else it was not needfull to præserve them: seeing they might bee generated againe by others, as the Mules, the Hyæna's and the like: the one begotten by Asses and Mares, the other by Foxes and Wolves.[46]

He estimated 130 kinds of herbivores and 32 carnivores, and that all other kinds of animals from the Americas and India (i.e., that differed from "Northerne" species), and so on, were hybrids or varieties caused to transmute by local conditions:

---

[41] For example, Lorenzo Valla's debunking of the *Donation of Constantine*, in 1440, using philological techniques. See also the influence of rabbinic ideas in Cohn 1999, 33ff and the discussion in Pleins 2009, chapter 5.

[42] According to Buteo 2008, he was so named due to there being a local hawk called *bourrel*; a Middle Latin name for hawk is *buteo*. See also Shannon 2013, 270–273.

[43] Buteo 1554; English translation in Buteo 2008, although this is a work of creationist scholars and should be treated cautiously.

[44] Farley 1977.

[45] Buteo 2008, 31.

[46] Raleigh 1614, 111–112, Bk I, Pt I, ch. 7, § 9.

TABLE 3.3
Buteo's List of Beasts on the Ark

Animals That Eat Forage and Grain	Animals That Eat Meat	Reptiles	Birds
Elephant	Lion	Vipers	Not enumerated
Wild ox	Leopard	Serpents	
One-horned ox	Leucrocota	Asps	
Arabian camel	Cala	Horned snakes	
Bactrian camel	Dragon	Hydras	
Dromedary	Boas	Basilisks	
Giraffe	Bulls with flexible horns	Lizards otherwise	
Rhinoceros	Manticore	green and tiny	
Unicorn	Panther	Geckos	
Bison	Tiger	Chameleons	
Buffalo	Wolf	Salamanders	
European bison	Dog wolf		
Oryx	Deer wolf (Jackal)		
Reindeer	Hellhound		
Pegasi	Cepha		
One-horned Indian ox	Lynx		
Ostrich	Sphinx		
Common cow	Hyena		
Deer	Wild axis		
Antelope	Sea otter		
Red deer	Beaver		
Common horse	Giant otter		
Wild horse	Satyr		
Spotted horse	Common otter		
Hippopotamus	Hyena (glavum)		
Donkey	African hunting dog		
Wild donkey	Fox		
One-horned Indian donkey	Cat		
Bear	Egyptian weasel		
Pig	Common weasel		
Boar	Sea calf		
Domestic goat			
Wild goat			
Mountain-goat			
Ibex			
Monkey			
Cebus			
Baboon			
Long-tailed monkey			
Other monkeys			
Hare			
Alpine white hare			

(Continued)

TABLE 3.3 (CONTINUED)
Buteo's List of Beasts on the Ark

Animals That Eat Forage and Grain	Animals That Eat Meat	Reptiles	Birds
Rabbit of the hare family			
Common rabbit			
Wild rabbit			
Badger			
Squirrel			
Dormouse			
Mice and mole (born of corruption, so not on Ark)			
Mule (not on Ark)			
Hedgehog			
Porcupine			

Wee also see it daily that the natures of fruits are changed by transplantation, some to better, some to worse, especially with the change of Climate. Crabs [crabapples] may be made good fruit by often grafting, and the best Mellons will change in a yeare or two to common Cowcummers by being set in a barren soile: Therefore taking the kindes præcisely of all creatures, as they were by God created, or out of the earth by his ordinance produced: the *Arke*... was sufficiently capacious to contain of all, according to the number by God appointed...

The next major work of Ark logistics was that of the German Jesuit, Athanasius Kircher (1602–1680), who had a reputation as a polymath. In a tract entitled *De arca Noë* (1675),[47] he expanded upon Buteo's work, and listed no more than 50 kinds of quadrupeds.[48] The rest were hybrids formed afterwards from these kinds, or geographical varieties from by the conditions in which they found themselves. The lower animals were formed, as with Buteo, by spontaneous generation. As Breidbach and Ghiselin note, however, he immediately came under criticism by Francesco Redi (1626–1697), after he carried out numerous experiments to test spontaneous generation.[49] Kircher held fast, in part because his entire argument relied upon these methods of generating new kinds, and in part because of his commitment to a universal taxonomy of all things which God had intended.

The next such work was offered by Bishop John Wilkins.[50] We shall encounter him again below, but for now it is sufficient to note that he listed only 58 kinds, even fewer than Buteo (Figure 3.3), although slightly more than Kircher. He notes that there is no problem with some species:

---

[47] Kircher 1675.
[48] Breidbach and Ghiselin 2006, 998.
[49] Breidbach and Ghiselin 2006, 999.
[50] Wilkins 1668. See the essays in Subbiondo 1992 for context and commentary.

Beasts Feeding on Hay				Beasts Feeding on Fruits, Roots, and Insects				Carnivorous Beasts			
Number.	Name	Proportion to Beeves.	Breadth of the Stalls (feet)	Number.	Name	Proportion to sheep.	Breadth of the stalls. (feet)	Number.	Name	Proportion to wolves.	Breadth of the stalls. (feet)
2	Horse	3	20	2	Hog	4		2	Lion	4	10
2	Asse	2	12	2	Baboon	2		2	Beare	4	10
2	Camel	4	20	2	Ape	2		2	Tigre	3	8
2	Elephant	8	36	2	Monky			2	Pard	3	8
7	Bull	7	40	2	Sloth			2	Ounce	2	6
7	Urus	7	40	2	Porcupine	7 }	20	2	Cat	2	6
7	Bisons	7	40	2	Hedgehog			2	Civet-cat		
7	Bonasus	7	40	2	Squirril			2	Ferret		
7	Buffalo	7	40	2	Ginny pig			2	Polecat		
7	Sheep	1		2	Ant-bear	2		2	Martin		
7	Stepciseros[1]	1	30	2	Armadilla	2		2	Stoat	3 }	6
7	Broad-tail	1		2	Tortoise	2		2	Weesle		
7	Goat	1				—	20	2	Castor		
7	Stone-buck	1	30					2	Otter		
7	Shamois	1						2	Dog	2	6
7	Antilope	1						2	Wolf	2	6
7	Elke	7	30					2	Fox		
7	Hart	4	30					2	Badger	2 }	6
7	Buck	3	20					2	Jackall		
7	Rein-deer	3	20					2	Caraguya		
7	Roe	2	36								
2	Rhinocerot	8									
2	Camelopard	6									
2	Hare	(2 Sheep)	30								
2	Rabbet										
2	Marmotto										
		92	514							27	72

FIGURE 3.3 Bishop Wilkins' Table of Animals on the Ark.[51]

The *Serpentine-kind, Snake, Viper, Slow-worm, Lizard, Frog, Toad*, might have sufficient space for their reception, and for their nourishment, in the Drein or Sink of the *Ark*, which was probably three or four foot under the floor for the standings of the Beasts. As for those lesser Beasts, *Rat Mouse, Mole*, as likewise for the several species of Insects, there can be no reason to question, but that these may find sufficient room in several parts of the *Ark*, without having any particular Stalls appointed for them.[52]

However, although he does not mention insects spontaneously generating (Francisco Redi's experiments against mice and insects spontaneously generating having had a bite[53]), he does exclude hybrids, as his predecessors had:

---

[51] A probable typographical error for "Strepciseros," which Gesner (1551–1587, vol. 1, 323) named as a goat, but Wilkins called a sheep. I am grateful to Glenn Branch for tracking that down.

[52] Wilkins 1668, 165.

[53] Farley 1977, 14f.

In this enumeration I do not mention the Mule, because 'tis a mungrel production, and not to be rekoned as a distinct species. And tho it be most probable, that the several varieties of Beeves, namely that which is stiled *Vrus, Bisons, Bonasus* and *Buffalo* and those other varieties reckoned under *Sheep* and *Goats*, be not distinct species from *Bull, Sheep*, and and *Goat*; There being much less difference betwixt these, then there is betwixt several Dogs: And it being known by experience, what various changes are frequently occasioned in the same species by several countries, diets, and other accidents: Yet I have *ex abundanti* to prevent all cavilling, allowed them to be distinct species, and each of them to be clean Beasts, and consequently such as were to be received in by sevens ...[54]

Here we see that the term *species* is being given a peculiarly zoological mean-ing. Given that his collaborator John Ray gives our first definition of *species* in a biological context (below), it is clear that the issue of Noah's Ark has contributed to the rank of living kinds being of one particular level. The very need for a lowest level of *kind* on the Ark generated the rank of living species. It is this question that Ray addressed. This also explains why later writers spoke of the *primum genus* or *species* (see, for example, Fuchs, below). These were the initial kinds from which other kinds (*secundum genera* or *species*) were formed, usually by hybridization or local climatic modifications. The influence of Conrad Gesner's *Historia animalium* on Kircher[55] and Wilkins is also evident, and to this novel tradition we now turn.

FUCHS AND GESNER: IMAGES, GENUS, AND SPECIES

Following the Renaissance, beginning in the sixteenth century after the decline in empirical work on animals and plants in the fourteenth century, modern biology begins to take root.[56] The period of the Reformation was largely lost to natural history, with the exceptions of the herbals of Leonhart Fuchs (1501–1566) and Otto Brunfels (1500–1534), and the bestiary of Conrad Gesner (1516–1665) and the naturalists that followed him such as Guillame Rondelet (1507–1556) and Pierre Belon (1517–1564), who is remembered for his study of the homology between bird and human skeletons.

While Brunfels is regarded as derivative, and Heironymus Fock as naïve, Fuchs has some novelties that were later influential.[57] He occasionally used binomials (with the "genus" name often *after* the "species" name), and organized his kinds (*genera* and *species* are used indifferently) under heads (*capita*). However, he has nothing resembling the Linnaean ranks, although his descriptions can be mapped to Linnaean species as Sprague and Nelmes have shown. Moreover, he did not divide his "genera" evenly. For example, Sprague and Nelmes list his classification of *Ranunculus* (a flowering plant genus that includes the buttercups), with the corresponding Linnaean species (Figure 3.4). Each of the species listed is listed in an apparently arbitrary order. There is no type species of the "genus."

---

[54] Wilkins 1668, 164f.
[55] Breidbach and Ghiselin 2006, 998.
[56] A good summary, with excellent reproductions of woodcuts of the period, is Pavord 2005, which also includes a nice discussion of the Arabic contribution to the herbalist tradition.
[57] Sachs 1890, Sprague and Nelmes 1928/31.

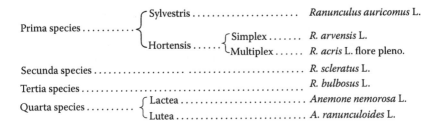

FIGURE 3.4 Fuch's classifications. A representation of Fuchs' capita and genera/species in his "new herbal" *De Historia Stirpium* of 1542.[58] In other cases, he would list the primum genus, and so on.[59]

Of the bestiary writers of the Reformation, Gesner is most interesting, as like Frederick and Albert, he studied the organisms for himself. Of equal note is that he engaged woodcuts to be made of his specimens, allowing readers to identify which actual species he was talking about. His *Historia animalium* ran to 3500 pages in four volumes, organized by Aristotelian categories of viviparous and oviparous organisms, then birds, fishes, and reptiles and insects.[60] In his *Historia plantarum* (1548–1566) he summarized twenty years' work of direct descriptions and drawings of 1500 plants, including dissections of flowers and fruit.

The formal distinction into genera and species in botany arises in the work of Gesner, according to Arber, when he employs the practice of giving genera substantive names, and Arber considers him the earliest to do so.[61] However, he was not consistent in that nomenclature, and his work was not widely known at first. Arber instead regards Fabius Columna (or Fabio Colonna, 1567–1650), in his *Ekphrasis* (1616), under the influence of Cesalpino, as having published the first views on the nature of genera in botany, relying on flower and seed rather than the older morphologies of leaf and stem to distinguish them.[62] Columna also used woodcuts for botanical illustration in his 1592 herbal *Phytobasanos*, for the first time.

Gesner's *Historia* was used as the basis for Edward Topsell's *History of Four-Footed Beasts*, in which Topsell wrote in the essay "Of the Crocodile":

> Because there be many kindes of Crocodiles, it is no marvel although some have taken the word *Crocodilus* for the *Genus*, and the several *Species*, they distinguish into the Crocodile

---

58 Redrawn from Sprague and Nelmes 1928/31, 557.
59 Classen 2001.
60 Nordenskiöld 1929, 93f.
61 Arber 1938, 166.
62 Arber herself, a botanist in the ideal morphology tradition, is a species "nominalist," writing:

> The progression from the vague concepts of the early writers to the sharp definition of genera and species to which we are now accustomed, has been in some ways a doubtful blessing. There is to-day, as a recent writer has pointed out, a tendency to treat these units as if they possessed concrete reality, whereas they are merely convenient abstractions, which make it easier for the human mind to cope with the endless multiplicity of living things. [Arber 1938, 168f]

The "recent writer" may be Senn 1925, whom she cites in her Appendix III [Arber 1938, 306].

of the earth and the water. Of the earth are sub-divided into the Crocodiles of *Bresilia*, and the *Scincus*: the Crocodiles of the water into this here described, which is the vulgar one, and that of *Nilus*, of all which we shall entreat in order, one successively following another. But I will not contend about the *Genus* or *Species* of this word, for my purpose is to open their several natures, so far as I have learned, wherein the works of Almighty God may be known, and will leave the strife of words to them that spend their wits about tearms and syllables only. … Thus much I finde, that the Ancients had three general tearms for all egge-breeding Serpents. Namely, *Rana, Testudo, Lacerta*: And therefore I may forbear to entreat of *Crocodilus* as a *Genus*, and handle it as a *Species*, or particular kinde.[63]

Here we see the word *species* used as a synonym for "kind" in English, and the use of the logical term *genus* for the enclosing class.

Although a Protestant and friend of Zwingli, Gesner was regarded well by all, except his rival Ulisse Aldrovandi, who plagiarized him and sneered when referring to him,[64] although Aldrovandi is regarded by Nordenskiöld as superior to Gesner in classification and methods, if not accuracy.[65] Stresemann notes that while his arrangement of species was novel, diverging from the alphabetical order Gesner and the bestiaries and etymologies used, Aldrovandi's was no improvement, based on beak structure.[66] One of Gesner's friends was Caspar Bauhin, to whom we now turn.

CESALPINO AND BAUHIN: THE BEGINNINGS OF MODERN TAXONOMY

True investigation of nature consists not only in deducing rules from exact and comparative observation of the phenomena of nature, but in discovering the genetic forces from which the causal connexion, cause and effect may be derived. In pursuit of these objects, it is compelled to be constantly correcting existing conceptions and theories, producing new conceptions and new theories, and thus adjusting our own ideas more and more to the nature of things. The understanding does not prescribe to the objects, but the objects to the understanding. The Aristotelian philosophy and its medieval form, scholasticism, proceeds in exactly the contrary way; it is not properly concerned with acquiring new conceptions and new theories by means of investigation, for conceptions and theories have been once and for all established; experience must conform itself to the ready-made system of thought; whatever does not so conform must be dialectically twisted and explained till it apparently fits in with the whole.

Julius Sachs[67]

The nineteenth-century historian of botany Julius Sach's rather cynical view of medieval and Renaissance botany does an injustice to the writers themselves. For example, Luca Ghini (1490–1556), who founded the botanical garden at Pisa for Cosimo de Medici, introduced the practice of herbariums, in which plants were dried and

---

[63] Topsell 1607, 682.
[64] Raven 1953, chapter 5.
[65] Nordenskiöld 1929, 95.
[66] Stresemann 1919, 22.
[67] Sachs 1890, 86.

pressed so that exact identification of plants could be made. He influenced a number of his contemporaries, including Gesner, Brunfels, Fuchs, and Andreas Cesalpino, who succeeded him as director of the garden. Although he did not write anything that has survived, he is credibly the founder of empirical botany.[68]

According to Lovejoy, the Florentine Andreas Cesalpino (1519–1603) was an enthusiast of Aristotelian classification, and Lovejoy notes that it was the fresh study of Aristotelian writings that set Cesalpino to producing his *De Plantis* (1583).[69] In this and his *Peripatetic Problems* of 1588, under the Latin title *Quaestionum peripateticarum, libri V,*[70] he worked on strict Aristotelian lines: "exhaustive comparative analysis of the forms, concisely worded theoretical definitions, and, based on these, abstract conclusions."[71] Sachs quotes him saying in chapter 13:

> That according to the law of nature like always produces like and that which is of the same species with itself.[72]

In chapter 14, Cesalpino stated:

> We seek similarities and dissimilarities of form, in which the essence ('substantia') of plants consists, but not of things which are merely accidents of them ('quae accidunt ipsis').[73]

Cesalpino therefore seems to be a mediate source of both the idea that species are fixed, and that species have an underlying essence, or substance in the Latin, that is distinct from other characters that may vary accidentally, which was the standard view in medieval logical writings. Whether this meant that he was a fixist, though, is open to doubt. Cesalpino, like all those of the time, was concerned to differentiate *genera et species* in *De plantis*, by identifying the botanical characters (*differentiae*) and he uses the fructification of the plants as his basis.[74] He says in the Preface to *De plantis*:

> Since science consists in grouping together of like and the distinction of unlike things, and since this amounts to the division into genera and species, that is, into classes based on characters (*differentiae*) which describe the fundamental nature of the things classified, I have tried to do this in my general history of plants, ...[75]

He specifically names Theophrastus as the authority for this methodology.

---

[68] Morton 1981, 121–124.

[69] Arber 1938, 142f, notes that Linnaeus' personal copy of *De plantis* is heavily annotated. She notes:

> Cesalpino's strength lay in the fact that he approached his subject with a trained mind; he had learned the lesson which Greek thought had then, and has now, to offer to the scientific worker—the lesson of how to think.

[70] Nordenskiöld 1929, 113.

[71] Nordenskiöld 1929, 192.

[72] Sachs 1890, 52.

[73] Quoted in *loc. cit.*

[74] Morton 1981, 135–140:

> Cum igitur scientia omnis in similium collectione & dissimilium distinctione consistat, hæc autem distributio est in genera & species veluti classes secundum differentias rei naturam indicantes, conatus sum id prestare in universa plantarum historia. [Caesalpini 1583, Book I, vii]

[75] Translated in Morton 1981, 135.

Independently of Cesalpino, Caspar (or Kaspar) Bauhin of Switzerland (1550–1624) organized known scientific names and descriptions of plants into two works (*Prodromus theatri botanici*, 1620 and *Pinax theatri botanici*, 1621), in which the genus–species arrangement was used, although he did not use it consistently as Linnaeus.[76] Even so, Bauhin used the common likeness of plant forms as his basis for classification, unlike Cesalpino's artificial system, which was based on the special characters of what he considered the "soul" or heart of the plants (hearkening to Aristotle's *De Anima*). In his *Phytopinax* (1596), Bauhin states in the preface that he has applied one name to each plant for clarity.[77] Cesalpino tended, according to Nördenskiold,[78] to focus on the fruits of plants in his classifications, a decision that also later influenced Linnaeus. Both Cesalpino and Bauhin were transitional between the older herbals tradition in which classification was either by alphabet or by utility in medicine and cooking.

THE UNIVERSAL LANGUAGE PROJECT

During the Renaissance, philosophers were more concerned with souls (especially the rational soul) and knowledge than species, but toward the end neo-Platonism was revived, particularly at Cambridge University under the general leadership of Ralph Cudworth and the later title of the "Cambridge Platonists." This movement was instrumental in the establishment of seventeenth-century science in England in particular,[79] and Bishop John Wilkins among others began discussion groups to review recent experiments and results. One of the movements they continued was that of the search for a logical system that could ensure a perfectly expressive and complete language.[80]

Mary Slaughter has discussed the universal language project at length, and she traces the influence of this approach from the medieval herbalists and the heritage of Aristotle. She noted that the sixteenth and seventeenth centuries inherited an "essentially Aristotelian" worldview that determined the way people thought about language.[81] The theory of language that they developed is founded on the epistemology of Aristotelian scholasticism—based on analytic differentiation of the *topoi*, the Categories. The "universal grammarians" Gerardus Vossius, Juan Caramuel, and Tommaso Campanella of the sixteenth century in turn gave way to the "universal language projectors" of the seventeenth. These began with the work of Francis Bacon's (1561–1626), *The Advancement of Learning* (1605),[82] in which, Slaughter says, he systematically

> described, analyzed and classified Renaissance provinces of knowledge, dividing learning into the arts of imagination, memory and reason.... Among the subdivisions of the arts of reason was rhetoric; among its parts Bacon included the Art of Transmission, or communication."[83]

---

[76] Although Sachs 1890, 33–35, thinks he was quite consistent and is dismissive of Linnaeus's "dry systematising manner," 40.
[77] Arber 1938, 168.
[78] Nordenskiöld 1929, 193f.
[79] See Mandelbrote 2007 for an account of how the Cambridge Platonists and the natural theologians like Wilkins and Boyle interacted.
[80] Slaughter 1982, Rossi 2000.
[81] Slaughter 1982, 87.
[82] Bacon 1913.
[83] Slaughter 1982, 89.

Bacon argued that letters of the Latin alphabet are conventional signs, but the Chinese characters recently reported to the Europeans[84] were "real characters, not nominal" which he said

represent neither letters nor words, but things and notions; insomuch that a number of nations whose languages are altogether different, but who agree in the use of such characters... communicate with each other in writing; to an extent that indeed any book written in characters of this kind can be read off by each nation in their own language.[85]

Words are signs that can represent the world, or they can be false and misleading, which he called "Idols of the Marketplace, ... idols which have crept into the understanding through the alliances of words and names."[86] Slaughter comments that for Bacon linguistic problems merely reflect conceptual problems. He wrote,

[t]here is no soundness in our notions whether logical or physical. Substance, Quality, Passion, Essence itself, are not sound notions; much less are Heavy, Light, Dense, Rare, Moist, Dry, Generation, Corruption, Attraction, Repulsion, Element, Matter, Form and the like; but all are fantastical and ill defined.[87]

Bacon goes on to say that we can be pretty sure that words for animal species and simple sensory perceptions are accurate enough:

Our notions of less general species, as Man, Dog, Dove and of the immediate perceptions of the sense, as Hot, Cold, Black, White, do not materially mislead us; yet even these are sometimes confused by the flux and alteration of matter and mixing of one thing with another. All the others which men have adopted are but wanderings, not being abstracted and formed by proper methods.[88]

So, there is a method (*organon*) one must adopt to do this properly, says Bacon:

There are and can be only two ways of searching into and discovering truth. The one flies from the senses and particulars to the most general axioms, and from these principles, the truth of which it takes for settled and immovable, proceeds to judgment and to the discovery of middle axioms. And this way is now in fashion. The other derives axioms from the senses and particulars, rising by a gradual and unbroken ascent, so it arrives at the most general axioms last of all. This is the true way, but as yet untried.[89]

Bacon, in effect, treats science as a bottom-up process of generalization, in opposition to the top-down classification practiced by the Aristotelians. As Slaughter comments, enumeration under Bacon's methodology is followed by classification, which is the "critical operation in the discovery and the definition of the essences of things. Natures could

---

[84] Porter 2001.
[85] Quoted in Slaughter 1982, 40.
[86] Bacon 1960, I.lix; cf. Slaughter 1982, 92.
[87] Bacon 1960, I.xv.
[88] Bacon 1960, I.xvi.
[89] Bacon 1960, I.xix.

be properly defined only when instances were properly classified into 'the true divisions' of nature."[90] For him and his contemporaries, real science meant that the labels of things were correct representations or signs of the way the world was, and that essences were found as axioms, or generalizations, made step by step from empirical inductions.

Bacon recognized that organisms and other things could deviate from the type. There is no necessity for things to be constrained by their species' essences:

> Among Prerogative Instances I will put in eighth place *Deviating Instances*, that is, errors, vagaries, and prodigies of nature, wherein nature deviates and turns aside from her ordinary course. Errors of nature differ from singular instances in this, that the latter are prodigies of species, the former of individuals. Their use is pretty much the same, for they correct the erroneous impressions suggested to the understanding by ordinary phenomena, and reveal common forms. ...
> ... we have to make a collection or particular natural history of all prodigies and monstrous births of nature; of everything in short that is in nature new, rare, and unusual.[91]

Bacon also thinks it worthwhile to discover which things are a mixture of species,[92] but it is clear that he considers the term in a much broader manner than the modern sense, and his notion of hybrids in the final phrase is quite arbitrary:

> ... I will put in the ninth place *Bordering Instances*, which I will also call *Participles*. They are those which exhibit species of bodies which seem to be composed of two species, or to be rudiments between one species and another. ...
> Examples of these are: moss, which holds a place between putrescence and a plant; some comets, between stars and fiery meteors; flying fish, between birds and fish; bats, between birds and quadrupeds; also the ape, between man and beast—
> Simia quam similes turpissima bestia nobis;[93]
> Likewise the biformed births of animals, mixed of different species, and the like.[94]

Elsewhere, in the *New Atlantis* (1627) he has his mythical sage enumerate the things his utopian society can do with animal breeding in Salomon's House:

> we make them differ in colour, shape, activity, many ways. We find means to make commixtures and copulations of divers kinds, which have produced many new kinds, and them not barren, as the general opinion is.[95]

Although he does not think non-sterile hybridization is impossible in forming new species, species are only mutable in a limited way.

---

[90] Slaughter 1982, 93.
[91] Bacon 1913, II.xxix.
[92] Bacon 1960, II.xxx.
[93] "How like us is that very ugly beast, the monkey," Ennius (239–169?BCE) as quoted in Cicero's "On the Nature of the Gods." Thanks to Tom Scharle for the reference and translation. *Simia* can mean "ape" or "monkey" and is most probably derived from the Greek *simos* (σιμός), "snub-nosed" [Liberman 2013]. In the first edition I stated wrongly that it was based on the Latin *similis*.
[94] Bacon 1960, 178f.
[95] Bacon 1913, 248.

Bacon was followed by René Descartes (1596–1650), who proposed that an artificial language could be constructed on the true divisions of nature (in his letters), but the work of the linguist John Amos Comenius (1592–1670) in *Janua linguarum researata* (1631) and *Janua linguarum researata vestibulum* (1633) made a start at a universal language answering to all things. In later work he expanded the scope of this project, and he had unrealized plans for

> an inductive history in which all things all things, which have ever been exactly observed and proved beyond all possibility of mistake to be true, are faithfully collected and set before our eyes; so that by an adequate examination of each one and by comparison of one with another the Universal Laws themselves of Nature may be brought within our knowledge.[96]

Descartes tried unsuccessfully to engage the botanist and polymath Joachim Jung (1587–1657) to edit his works. Other universal language projectors included Samuel Hartlib, Theodore Haak, Francis Lodowyck, Cave Beck, Francis Van Helmont, George Dalgarno, Athanasius Kircher, Johann Joachim Becher, Seth Ward, and various others involved in the nascent Royal Society at Cambridge.[97]

One contributor, one might say the zenith, of the universal language project was Bishop John Wilkins (1614–1672). My namesake was the brother-in-law of Oliver Cromwell and is widely regarded as the founder of the Royal Society,[98] of which he was the first Secretary. He produced *An essay towards a real character and a philosophical language* (1668), the only long-standing influence of which is said to have been the foundation of the scheme for Roget's *Thesaurus* (and the occasion for a Borges essay). In fact, Wilkins' scheme was not so much a universal language as a kind of indexical classification of concepts of his time (which, in typical style for the time and since, he thought to be universally true of the entire human condition), where words were in fact unique keys to each conceptual "address." He managed to engage (for money) a naturalist by the name of John Ray to produce a table of species according to his own a priori categories. Ray felt that this a priori scheme was too constrictive, but according to Raven it did provide him with a motivation to do better in his later publications.[99] Ray also translated the *Essay* into Latin, and remained a close friend to Wilkins until his untimely death from kidney stones, despite his objections to the underlying philosophy of the project. The Latin edition was never published.[100]

Out of the neo-Platonic resources of the Cambridge Platonists and the Universal Language Project, the transition was begun to an autonomous program of biological classification. This heritage had two opposing aspects. First, as we have seen, there was the history of species as the sharp categories of a top-down classification. The other, via Descartes and Leibniz, was the Great Chain of Being notion of continuous gradation from simpler forms to more complex. Ironically, both of these develop ideas nascent or explicit in Aristotle's writings. Lovejoy writes:

---

[96] Quoted by Slaughter 1982, p102f.
[97] Rossi 2000.
[98] Wright Henderson 1910, Wilkins 1970, Slaughter 1982.
[99] Raven 1986, 182f, 192.
[100] Clauss 1982.

The first [aspect] made for sharp divisions, clear-cut differentiations, among natural objects, especially among living beings. To range animals and plants in well-defined species, presumably (since the Platonic dualism of realms of being was still influential) corresponding to the distinctness of Eternal Ideas, was the first business of the student of the organic world, The other tended to make the whole notion of species appear a convenient but artificial setting-up of divisions having no counterpart in nature. It was, on the whole, the former tendency that prevailed in early modern biology.[101]

In the tension between sharp classification and gradual variation from one form to another in the Great Chain, we see the early stages of natural species realism, based on typological definitions, and species nominalism, based on the unreality of any divisions between them. Much of the early biological debate over species is an attempt to deal with this tension, and indeed it continues to the present day. This is made more complicated by the second main developing tradition of the distinction between "natural" and "artificial" classification.

LOCKE AND LEIBNIZ ON REAL AND NOMINAL ESSENCES

For the natural tendency of the mind being towards knowledge; and finding that, if it should proceed by and dwell upon only particular things, its progress would be very slow, and its work endless; therefore, to shorten its way to knowledge, and make each perception more comprehensive, the first thing it does, as the foundation of the easier enlarging its knowledge, either by contemplation of the things themselves that it would know, or conference with others about them, is to bind them into bundles, and rank them so into sorts, that what knowledge it gets of any of them it may thereby with assurance extend to all of that sort; and so advance by larger steps in that which is its great business, knowledge. This, as I have elsewhere shown, is the reason why we collect things under comprehensive ideas, with names annexed to them, into genera and species; i.e. into kinds and sorts.

Locke, *Essay on human understanding*[102]

A friend of the Cambridge Platonists and universal language projectors was John Locke (1632–1704). But Locke was also a friend of Robert Boyle, the atomistic ("corpuscularean") chemist, whose philosophy demoted sensory perceptions from immediate empirical experiences of things to secondary qualities, since we do not observe the corpuscles of which things are composed. Locke followed Boyle rather than Bacon in this, and so our knowledge of things must be nominal rather than essential.[103] Locke is sometimes called a nominalist, and there is a sense in which this is true, but it relates to his views about names, and not about the underlying things the names refer to.[104] Names denote abstractions, and some names denote the

---

[101] Lovejoy 1936, 227.
[102] Book II, chap. 32, §6.
[103] Slaughter 1982, 193–207.
[104] See Mackie 1976, chapter 3, Fales 1982, who hold that Locke did not deny real essences, but that they were the underlying substratum, or microstructure; but see also Stanford 1998 for a rejection of the substratum argument for real essences.

essences of ideas. Locke does not deny that there are what he calls "real essences," but rejects only that the essences of kinds, or "sortals," as he calls them, agree to anything else but "nominal essences."

Locke is not particularly remembered by biologists for his contribution to the species debate, but he should be.[105] His views, expressed especially in chapters 3 to 7 of Book III of the *Essay*, are the first statement of a position one may call *species conventionalism*, and which is held today even by those who reject his essentialistic notion of names. He triggered a response by Leibniz and was influential on French naturalists, including Buffon and Lamarck.

> The learning and disputes of the schools having been much busied about genus and species, the word essence has almost lost its primary signification: and, instead of the real constitution of things, has been almost wholly applied to the artificial constitution of genus and species. It is true, there is ordinarily supposed a real constitution of the sorts of things; and it is past doubt there must be some real constitution, on which any collection of simple ideas co-existing must depend. But, it being evident that things are ranked under names into sorts or species, only as they agree to certain abstract ideas, to which we have annexed those names, the essence of each genus, or sort, comes to be nothing but that abstract idea which the general, or sortal (if I may have leave so to call it from sort, as I do general from genus), name stands for. And this we shall find to be that which the word essence imports in its most familiar use.[106]

Locke considers "species" to be merely the Latinized version of the good English word "sort" or "kind,"[107] and held that species are conventional names used mainly for specialists to communicate. In fact, most of the then-current species names were based on what we would now call "folk taxonomy":

> *This shows Species to be made for Communication.*—The reason why I take so particular notice of this is, that we may not be mistaken about *genera* and *species*, and their *essences*, as if they were things regularly and constantly made by nature, and had a real existence in things; when they appear, upon a more wary survey, to be nothing else but an artifice of the understanding, for the easier signifying such collections of *ideas* as it should often have occasion to communicate by one general term; under which divers particulars, as far forth as they agreed to that abstract *idea*, might be comprehended. And if the doubtful signification of the word *species* may make it sound harsh to some, that I say the species of mixed modes are "made by the understanding"; yet, I think, it can by nobody be denied that it is the mind makes those abstract complex *ideas* to which specific names are given. And if it be true, as it is, that the mind makes the patterns for sorting and naming of things, I leave it to be considered who makes the boundaries of the sort or *species*; since with me *species* and *sort* have no other difference than that of a Latin and English *idiom*.[108]
>
> But supposing that the *real essences* of substances were discoverable by those that would severely apply themselves to that inquiry, yet we could not reasonably think that the *ranking of things under general names was regulated by* those internal real

---

[105] Cain 1997, Jones 2007 have taken some steps in this direction.

[106] Book III, chap. III, §15.

[107] Book III, chap. I, §6.

[108] Book III, chap. V, §9.

constitutions, or anything else but *their obvious appearances*; since languages, in all countries, have been established long before sciences. So that they have not been philosophers or logicians, or such who have troubled themselves about *forms* and *essences*, that have made the general names that are in use amongst the several nations of men: but those more or less comprehensive terms have, for the most part, in all languages, received their birth and signification from ignorant and illiterate people, who sorted and denominated things by those sensible qualities they found in them; thereby to signify them, when absent, to others, whether they had an occasion to mention a sort or a particular thing.[109]

Locke here recognizes a distinction between folk taxonomy and a proper (philosophical) enquiry into the scientific issues. Leibniz in the *New Essay on Human Understanding*[110] paraphrases this more succinctly (Philolethes is Locke, Theophilus is Leibniz):

PHIL. §25. Languages were established before sciences, and things were put into species by ignorant and illiterate people.

To which *he* responds,

THEO. This is true, but the people who study a subject-matter correct popular notions. Assayers have found precise methods for identifying and separating metals, botanists have marvelously extended our knowledge of plants, and experiments have been made on insects that have given us new routes into the knowledge of animals. However, we are still far short of halfway along our journey.

Leibniz is more of an optimist about the causal powers that form species being available to investigators than Locke, who seems to propose a permanent conventionalism based on the current inability to define species according to their "internal real constitutions." Nevertheless, Leibniz held to a view that species were not real as discrete objects. He was an adherent of the *scala naturae* as the Great Chain of Being was known in Latin, and popularized the *lex completio*—that there could be no incompleteness in the world as made by a beneficent God.

Locke effectively argued that *species* were simply names we gave to similarities in perceived phenomena:

Nature makes the similitudes of substances. This, then, in short, is the Case: *Nature makes many particular Things, which do agree one* with another in many sensible Qualities, and probably too in their internal Frame and Constitution: but it is not this *real Essence* that distinguishes them into *Species*; it is *Men* who, taking occasion from the Qualities they find united in them, and wherein they observe often several Individuals to agree, *range them into sorts, in order to their Naming*, for the convenience

---

[109] Book III, chap. 6, §25.
[110] Leibniz 1996, 319.

of comprehensive Signs; under which Individuals, according to their Conformity to this or that abstract *Idea*, come to be ranked as under Ensigns: so that this is of the Blue, that the Red Regiment; this is a Man, that a Drill:[111] and in this, I think, consists the whole business of *Genus and Species*.[112]

Locke discussed, rather interestingly, the presence in species of divergences from the type.[113] In this he was preceded by Cusa, as we have seen, and also by Francis Bacon in the *New Organon*,[114] but Locke's discussion is surprisingly modern in a way that Bacon's is recognizably medieval. He greatly influenced, among others, Buffon, through the writings of French admirers of Locke's empiricism and new way of ideas (from the Greek *ideai*, a form of *eidos*). He is often thought to be solely a species nominalist, as Buffon transitorily was, but it seems more accurate to say that he believed our ideas and associated names were conventional, but that, as Bacon thought, there was some underlying essence that was likely to remain out of our reach. This idea was famously given its canonical expression in the doctrine of the *noumenal* and the *phenomenal* of Kant.

Locke's views on relations[115] are also significant: Although he rejects the idea that relations are outside the understanding, he nevertheless has a category of "natural relations" that apply to genealogical relationships between organisms. Although we only have access to the ideas of these simple and mixed modes, they are nevertheless something in the natural world. He does assert here that the *words* used for the relations, such as "father," "brother," and "cousins-germaine," are conventional, though, even if the relations are not. Relations resurfaced in the writings of William Hamilton,[116] Russell and Whitehead,[117] and throughout the middle of the twentieth century.

WILKINS AND RAY: PROPAGATION FROM SEED

John Ray (1627–1705) was a seventeenth-century naturalist who, in conjunction with nobleman Francis Willughby (1635–1672), prepared the first systematic flora for a region—at first of Cambridgeshire, and later of Britain.[118] In the *Historia plantarum* (1686–1704), he and Willughby attempted to describe all known species of plants. He also collaborated with Willughby, before his untimely death, on a treatment of insects, animals, and fishes. To this end, he needed to define "species," and he was the first to do so entirely in a biological context. This strongly influenced Linnaeus' conceptions of species and other ranks.

---

[111] That is, a mandrill, probably referring to baboons.
[112] *Essay* III.iv.36.
[113] Book II, chap. VI, §§16–17, 26–27.
[114] Book II, §29 [cf. Glass 1959, 36].
[115] *Essay*, Book II, chapter XXVIII, §2.
[116] Hamilton et al. 1874, Vol. 2.
[117] Whitehead and Russell 1910.
[118] Raven 1986.

In the *Historia plantarum generalis*, in the volume published in 1686, Ray defined a species thus:

> So that the number of plants can be gone into and the division of these same plants set out, we must look for some signs or indications of their specific distinction (as they call it). But although I have searched long and hard nothing more definite occurs than distinct propagation from seed. Therefore whatever differences arise from a seed of a particular kind of plant either in an individual or in a species, they are accidental and not specific. For they do not propagate their species again from seed; thus, for example, we do not have Caryophylli with a full or multiple flower distinct in species from Caryophylli with a simple flower, because they derive their origin from their seed, and when sown from seed produce simple Caryophylli again. But those which never arise with the same appearance from seed, are indeed to be considered specific; or if comparison is made between two kinds of plant, those plants which do not arise from the seed of one or the other, nor when sown from seed are ever changed one into the other, these finally are distinct in species.
>
> For thus in animals a distinction of sexes does not suffice for proving a diversity of species, because both sexes arise from the same kind of seed and frequently from the same parents, although by many striking accidents they differ among themselves. It requires no other proof that a bull is of the same species as a cow, and a man as a woman, than that both have very often arisen from the same parents or from the same mother. So, equally in plants, there is no more certain indication of a sameness of species than to be born from the seed of the same plant either specifically or individually. For those which differ in species keep their own species for ever, and one does not arise from the seed of the other and vice versa.[119]

Mayr notes, "Here was a splendid compromise between the practical experience of the naturalist, who can observe in nature what belongs to a species, and the essentialist definition, which demands an underlying shared essence."[120] However, it seems to me the influence of the generative conception is more apparent here. Ray is dealing with the Aristotelian problem of accidental variation—like Locke he believes there must be a real essence, and not merely a nominal one. He conjectures that the real essence is, or rather defines it to be, based upon descent by generation. But in most cases, as many have observed since (see below, Buffon), descent is *not* observable; and he clearly did not observe the fact that one species never springs from the seed of another. In fact, he wrote, in the *Methodus plantarum* of 1682:

> ... I would not have my readers expect something perfect or complete; something which would divide all plants so exactly as to include in positions anomalous or peculiar; something which would so define each genus by its own characteristics that no species be left, so to speak, homeless or be found common to many genera. Nature does not permit anything of the sort. Nature, as the saying goes, makes no jumps and passes from extreme to extreme only through a mean. She always produces species

---

[119] Lazenby 1995, 1157, Vol. III, chapter 20.
[120] Mayr 1982, 257.

intermediate between higher and lower types, species of doubtful classification linking one type with another and having something common with both—as for example the so-called zoophytes between plants and animals.[121]

According to Glass, Ray did allow some limited transmutation between related species, especially hybridization, a problem that Linnaeus later also had to accommodate. Ray also treated *species* as the Aristotelian logic did, as a subordinate kind to a superordinate genus, but he was not too strict. Raven says that he attempted

> not only to formulate a correct definition and arrangement of the 'genera' (that is, the large groups or orders), but to marshal correctly the 'species subalternae' (or genera) and the 'species infimae' (or species).[122]

An originator of the British natural theology tradition—he wrote the book *Wisdom of God Manifested in the Works of the Creation*[123]—Ray also had a problem with extinction. It was not that he did not find it an operational concept; it was ruled out in terms of the principle of plenitude.[124] Initially he thought that fossil forms would be found alive elsewhere, but later he was forced into denying that fossils were even the remnants of living forms, but had instead grown within the rocks.

Ray is responsible for formulating the first explicit entirely *biological* notion of species, but this is not the same as saying that he presented what we would now consider a Biological Species Concept. For him, this was a sense of "species" that applied to reproducing *forms*, that is, living things. It was the first time a concept was proposed that applied *only* to the classification of living things—prior to this, all such concepts were general-duty concepts of classification that were then applied equally to, say, books or rocks as to life. Ray clearly saw his classificatory logic as the continuation of the Aristotelian, scholastic, tradition, but his adherence to the Great Chain of Being, its neo-Platonist heritage, and to the idea that there needed to be a real essence, led him to propose not so much an operational concept as Mayr would have it, but a metaphysical one.

---

[121] Ray 1682, 2, Prefatio ad Lectorum. Quoted in Glass 1959, 35. The provenance of the saying Ray adopts here—*natura non facit saltum*—is interesting. Usually associated with Linnaeus, it can be found also in Leibniz and even, in a form, in Albertus Magnus: "nature does not make [animal] kinds separate without making something intermediate between them, for nature does not pass from extreme to extreme *nisi per medium*" [quoted in Lovejoy 1936, 79]. The idea can be traced back to the views of Plotinus and Porphyry, and probably also to the Gnostic idea of emanation. Ray also used similar phrases: *Natura nihil facit frustra* (nature makes nothing in vain) and *Natura non abundant in superfluis, nec deficit in necessarius* (Nature abounds not in what is superfluous, neither is [it] deficient in necessaries)—in the *Wisdom of God* quoted in Cain 1999, 233. This saying, adopted much later by Darwin, is also found in a similar form in Leibniz's *New Essays*: "In nature everything happens by degrees, and nothing by jumps," [Leibniz 1996, 473, Book IV, chapter xvi]. It is an expression of the Great Chain of Being.
      A recent English translation of Ray's *Methodus* has been published by the Ray Society [Ray 2015] but I have been unable to access it.

[122] Raven 1986, 219; Quoted in Lazenby 1995, 49–50, Vol. I.

[123] Ray 1691, Gould 1993, 140. Ray says a lot about species in this work, mostly defending the constancy of propagation from seed.

[124] Bowler 2003, 37f.

As mentioned above, Ray had prepared a table of species before the *Historia* for Bishop John Wilkins' magnum opus *Essay Toward a Real Character and Philosophical Language*, published by the Royal Society in 1668.[125] He wrote to Martin Lister (1639–1712) complaining of the limitations of Wilkins' schemata:

> This next week we expect the Bishop of Chester [Wilkins] at Middleton, who desires our assistance in altering and amending his tables of natural history. To make exact philosophical tables, you know, is a matter very difficult, not to say impossible; to make such as are tolerable requires much diligence and experience, and is work enough for one man's whole life, and therefore we had need call in all the assistance we can from our friends, especially being not free to follow nature, but forced to bow and strain things to serve a design according to the exigency of the character.[126]

Wilkins began his *Essay* with the logic of division:

KIND. I. That common Essence wherein things of different natures do agree, is called
    GENUS, general, common Kind.
    That common nature which is communicable to several Individuals, is called
    SPECIES, Sort or special kind, specifie, specifical. Breed.[127]

In his table of his General Scheme, he explicitly describes living kinds:[128]

All kinds of things and notions, to which names are to be assigned, may be distributed into such as are either more

> *General*; namely those Universal notions, whether belonging more properly to
>   *Things*; called TRANSCENDENTAL
>
> …
>
>   *Words*;   DISCOURSE. IV
> *Special*; denoting either
>   CREATOR. V
>   *Creature*; namely such things as were either created or concreated by God,
>   not excluding several of those notions, which are framed by the minds of
>   men, considered either
>   *Collectively*; WORLD. VI
>   *Distributively*; according to the several kinds of Beings, Whether such as do
>   belong to
>   *Substance*;
>   *Inanimate*; ELEMENT. VII
>   *Animate*; considered according to their several
>   *Species*; whether
>   *Vegetative*
>   *Imperfect*; as Minerals.
>   STONE. VIII
>   METAL. IX
>   *Perfect*; as Plant,

---

[125] Wilkins 1668; a reprint is available [Wilkins 2002]. An excellent discussion of the social and intellectual debate of the time is in DeLacy 2016.

[126] April 28, 1670, in Lankester 1848, 55f. On the arachnologist Lister, see Roos 2011.

[127] *Essay*, 26.

[128] *Essay*, 23. I have recast his bracketed table as an indented list for clarity.

HERB consid. accord. to the
LEAF. X
FLOWER. XI
SEED-VESSEL. XII
SHRUB. XIII
TREE. XIV

Sensitive;

EXANGUIOUS. XV
Sanguineous;
FISH. XVI
BIRD. XVII
BEAST. XVIII

Parts.

PECULIAR. XIX.
GENERAL. XX

Wilkins was connected to the so-called "Cambridge Platonists," a neo-Platonic school that was directly and indirectly responsible for inspiring much scientific work done in Britain at the time, and by whom the Royal Society was influenced. The neo-Platonic element is obvious in Wilkins' use of privative classes ("imperfect," "exanguious") and yet he classes organisms on the Aristotelian theory of vegetative and sensitive properties. Overall he changes the ten Aristotelian categories to "forty *Genus's*."

Like Wilkins, Ray was also influenced by Ralph Cudworth, whose ideas were consciously in the school of Plotinus and Porphyry. In addition, Ray had received the standard *Trivium* education (Grammar, Logic, and Rhetoric) that was presented at the grammar schools of the time, and which at sixteen and a half he began also to receive from Cambridge. With his friend and mentor Henry More, Ray was thoroughly inculcated in the techniques and terminology of the older scholastic logic. Although he accepted the *scala naturae*, he nevertheless held to the fixity of species since creation, a view that he bequeathed to Linnaeus. That Ray held to fixity of species as created by God is illustrated by this comment, made in a letter:

... the number of species being in nature certain and determinate, as is generally acknowledged by philosophers, and might be proved also by divine authority, God having finished his works of creation, that is, consummated the number of species in six days.[129]

NEHEMIAH GREW: THE ESSENCE OF SPECIES

Almost contemporaneous with Ray was Nehemiah Grew (1641–1712), who undertook to bring the anatomical studies of plants under the scope of microscopic studies. Grew knew Ray, and also John Wilkins (to whom his *Anatomy* is dedicated). His notion of species was fairly standard, but what is remarkable, apart from the exquisite figures of cell structure, in his treatment is his justification of the use of classification as a way of making inductive generalizations. *The anatomy of plants*

---

[129] Quoted in Greene 1959, 131. See also "Mr Ray on the Number of Plants" in Derham 1718, 344–351.

is a series of lectures given before the Royal Society, including his general overview of botany ("vegetables"—Grew is primarily interested in crop plants), *An idea of a philosophical history of plants*, given on January 8, and January 15, 1672.

[12. §.] For in looking upon divers *Plants*, though of different *Names* and *Kinds*; yet if some affinity may be found betwixt them, then the *Nature* of any one of them being well known, we have thence ground of conjecture, as to the *Nature* of all the rest. So that as every *Plant* may have somewhat of *Nature individual* to it self; so, as far as it obtaineth any *Visible Communities* with other *Plants*, so far, may it partake of *Common Nature* with those also.[130]

In short, Grew is arguing that if we can identify the nature of one kind, we can expect that others will share in it to the extent that they share an "affinity" with it. Nature is here a kind of inductive warrant, because genera share their own properties, and kinds will therefore also share the generic properties. He expands on this later, noting that if we know the essence of a plant (the causes of its being and becoming what it is) we can know what is necessary to it, and what is accidental— based on its internal structure as well as its outward appearance:

20. §. From all which, we may come to know, what the *Communities* of *Vegetables* are, as belonging to all; what their *Distinctions*, to such a Kind; their *Properties*, to such a Species; and their *Peculiarities*, to such Particular ones. And as in *Metaphysical*, or other Contemplative Matters, when we have a distinct knowledge of the *Communities* and *Differences* of Things, we may then be able to give their true *Definitions*: so we may possibly, here attain, to do likewise: not only to know, That every *Plant* Inwardly differs from another, but also wherein; so as not surely to Define by Outward *Figure* than by the Inward *Structure*. What that is, or those things are, whereby any *Plant*, or Sort of *Plants*, may be distinguished from all others. And having obtained a knowledge of the *Communities* and *Differences* amongst the *Parts* of *Vegetables*; it may conduct us through a *Series* of more facile and probable *Conclusions*, of the ways of their *Causality*, as to the *Communities* and *Differences* of *Vegetation*.[131]

Thus, essence for Grew is a matter of causally efficacious internal microstructure, which we would now regard as a natural kind. This makes him one of the few actual material essentialists we shall encounter in the history of biology before the nineteenth century.[132] He employs the Aristotelian definitional approach, naturally, but treats the definitions as identifying the true causes, the "active capacitating Causes," of the nature of the plants.

53. §. The prosecution of what is here proposed, will be requisite, To a fuller and clearer view, of the *Modes* of *Vegetation*, of the *Sensible Natures* of *Vegetables*, and of their more Recluse *Faculties* and *Powers*. First, of the *Modes* of *Vegetation*. For suppose we were speaking of a *Root*; from a due consideration of the *Properties*

---

[130] Grew 1682, 6.
[131] Grew 1682, 10.
[132] Material but not taxonomic, as he is not concerned with plant taxonomy but physiology (anatomy).

of any *Organical Part* or *Parts* thereof; 'tis true, that the real and genuine *Causes* may be rendred, of divers and other dependent *Properties*, as spoken generally of the whole *Root*. But it will be asked again, What may be the *Causes* of those *first* and Independent ones? Which, if we will seek, we must do by inquiring also, What are the *Principles* of those *Organical Parts*? For it is necessary, that the *Principles* whereof a Body doth consist, should be, if not all of them the *active*, yet the *capacitating Causes*, or such as are called *Causae sine quibus non*, of its becoming and being, in all respects, both as to *Substance* and *Accidents*, what it is: otherwise, their Existence, in that Body, were altogether superfluous; since it might have been without them: which if so, it might then have been made of any other; there being no necessity of putting any difference, if neither those, whereof it is made, are thought necessary to its Being.[133]

Pretty clearly, Grew thinks of these causes as developmental, and hence generational. The outer, or morphological as we would now say, appearances may lead us to uncover those developmental properties. Grew thought these were to be found in the microscopic anatomy and physiology of the organisms, and his illustrations of the microstructures of the leaves, stalk, and so on remain exceptionally accurate.

TOURNEFORT: NAMES FOR SENSIBLE DIFFERENCES

In his *Botanical institutions* (*Institutiones rei herbariae*, 1700, earlier published in 1694 as *Elemens botanique*) Joseph Pitton, de Tournefort (1628–1708) held that it did not really matter whether one diagnosed a species or a variety:

> I not only enumerate the several Species of Plants, but often mention what the Botanists call Varieties; not at all solicitous whether they be really the same Species only varied and somewhat diversified, for as they differ in some sensible Qualities, they ought to be distinguished by peculiar Titles.[134]

He used both flowers and fruits to define species, in contrast to the Linnaean artificial account based on just the sexual apparatus of the plant. So far as I can find, Tournefort did not define species, but he did use many different characters—the root, leaves, stalk, branch—to identify the differentia between species of a genus. However, he also used, but did not rely upon, accidents such as color, taste, scent, size, site, and similarity to popularly known objects.[135] Tournefort clearly thought only of species as definitional and nomenclatural items. However, it has been said that he treated species, genera, and classes as objectively real objects.[136]

---

[133] Grew 1682, 20.
[134] Tournefort 1716–30, 2.
[135] Tournefort 1700, 63.
[136] Leroy 1956, 327:

> Pour TOURNEFORT il y a une réalité objective *des espèces, des genres, des classes*, et d'autre part la possibilité d'échafauder une *classification naturelle* vraiment scientifique puisque indépendante de l'observateur.

LINNAEUS: SPECIES AS THE CREATOR MADE THEM

Like Ray, the traditional logic was also taught to a poor Swedish student named Carl Linnaeus.[137] Born in 1707, Linnaeus died in 1770, the most celebrated Swede of his day. In 1761, he was knighted with the vernacular name Carl von Linné, and took the Latinized name Carolus Linnaeus. Linnaeus was a botanist, and trained in Holland where he published his first botanical works.

Before Linnaeus, species were given all kinds of descriptive names, usually in Latin, up to ten words or so long. Each author made up their own terms, and there was no real convention for referring to species. On Linnaeus' account, both species and genera were fixed, real and known by definitions. He apparently believed that the genus was more real than the species, and he allowed late in life that species may occasionally arise, but only within genera, through hybridization.

Some consider Linnaeus to be an essentialist regarding species.[138] This was due to the fact that, unlike the medieval logical conception, for Linnaeus all species (at least in botany, zoology, and mineralogy) were infimae species. He attempted to provide a diagnostic definition for each species, although his practice and adopted motto "*In scientia naturali principia veritatis observationibus confirmari debent*" ("in natural science, the principles of truth ought to be confirmed by observation")[139] suggests that he was not firmly wedded to a priorism.

In the 1735 first edition of the *Systema Naturae*[140] Linnaeus proposed a system of five ranks under the *summum genus* of the Empire of Nature (*Imperium Naturae*):[141]

Methodical arrangement, which is the soul of science, indicates every natural body at first sight, so that it may be known by its own name; and this name points out whatever the industry of the age has discovered concerning the body to which it belongs: Thus, amidst the greatest apparent confusion of things, the order of Nature is seen to retain the highest degree of exactness. This systematic arrangement is most conveniently divided into branches, subordinate to each other, which have received various appellations; thus,

Class,	Order,	Genus,	Species,	Variety.
Highest genus,	Intermediate genus,	Proximate genus,	Species,	Individual.
Province,	District,	Parish,	Ward,	Hamlet.
Legion,	Battalion,	Company,	Mess,	Soldier.

---

[137] I am informed by Staffan Müller-Wille (*pers. comm.*) that Linnaeus, being from a relatively poor district of Sweden, Småland, known (presumably by an Englishman) as the "Scotland of Sweden," was taught from old standard textbooks, and not out of the neo-Platonists early or late, as far as is recorded [see also Goerke 1973, Frängsmyr 1983, Koerner 1999]. According to Hagberg 1952, 44ff, Linnaeus was greatly influenced by Aristotle's *Historia Animalium* as a young student.

[138] For example, Mayr 1969, Stafleu 1971, Mayr 1982.

[139] Stafleu 1971.

[140] Linné 1735, Linné 1956, reprint of 10th edition, Linné 1758, 7.

[141] Linné 1792.

Linnaeus compares his hierarchical branches to the logical, regional, and military equivalents. It is also telling that he uses political terms for his own ranks; following on from the Imperial role God plays over nature, but also echoing his hope that a good taxonomy will serve the Swedish empire.

Later taxonomic conventions added the ranks of *Phylum* between Kingdom and Class, and *Family* between Order and Genus, giving seven ranks.[142] The philosophical notion of species was not entirely helpful in botany, so Linnaeus changed it a little. Instead of any number of subaltern genera, he made the scale of classes absolute, and instead of working downwards, he started in the middle (at the genus). Linnaeus' ranks began at species, and these existed in genera. Hence, to name a species you needed to give the generic name and the species name. Humans are members of the genus *Homo* (or Man; according to Linnaeus, one of several[143]) and our species is called *sapiens* (the wise one). So in Latin our "name" is "the wise man." Humans, under his initial system, are:

Animals (*Regnum Animale*)
  Mammals (*Classis Mammalia*)
    Primates (*Ordo Primates*)
      Man (*Genus Homo*)
        Wise or rational (*Species sapiens*)[144]

The "rational animal" definition of medieval logical taxonomies is evident. What Linneaus did differently was to make species and genera fixed ranks. He established this universal system for the naming and classification of all organisms. There were, for example, various kingdoms—plants (Plantae) and animals (Animalia). Each species had a street address (its generic name, or *genus*) and a street number (its species name, or *epithet*).[145] Now, taxonomists (those who classify taxa, or groups of organisms) could use a single and relatively simple system for their organisms, and all could agree on how to name them, and what to name.

Linnaeus was most definitely a special creationist—that is, he believed that each species was created specially by God, and Haller famously said of him that he thought himself a "second Adam."[146] His project of classification relied heavily on the logic of scholastic philosophy, which he employed in ways reminiscent of the Universal Language Project, to impose order upon the world. He said once to his

---

[142] Later, further ranks in the so-called Linnean System, also known during the early nineteenth century as the "natural system," were added; see Appendix A.

[143] See Broberg 1983 for a comprehensive account of Linnaeus' treatment of humans in relation to apes and the reaction he received from the religious, both naturalists and theologians.

[144] Initially he referred to humans as *Homo diurnis*, in contrast to *Homo nocturnis* for the orangutan. Some have held that *sapiens* should be read as "the knowing man" [Broberg 1983, 176]. The classical meaning of *sapiens* is "wise or discerning" or "sage," and it is used this way in Ovid, for example. Linnaeus, who quotes the Oracle's advice "know thyself" (*Nosce te ipsum*) on which Socrates based his investigations as the definition of *Homo sapiens*, may have been alluding to this. However, I still think *wise* is a better translation. Thanks to Polly Winsor for pointing this out.

[145] In modern practice, the genus name is always capitalized and the species epithet is always lower case, and both are always italicized. Other taxonomic ranks are capitalized but not italicized.

[146] Ramsbottom 1938, 195n.

friend Sauvages that he was unable "to understand anything that is not systematically ordered,"[147] and he was obsessive in this regard. To this end, he established a system that did impose that order on the living and nonliving world.

Linnaeus' widely cited "definition" of *species*, repeated throughout his writings in various wordings, is:

> There are as many species as the diverse forms created in the beginning (*Species tot sunt numeramus, quot diversae formae in principio sunt creatae*).[148]

This is not really a definition, and is more of a statement of piety. However, in 1744 he was forced to allow that some species are the result of hybridization, at least in plants, because (he thought) he observed it happening. A species of plant he placed in a genus *Peloria* (from the Greek *pelor*, meaning monstrosity) was in stem and leaf structure part of the *Linaria* genus, but the flower was clearly different.[149] This admission was widely known by subsequent botanical writers.[150] Still, he thought that genera were real and the possibilities for change limited. Per Larson, Linnaeus imagined in the *Fundamenta fructifications*

> that God created one species for each natural order of plants differing in habit and fructification from all others. These species, mutually fertile, gave birth to as many genera as there were different parents, their fructification somewhat changed.[151]

In the *Pralectiones*,[152] Linnaeus went further:

> The principle being accepted that all species of one genus have arisen from one mother through different fathers, it must be assumed:
>
> 1. That in the beginning the Creator created each natural order only with one plant with reproductive power.
> 2. That by their various mixings different plants have arisen which belong to the mother's natural order as they are similar to the mother with regard to their fructifications, and are, as it were, species of the order, that is, genera.
> 3. We may assume that plants have arisen within the orders, that is, by genera of one order, may mix with each other. In this way there will arise species that should be referred to the mother's genus as her daughters.[153]

Linnaeus thus employed the Great Chain of Being in a rather unusual way. Most "chainists" accepted what was later called the Principle of Plenitude (the *lex completio*), which stated that God would create everything that could be created, since he would not make an incomplete creation. This usually meant that species graded into each other in a series of varieties. Linnaeus instead represented species using the

---

[147] Lindroth 1983, 23.
[148] *Fundamenta botanica* No. 157, [Linné 1736, 18].
[149] Hagberg 1952, 196f, Glass 1959.
[150] For example, Lee 1810, Gray 1821.
[151] Larson 1967, 317.
[152] In 1744. A version may be found in Linné 1792, 16–18.
[153] Quoted in Larson, *loc. cit.*

metaphor of countries adjoining each other.[154] In his early writing, all the territory is regarded as pretty much filled—as he said, nature does not make jumps—but the countries are discrete and distinct from one another. In the later work, this strict fixism of the first edition of the *Systema Naturae* has been modified. All hybrids did was fill in a rare empty bit of territory in God's time and plan. The borders were set by the genera, and all genera arose from a single species created by God.

At the end of the 1750s, says Hagberg,[155] Linnaeus was in a state of perplexity with respect to species. In 1755, his student Nils Dahlberg published *Metamorphoses plantarum*,[156] dealing primarily with the development of plants, but also with monstrosities and varieties. Such later hybrids he called the "children of time" in an anonymous entry in a competition at St. Petersburg in 1759,[157] and also in the *Species plantarum*, where he speculated that a species of *Achillea* (yarrow, or staunchweed), *alpina*, might have formed from another, *ptarmica*, "*[a]n locus potuerat ex prae-cedenti formasse hanc?*" ("Could this have been formed from the preceding one by the environment?"[158]). Hagberg says, "Linnaeus never succeeded in pin-pointing his new conception of species. But the old one, that formed the basis of *Systema Naturae*, was utterly and irrevocably abandoned." In fact, reports Lindroth,[159] he started seeing hybrids everywhere, even where they were not.

Linnaeus also noted that species grew differently according to the conditions of their locale. Of the genera *Salix*, *Rosa*, *Rubus*, and *Hieracium* (willows, roses, brambles, and hawkweeds), Linnaeus said that their description was problematic because of variability ("metamorphosis") of form in different soils and climates.[160] Linnaeus also experimented on propagating a hybrid geranium, with success, in 1759; he believed that maternal influences of hybrids affected the "medullary substance" and fructification of plants, but the leaf structure was due to the paternal species.[161] As time went on, he removed the statement that there were no new species from his 1766 edition of the *Systema Naturae*, and crossed out the statement *natura non facit saltum* from his own copy of his *Philosophia Botanica*. A full list of Linnaeus' various pronouncements on species can be found in Ramsbottom.

When Linnaeus was working, European trade and exploration were limited. Linnaeus himself classified around 6000 species of mainly Mediterranean and northern European plants, and later animals.[162] This was more than had been done before, but still it was a fraction of what we know today. His students and adherents sent him specimens from around the world, and there was a steady "trade" in

---

[154] *Philosophia botanica* §77 [Linné 1751]:

    Nature does not make leaps.
    All plants exhibit their contiguities on either side, like territories on a geographical map [translated in Linné and Freer 2003].

[155] Hagberg 1952, 199.
[156] Dahlberg and Linné 1755.
[157] Hagberg 1952, 201f.
[158] Volume II, 1266 of the second edition, quoted in Greene 1959, 134.
[159] Lindroth 1983, 95.
[160] Ramsbottom 1938, 200f.
[161] Ramsbottom 1938, 210f.
[162] Stafleu 1971.

specimens between him and other taxonomists and collectors.[163] Linnaeus hoped
that his system would enable taxonomists to list all actual species, but he also knew
that his system was artificial—that is, not the pure result of studying the actual characters of organisms, but also imposing an a priori scheme on them for convenience.
He hoped there *would* be a "natural" scheme developed on the basis of an aggregation of characters, but he was never able to do more than a partial sketch of one. In
his later work, he set up a "rational" system that allowed for there to be 3600 genera
in plants, each of which could generate species through hybridization. Although this
was supposed to be a "natural" system (one based on the closeness of resemblance
of all traits of the organisms and not just a single character), in fact he chose just
three features of plants and restricted the varieties to 60 types of each (hence $60^3 =$
216,000 maximum of plant species). However, this was fragmentary and in an appendix, and was not developed further.[164]

In summary, Linnaeus proposed a five-rank taxonomic system, and there were
only a set number of species possible, although later he was forced by various observations, including his own, to accept that new species could be created through hybridization. All that remains of his taxonomic enterprise now are the names and general
ranks of his system, but even this has been dramatically modified, with such groups
as tribes, sub-families, and so on being added to deal with the massive increase in
species discovered since. There are now as many as 18 ranks or more, each with
supra- and sub-ranks, in the taxonomic hierarchy that is called "Linnaean."[165]

Linnaeus distinguished between the diagnostic characters (*characters*) and actual
traits (*notae*) of organisms, but it seems not much came of this distinction. He appears
to have despaired of a natural system in his foreseeable future, and so promoted a
purely diagnostic and hence conventional taxonomy, even though he believed that
species were themselves natural, along with genera. This tension underlies much of
later taxonomy.

Strictly speaking, Linnaeus did not *have* a "species concept," despite recent arguments that he held to a "biological" species concept.[166] If he "had" a "concept" it was
Ray's generative concept from seed, and merely involving generation is insufficient
to make it a reproductive isolation conception of species, which is what "biological"
implies in this context. However, it is clear that he regarded interfertility as a good
test of species.

BUFFON: DEGENERATION, MULES, AND INDIVIDUALS

Georges-Louis Leclerc, Comte de Buffon (1707–1788), referred to simply as Buffon,
was one of the last naturalists with an encyclopedic knowledge of zoology. Indeed,
he effectively defined that discipline. He was a French aristocrat (a count) who
superintended the "King's Garden" (*Le Jardin du Roi*, later *Le Jardin des Plantes*).

---

[163] Müller-Wille 2003.
[164] Linné 1787.
[165] Cf. Mayr and Ashlock 1991. See Appendix A.
[166] Müller-Wille and Orel 2007. Linnaeus did use interfertility as a criterion for conspecificity, but that
     was neither operational nor novel at the time.

His pupil and later associate was the famous early evolutionist Lamarck, but Buffon was not what we would understand to be an evolutionist himself.

Buffon strongly disapproved of Linnaeus' binomial system and particularly of his use of sexual characters in discriminating plants. As a result, he and his followers were often in argument and political maneuverings against the Linnaeans. He was the primary author of the 44-volume *Natural history, with particular reference to the Cabinet of the King (Histoire naturelle[167])* of which he issued 36 volumes, and in the course of this stylistically elegant but often confusing and sometimes contradictory series he made a number of passing comments regarding species, which influenced many later ideas on the subject.

Buffon was a relatively standard adherent to the Great Chain—he adopted the "law of continuity" (*lex continua*) of Leibniz and his followers, but he did not necessarily accept the Principle of Plenitude, and so did not expect that every possible kind of species would necessarily exist. He wrote that it was an error in metaphysics trying to find a natural definition of species.

> The error consists in a failure to understand nature's processes (*marche*), which always take place by gradations (*nuances*). … It is possible to descend by almost insensible degrees from the most perfect creature to the most formless matter. … These imperceptible shadings are the great work of nature; they are to be found not only in the sizes and forms, but also in the movements, the generations and the successions of every species. … [Thus] nature, proceeding by unknown gradations, cannot wholly lend herself to these divisions [into genera and species]. … There will be found a great number of intermediate species, and of objects belonging half in one class and half in another. Objects of this sort, to which it is impossible to assign a place, necessarily render vain the attempt at a universal system. …
>
> In general, the more one increases the number of one's divisions, in the case of the products of nature, the nearer one comes to the truth; since in reality individuals alone exist in nature.[168]

In his opinion, the boundaries between species were arbitrarily drawn—species grade into each other. However, this was not a statement about transmutation. He merely thought that the variation between species was continuous, not that species continuously arose from prior species. That idea was left to his student Lamarck to elaborate (although Darwin's grandfather Erasmus independently trumped Lamarck by several years with a similar approach). However, Buffon did allow that some change was possible. He thought that there were "types" of organisms roughly equivalent to Linnaeus' genera. The original type was the "true" form, and various species could degrade from that type to become a kind of "monster." Again, the influence of Aristotle is apparent, but there were limits to the sort of change species could undergo.

Two animals are of the same species, he wrote in the second volume of the *Histoire naturelle,*

---

[167] Buffon 1749–1789. See Sloan 1979, Eddy 1994, Roger 1997.

[168] *Histoire naturelle*, Tome I (1749, 12, 13, 20, 38) quoted in Lovejoy 1936, 230. For a complete translation of the "Premiere Discours," see Lyon 1976.

... if, by means of copulation, they can perpetuate themselves and the likeness of the species; and we should regard them as belonging to different species if they are incapable of producing progeny by the same means. Thus the fox will be known to be a different species from the dog if it proves to be a fact that from the mating of a male and female of these two kinds of animals no offspring is born; and even if there should result a hybrid offspring, a sort of mule, this would suffice to prove that fox and dog are not of the same species—inasmuch as this mule would be sterile (*ne produirait rien*). For we have assumed that, in order that a species might be constituted, there was necessary a continuous, perpetual and unvarying reproduction (*une production continue, perpétuelle, invariable*)—similar, in a word, to that of other animals.[169]

Intriguingly, Mayr in his history omits the last sentence, perhaps from a desire to find forerunners for his own biological species concept and downplay the morphological aspect of *species* in predecessors.[170] Hence, we again find in Buffon the generative notion of species that, like Ray and Linnaeus and others before and after, includes both form and reproduction.

Buffon had a mechanistic story for what kept organisms reproducing according to type—it involved what he called the *moule interieur*, or interior mold. This was an epigenetic, but particulate, hereditary factor that was held constant, or at least not deformed too much. It was derived from the *premier souche*, the primary stock from which all species of a type degenerated, and because it was shared, he was convinced that all species within the type were actually one species and interfertile, and he undertook experiments to prove this, with limited success.[171]

His was not the first reproductive concept of biological species but he was first to make reproductive isolation the test of whether two organisms should be included in the same species. At other times, he seemed to claim that only the organisms themselves were real, and that species were just convenient fictions or names of biologists.[172] Inconsistent in his definitions over the course of the *Natural History*, he had denied that species are real, asserting that only individuals exist (the first example of biological species "nominalism"). Farber and Eddy delimit two stages in Buffon's intellectual development.[173] At first, around 1749, in his "Premier discours" (volume 1 of the *Histoire*) he declared that the reality of species cannot be determined using the artificial methods of naturalists. Later, in 1753, he stated that he considered species to be the "constant succession of similar individuals that reproduce," and that "[t]he term *species* is itself an abstraction, which in reality corresponds only to the destruction and renewal of beings through time."[174]

In volume 2 of the *Histoire* (1749), Buffon adopted the notion of the *moule interieur*; generation was entirely a physical process akin to Newtonian forces, in many ways similar

---

169 Lovejoy 1959, 93f.
170 Mayr 1982, 334.
171 Given Buffon's heterodoxy, it is unlikely that he was directly influenced by the Noachian logistics of Buteo, Kircher, and Wilkins, but indirectly the notion that a primary species might diverge into local varieties due to the influence of geographical conditions which derived from it must have.
172 He called them naturalists, of course, because the term *biology* was not coined until the end of the eighteenth century.
173 Farber 1971, Eddy 1994.
174 Full citations and page numbers to be found in Eddy 1994, 646n.

to Darwin's later pangenesis hypothesis. As Eddy recounts the hypothesis, unused molecules in an organism are brought together and reassembled in the form of the organism in the seminal fluids. This notion of form being physical is in itself very Aristotelian, as Sloan observes, but Buffon is also directly influenced by Leibniz (which Sloan also notes) and so there is a tension in his thought.[175] If species are forms, then forms must be distinct, but if they are arbitrary, and grade into each other, as Leibniz taught the *scala naturae*, then they are not distinct. Eddy discusses this and points out that Buffon did not, in his view, propose at this stage a historical or biological conception of species so much as a logical one,[176] but whether one adopts Sloan's or Eddy's view, he later came to adopt both, and it is in this that he most influenced his pupil Lamarck.

In the period of 1764–1765, Buffon moved to a temporal and physical conception, beyond question. But he did not think that form changed; it was fixed eternally, and at best change could be a process he came to call *degeneration* in an essay in 1766 entitled "De la dégénération des animaux." Form could be modified, he thought, by external environmental changes, but should the "species" be brought back into their ancestral environment, those changes would be reversed.[177] Such change from the *premier souche* was primarily due to damage; Eddy says, "[f]ar from seeing degeneration in terms of organic history, Buffon saw it as the death of the organic past; through degeneration, organisms lose their organic identity, become weak and vitiated, and in extreme cases, have trouble reproducing."[178] Roger reports that Buffon carried out experiments on hybridization to test this theory, and had surprising success, but that eventually he accepted that some animals, particularly domesticated ones, do not revert to the purity of the wild type.

He did, however, allow that the natural unit of biology was not fixed at a single level. What the Linnaeans called genera were roughly more like the natural families Buffon thought set the limits of variety. He noted in volume 6 of the *Histoire* (1779), in the essay on sheep, that human intervention had caused many "species" of domesticated animals to degenerate from a single stock:

> These physical genera are, in reality, composed of all the species, which, by our management, have been so greatly variegated and changed; as have all those species, so differently modified by the hand of man, have but one common origin in Nature, the whole genus ought to constitute but a single species.[179]

In his *Histoire naturelle des oiseaux* (1770), Buffon claims that the size of the species of bird directly correlates with the number of species that degenerate from the *premier souche*:

> A sparrow or warbler has perhaps twenty times as many relatives as an ostrich or a turkey; for by the number of relatives I understand the number of related species that are sufficiently alike among themselves to be considered side branches of the same stem, or at least ramifications of stems that grow so closely together that one can suspect that they

---

[175] Sloan 1985.
[176] Eddy 1994, 648n.
[177] Roger 1997.
[178] Eddy 1994, 652.
[179] "The Sheep," *Histoire Naturelle*, Quadrupèdes, vol 33, tome 6, 121f quoted in Greene 1959, 147.

have a common root, and can assume that originally they all sprang from this root, of which one is reminded by the large number of their shared similarities; and these related species probably have separated only through the influence of climate, food, and the procession of years, which brings into being every realizable combination and allows every possibility of variation, perfection, alteration, and degeneration to become manifest.[180]

He held at the end of his career the view that species were definite categories, and that one might determine the boundaries by seeing experimentally whether reproduction was possible; he reported such in his essay "Mulets" (1776) of the sterility of hybrids as a mark of species boundaries.[181] But as what he meant by *espèces* was more like Linnaeus' genera, or even the later *family* rank, he was not fussed about using the Linnaean terminology inconsistently, occasionally shifting from one term to another, perhaps deliberately to annoy the Linnaeans, or perhaps doing exactly what Theophrastus had done, using the terms informally. Local variant forms for him were more geographical varieties than they were species, which was the *premiere souche* (first stock). Moreover, he believed that these varieties were closely related to similar forms in the Old as in the New World. It was a kind of vicariance definition of species; Stresemann gives the example of Buffon putting shrikes (gen. *Lanius*) into a single species with species from Senegal (*Tchagra senegala*), Madagascar (*Leptopterus madagascarinus*), and Cayenne (*Thamnophilus doliatus*) as climatic variations.[182]

ADANSON: MANY CHARACTERS ARE NEEDED

Michel Adanson (1727–1806) was a student and intimate of Antoine-Laurent de Jussieu's uncle, Bernard, and for a time lived with the de Jussieu family. He laid out a taxonomic procedure that derived from Bernard's ideas in his *Family of plants* (*Familles des Plantes*[183]) in which he attempted a natural system in which all characters are to be used equally to uncover groups. He wrote

> It was necessary to seek in nature for nature's system, if there really was one. With this aim, I examined plants in all their parts, without omitting one, from roots to embryo, folding of leaves in the bud, manner of sheathing, development, position, and folding of the embryo and radicle in the seed relative to the fruit; in a word, a number of features to which few botanists pay attention.[184]

He allowed, where Ray and Linnaeus had not, the use of microscopic characters in classification. Adanson was later claimed, somewhat illicitly, as a precursor to the so-called "phenetics" school of classification,[185] but unlike the pheneticists of the 1960s, he did not hold that all characters have equal weight in classification, but

---

[180]   Stresemann 1975, 56 translating page 75 in the original.
[181]   Roger 1997, 327.
[182]   Stresemann 1975, 57.
[183]   Adanson 1763.
[184]   Morton 1981, 303.
[185]   Nelson 1979, Winsor 2004.

only that there should be no a priori specification of what characters were to be used, contrary to Linnaeus.[186] He noted that

> [w]hat is sufficient to constitute the genera of certain families is not sufficient for other families, and neither the same parts nor the same number of these parts invariably furnish these [constituent] parts in each family.[187]

Although he was primarily concerned with higher taxa than the species level, he did give a definition of species in the second part of the Preface to the *Familles des plantes*. A definition of species founded only on sexual reproduction fails to apply for all organisms, but applies only for some plants, and most animals. So, such a definition is artificial and arbitrary, and if all possibilities are considered, the term becomes hard to define.

> Although it is very difficult, not to say impossible, to give an absolute and general definition of any object of natural history whatever, one could say rather exactly that there are as many species as there are different individuals among them, different in any (one or more) respect, constant or not, provided they are definitely perceptible and taken from parts or qualities where those differences appear to be most naturally placed in accordance with the particular character of each family.[188]

He rejected the definition of Buffon, in part because he wanted still to include mineral species under the rubric,[189] but also because he recognized the existence of asexual plants and even some animals. He notes that although there may be, as Buffon had said, only individuals in nature, and that for God all is one, for us it is divided, and *"et cela sufit"* (and that suffices), the differences are real even if to divine inspection the species are not. He does think, though, that there is continuity in nature, and that we see gaps between species only because some have become extinct. In a handwritten definition in the fifth volume of his copy of the *Encyclopédie* of Diderot,[190] he wrote:

> *Species*: collection of all objects which nature separates individually from each other as so many isolated entities existing separately and which the imagination or the free and creative opinion of man unites *idéalement* each time that he finds an almost complete resemblance or a resemblance at any rate greater than with any other group, a collection to which he gives the name species.[191]

---

[186] Stafleu 1963, 184.

[187] Morton 1981, 305.

[188] Translation from Stafleu 1963, 185. The French is:

Definition de l'Espèce.
Ainsi, quoiqu'il soit très-dificil, pour ne pas dire impossible, de donor une définition absolue & générale d'aucun objet de l'Hist. nat. on pouroit dire assez exactemant qu'il existe autaunt d'Espèces, qu'il i a d'Individus diférans entreaux, d'une ou de plusieurs diférances quelkonkes, constantes ou non, pourvu qu'eles soient très-sensibles, & tirées des parties ou qualités où ces diférances paroissent plus naturelement placées, selon le génie ou les moeurs propres à chaque Famille; ... [Adanson 1763, clxviij]

[189] Stafleu 1963, 182f. The notion that minerals "grew" in the earth was still current at this time.

[190] Diderot himself was a transmutationist of the Great Chain variety. See Gregory 2007.

[191] Translated in Stafleu 1963, 186.

Although it was not published, this passage indicates, as Stafleu notes, that Adanson thought that it was resemblance for the observer that "made" species, and that varieties were the subspecific changes that occurred in species. So he was a kind of taxonomic essentialist; but was he a species fixist? Adanson sought to answer the question whether natural classification is meaningful if species change, which he believed they had done, and that he had observed it. He cited both Linnaeus' example of *Peloria* and J. Marchant's discovery in 1715 of a mutation in *Mercurialis*.[192] His own work with lettuce and basil led him to conclude that varieties had become fixed in each generation. Morton considers that Adanson in fact had a kind of genealogical approach to classification, given that plants were mutable (he is the first to use the term "mutation" in connection with taxonomic transmutation). However, in 1769 he re-examined the question and concluded that the changes he had observed before and since were neither due to hybridism nor did they breed true, and so "the transmutation of species does not take place in plants, any more than in animals." Thus, he ended up a fixist in practice, if not in principle.

Adanson's natural method approach to classification was in one way quite standard. He focused on the shared characteristics for larger groups and distinguished the smaller groups by their differences. Gareth Nelson notes

It is evident that Adanson acquired his notions of natural families not by study of his artificial systems, but by independent, and unspecified means ...:

"These diverse remarks, in showing the utility of botanical exploration [discovery of new species in the tropics], explain why I became more and more convinced of the necessity to consider plants in a totally new fashion. I believed it necessary to abandon old prejudice in favor of [artificial] systems, and the ideas on which the systems are based and which limit our knowledge; and to search in nature herself for her system, if it is true that she really has one. In this belief I examined all parts of plants, without exception, from the roots to the embryo ... First I made a complete description of each plant species, considering each part in detail in a separate article; and to the extent that I encountered new species with relations to species already described, I described the new species separately, omitting the similarities and noting only their differences. It was by the *ensemble* of these comparative descriptions that I perceived that the plants placed themselves naturally in classes, or families, that could be neither artificial nor arbitrary—not based on one or a few parts that must vary within certain limits, but on all parts; such that the absence of one part [in a given species] would be replaced and balanced by the addition of another part that would restore the equilibrium."[193]

JUSSIEU: SPECIES AS SIMPLES

Although he was mainly concerned with the nature of a natural system at much higher levels of classification than the species level, Antoine-Laurent de Jussieu (1748–1836) did also make some comments on species that indicate that at the time he published his massive *Genera plantarum secundum ordines naturalis disposita*,[194] two distinct approaches to species had developed. One was due to Ray, while the other was due to Linnaeus, and Jussieu was more in line with Ray than Linnaeus.

---

[192] Morton 1981, 309.
[193] Nelson 1979.
[194] Jussieu 1964.

Jussieu influenced many, if not most, of those who followed. He directly influenced botany in general—for instance Samuel Gray's 1821 *A natural arrangement of British Plants*[195] shows how he had become the standard even at the popular level. He influenced also the young Cuvier[196] and through him systematics as a whole.

In his early works, he treated species as having more restricted affinity than the affinity of the genus. He defines species in terms of the sharing of characters, noting,

> [j]ust so many plants agreeing in all their parts, or being consistent in their universal character, and born from and giving birth to those of like nature, are the individuals together constituting a *species*, [a term] wrongly used in the past, now more correctly defined as *the perennial succession of like individuals, successively reborn by continued generation.*[197]

So here Jussieu is treating the diagnostic aspect of species, which are formed from individual organisms by abstraction of characters, from the causal aspect of species, which are, again, a generative reality of like producing like, as he does elsewhere.[198]

But it is the *Genera plantarum* that had the greatest influence, and in the Introduction he explicitly defines species, the concept, thus, in a section entitled "The sure knowledge of species":

> … the species must first be known, and defined by its proper signs: [it is] a collection ["adhesio"] of beings that are alike in the highest degree, never to be divided, but simple by unanimous consent [and] simple by the first and clearest law of Nature, which decrees that *in one species are to be assembled all vegetative beings or individuals that are alike in the highest degree in all their parts, and that are always similar* ["conformia"] *over a continued series of generations*, so that any individual whatever is the true image of the whole species, past, present, and future. [italics original][199]

He then continues that genera are analogously formed from species that "conform in a large number of characters." Species are the only natural *rank* he recognizes. Species are simples from which the composite groups are formed. He says, in a section entitled "General observation of species affinities":

> As a species unifies individuals similar to each other, congeneric species are expected to follow similar patterns. Yet nobody admits the need to *unify* [*species*] *appearance with more numerous marks of conformity*, and therefore [those patterns] should [co-]occur with several different norms. This primary reason, before first assessment, should demonstrate affinities by classifying [similar] species in

---

[195] Gray 1821.

[196] Stevens 1994, 2, chapter 4.

[197] Stevens 1994, 292, page 340 in the original, emphasis added. Quotations from Jussieu's works are from the translations in this book.

[198] Jussieu 1964, 311, 313–314.

[199] Stevens 1994, 356f. The Latin of the definition is:

> … *in unam speciem colligenda sunt vegetantia seu individua omnibus suis partibus simillima & continuatâ generationum serie semper conformia…* [Jussieu 1964, xxxvij]

the same founding group. Genera are thus established with closely similar species, and are given with no formal and precise definition of classifying rules, especially in the natural method.[200]

Of the Linnaean scheme, he says merely that the nomenclature is useful, a matter of convenience: "the species name must be both simple and easy, but it must also signify,"[201] and "It is one thing to name a plant, another to describe it."[202]

Like Linnaeus, Jussieu held that plants have affinities like regions on a geographical map. The natural method

> links all kinds of plants by an unbroken bond, and proceeds step by step from simple to composite, from the smallest to the largest in a continuous series, as a chain whose links represent so many species or groups of species, or like a geographical map on which species, like districts, are distributed by territories and provinces and kingdoms.[203]

However, as Stevens observes, the geography was becoming sparser in the regions occupied.[204] Where for Linnaeus, the entire territory was more or less filled, for Jussieu, there were large unoccupied regions, and for Charles-François Brisseau de Mirbel (1776–1854)[205] and Alphonse de Candolle the groupings become discrete and separate.

CHARLES BONNET AND THE IDEAL MORPHOLOGISTS

Bonnet (1720–1793) was a Swiss (Genevan, as Switzerland did not then exist as such) zoologist who was greatly influenced by Leibniz's ideas about the continuity of nature (the *lex continui*) and he produced the classical ladder of nature as a result, first in his *Traité d'insectologie*,[206] and then in his *Contemplation de la nature*.[207] His ladder was envisaged as an artificial system of ranking things in terms of their progression, although he did so, he said, without "presuming to establish the progressive order of Nature." The ladder ran from fish to birds to quadrupeds, and each division itself was further divided: birds into aquatic, amphibious, and terrestrial, and so forth (Figure 3.5).

Bonnet's *Échelle des êtres naturels*, as he called it, implied that extant species were the current forms of "extinct" species, but that we can only diagnose the modern forms in terms of their reproduction of like forms. Like the early Buffon, he effectively denied the reality of species considered as essences, and plumped for a nominalistic individualist conception:

---

[200] Jussieu, *loc cit*. Italics original. Translation by Laurent Penet, and revised by the author.
[201] Jussieu 1964, 344.
[202] Jussieu 1964, 510n115.
[203] Jussieu 1964, 355.
[204] Stevens 1994, 74ff.
[205] Mirbel 1815.
[206] Bonnet 1745.
[207] Bonnet 1764, cf. Stresemann 1975, 172.

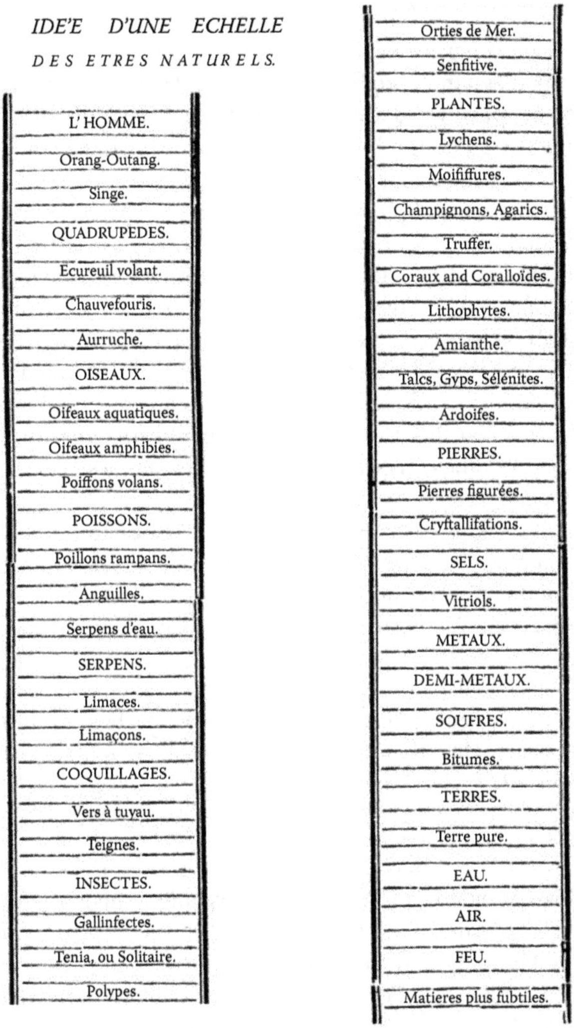

IDE'E   D'UNE   ECHELLE

DES ETRES NATURELS.

L' HOMME.	Orties de Mer.
Orang-Outang.	Senfitive.
Singe.	PLANTES.
QUADRUPEDES.	Lychens.
Ecureuil volant.	Moififfures.
Chauvefouris.	Champignons, Agarics.
Aurruche.	Truffer.
OISEAUX.	Coraux and Coralloïdes.
Oifeaux aquatiques.	Lithophytes.
Oifeaux amphibies.	Amianthe.
Poiffons volans.	Talcs, Gyps, Sélénites.
POISSONS.	Ardoifes.
Poillons rampans.	PIERRES.
Anguilles.	Pierres figurées.
Serpens d'eau.	Cryftallifations.
SERPENS.	SELS.
Limaces.	Vitriols.
Limaçons.	METAUX.
COQUILLAGES.	DEMI-METAUX.
Vers à tuyau.	SOUFRES.
Teignes.	Bitumes.
INSECTES.	TERRES.
Gallinfectes.	Terre pure.
Tenia, ou Solitaire.	EAU.
Polypes.	AIR.
	FEU.
	Matieres plus fubtiles.

FIGURE 3.5   Bonnet's scale of nature. Bonnet's chain of being from his *Traité d'Insectologie*[208] was enormously influential, even though it was a reworking of the medieval notions derived originally from Aristotle. It was a static, rather than temporal, ladder, but Lamarck transformed it into a temporal sequence.

If there are no cleavages in nature, it is evident that our classifications [*Distributions*] are not hers. Those which we form are purely nominal, and we should regard them as means relative to our needs and to the limitations of our knowledge. Intelligences higher than ours perhaps recognize between two individuals which we place in the same species more varieties than we discover between two individuals of widely separated genera. Thus these intelligences see in the scale of our world as many steps as there are individuals.[209]

---

[208] Bonnet 1745, 42.
[209] Bonnet 1769, I, 28. Quoted in Lovejoy 1936, 231.

Lovejoy notes wryly, "[t]hus the general habit of thinking in terms of species, as well as the sense of separation of man from the rest of the animal creation, was beginning to break down in the eighteenth century." However, in the same century in which the notion of *species* had acquired a biological sense distinct from the philosophical and logical tradition, it is highly significant that this "nominalism" occurs so early, and remains contentious from that day until this.

It is sometimes held that Bonnet would permit changes to species over time as germ lines ascended the scale, but in fact, he seemed to waver on this. To understand why, it is important to understand his preformationist doctrine of *emboîtement*. Preformationism in Bonnet's view was the theory (he explicitly called it a hypothesis) that all the seeds needed for future generations were created in the first organism of the species by God, and that these seeds were activated at reproduction. Bonnet held that while conditions, mostly the environment, could produce variation within a species so great that two conspecifics could be more different than they were from members of other species, nevertheless, the preformed seed would prevent the species from changing.

Nevertheless, he seems also to say in places that as the higher species were removed, other species would move into that level of sophistication. Again, though, this was chosen and preformed at creation by the good God.

It is significant that Bonnet did not see any gap between humans and apes. Anderson writes,

> The ape is brought so near to man in order to leave no doubt that they touch within their adjacent space. Thus Bonnet prefaced his remarks on the orang by writing: "It is here especially that one cannot fail to recognize the graduated progression and find verification for the famous Platonic axiom that nature goes nowhere by leaps."[210]

Donati in 1750 extended this view, says Stresemann, and "therefore conceived the notion that every being is a knot in the web of nature, and its resemblance to other forms may be compared to the threads between the knots,"[211] which may have influenced Linnaeus' conception of analogies between orders. Such ideas recur in the work of Johann Hermann,[212] and Jean Baptiste Robinet,[213] who produced a three-dimensional lattice in which species were nodes in the lattice. Stresemann traces this view through Schelling, Spix, Oken, and others through to William Swainson.[214] Swainson's view on classification led to a particular account of species as formal ideas. For him, following the ideas of William Macleay published in 1819, all taxa had to be organized in circles that touched ("osculating circles," Figure 3.6A),[215] and species were arranged at the rim of these circles and were analogous to the species in the adjacent circle. Swainson also had an

---

[210] Anderson 1976, 46. The quote is from the *Contemplation* III, chapter XXX [Bonnet 1764].
[211] Stresemann 1975, 172f.
[212] Hermann 1783.
[213] Robinet 1768.
[214] Stresemann 1975, 174–177.
[215] Macleay 1819; cf. Hull 1988, 92–96.

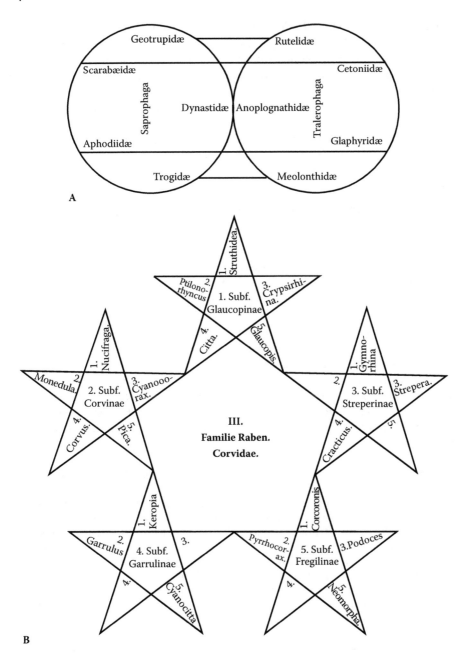

FIGURE 3.6 Quinarian schemes. **A.** Macleay's 1819 diagram of osculating circles. (Redrawn from page 29 of *Horae Entomologicae*.) **B.** Kaup's 1855 pentagonal "Druids' or Witches-foot" diagram. (Redrawn from scan of original.)

extended discussion in his *Preliminary Discourse*[216] on what counted as a "natural classification." This debate in part inspired Darwin to think through the reason for systematic classification and the nature of a natural system and it survived for some time after the *Origin*.[217] In Oken's native Germany the tradition he had begun continued in the work of Johann Jakob Kaup (1803–1879), who in turn influenced the Austrian Leopold Fitzinger (1802–1884). Kaup offered a pentagrammatic system,[218] but otherwise was much in the tradition of Macleay and Swainson (Figure 3.6B). However, such claims of the "end" of morphology depend upon a rather rigid definition of morphological biology, which as Amundson notes is neither historically plausible nor accurate.

IMMANUEL KANT AND THE CONTINUITY OF SPECIES

Kant's views on biological species are interesting primarily because they influenced the work of Blumenbach[219] whose own work, published in 1781, established the notion of races as distinct subspecific groups within the human species,[220] and later influenced the *Naturphilosophen*.[221] While Blumenbach worked mainly with skull morphology, his views on teleology were influenced deeply by Kant, and Kant regarded him as the scientist who best understood his ideas. So, we may consider Kant's view to be influential on some aspects of his contemporary biology, and also on later biology through Goethe and Oken.[222]

In his early lectures on geology, Kant made a similar distinction to that of Locke, between species defined by similarity relations, and genealogical classification, although unlike Locke he specifically named the genealogical groups "biological species" rather than simply "natural relations" as Locke had. He disparages species that are simply similar as "academic" or "scholastic," and as dry unproductive logical species:

> In the animal kingdom, the natural division into genera and species [*Natureinteilung in Gattungen und Arten*] is based on the law of common propagation and the unity of the genera [*Gattungen*] is nothing other than the unity of the reproductive power [*zeugenden Kraft*] that is consistently operative within a specific collection of animals. For this reason, Buffon's rule, that animals that produce fertile young with one another belong to one and the same physical genus [*physisches Gattung*] (no matter how dissimilar in form they may be), must properly be regarded only as a definition of a natural genus [*Naturgattungen*] of animals in general. A natural genus may, however, be distinguished from every scholastic genus [*Schulgattungen*]. A scholastic division [*Schuleintheilung*] is based upon classes and divides things up according to

[216] Swainson 1834.
[217] Coggon 2002.
[218] Kaup 1855.
[219] Lenoir 1980, Sloan 2002. Sloan 2006 argues that Kant's views on species, races and generation were strongly influenced by Buffon.
[220] Osborne 1971, 164.
[221] Nyhart 1995, Amundson 1998.
[222] Moss 2003, 10–12.

similarities, but a natural division [*Natureintheilung*] is based upon the common stem [*Stamme*], which divides animals according to kinship from the standpoint of generation. The first of these creates an academic system for memorization, the latter a natural system for the understanding [ein *Natursystem für den Verstand*]. The first has only the intent of bringing creatures under names [*Titel*]; the second has the intent of bringing them under laws [*Gesetze*].[223]

Kant has rightly identified the reason for the Linnaean system, which was at the time gaining ground among naturalists. It is unfortunate that Kant did not follow this up further, so far as I know. Natural classification is here based on similar causal processes of fertility and reproduction, a view he based on what turns out to be a misreading of Blumenbach's *Bildugstrieb*, or "formative force."[224] Kant returned to the question of the human races in an essay "Determination of the concept of human races" in 1785, during the course of which he addressed the reproductive aspect of species:[225]

---

[223] Translation by Mikkelson and Sloan from Sloan 2006. Immanuel Kant, *On the Different Races of Man*, [Kant 1775]; also in his *Werke* II [Kant 1969, 429] Another recent translation by Holly Wilson and Günter Zöller is Kant 2007, 82–97. A version of this passage can be found also in Dobzhansky 1962, 93.

Im Thierreiche gründet sich die Natureintheilung in Gattungen und Arten auf das gemeinschaftliche Gesetz der Fortpflanzung, und die Einheit der Gattungen ist nichts anders, als die Einheit der zeugenden Kraft, welche für eine gewisse Mannigfaltigkeit von Thieren durchgängig geltend ist. Daher muß die Büffonsche Regel, daß Thiere, die mit einander fruchtbare Jungen erzeugen, (von welcher Verschiedenheit der Gestalt sie auch seien mögen) doch zu einer und derselben physischen Gattung gehören, eigentlich nur als die Definition einer Naturgattung der Thiere überhaupt zum Unterschiede von allen Schulgattungen derselben angesehen werden. Die Schuleintheilung geht auf Klassen, welche nach Ähnlichkeiten, die Natureintheilung aber auf Stämme, welche die Thiere nach Verwandtschaften in Ansehung der Erzeugung eintheilt. Jene verschafft ein Schulsystem für das Gedächtniß; diese ein Natursystem für den Verstand: die erstere hat nur zur Absicht, die Geschöpfe unter Titel, die zweite, sie unter Gesetze zu bringen.

[224] Richards 2000.

[225] In "Bestimmung des Begriffs einer Menschenrace," *Gesammelte schriften, band 8*, 102 (italics indicate emphatic spacing):

Denn Thiere, deren Verscheidenheit so groß ist, das zu deren Existenz eben so veil verscheidene Erschaffungen nöthig wären, können wohl zu einer *Nominalgattung* (um sie nach gewissen Ähnlichkeiten zu klassificiren), aber niemals zu einer *Realgattung*, als zu welcher durchaus wenigstens die Möglichkeit der Abstammung von einem einzigen Paar erfodert wird, gehören. ... Aber auch alsdann würde *zweitens* dochimmer der sonderbare Übereinstimmung de Zeugungskräfte zweier verscheidenen Gattungen, die, da sie in Ansehung ihres Ursprungs einander ganz fremd sind, dennoch mit einander fruchtbar vermischt werden können, ganz umsonst und ohne einem anderen Grund, daß es der Natur so gefallen, angenommen werden. Wenn man, um dieses letztere zu beweisen, Thiere anführen, bei denen dieses ungeachtet der Verschedienheit ihres ersten Stammes dennoch, geschehe: so wird ein jeder in solchen Fällen die letztere Voraussetzung leugnen und vielmehr eben daraus, das eine solche fruchtbare Vermischung statt findet, auf die Einheit des Stammes schleißen, wie aus der Vermischung der Hunde und Füche, u. s. w. Die *unausbleibliche Unartung* beiderseitiger Eigenthümlichkeiten der Eltern ist also der einzig wahre und zugleich hinreichende Probirstein der Verschedienheit der Racen, wozu sie gehören, und ein Beweis der Einheit des Stammes, woraus sie entsprungen sind: nämlich der in diesem Stamm gelegten, sich in der Folge der Zeugungen entwickelden ursprünglichen Keime, ohne welche jene erblichen Mannigsaltigkeiten nicht würden entstanden sein und vornehmlich nicht hätten *nothwendig erblich* können.

For animals whose variety is so great that an equal number of separate creations would have been necessary for their existence could indeed belong to a *nominal family grouping* [*Nominalgattung*, lit. nominal genus] but never to a *real one*, other than one as to which at least the possibility of descent from a single common pair is to be assumed. ... [Otherwise] the singular compatibility of the generative forces of *two* species (which, although quite foreign as to origins, yet can be fruitfully mated with each other) would have to be assumed with no other explanation than that nature so pleases. If, in order to demonstrate this latter supposition, one points to animals in which crossing can happen despite the [supposed] difference of their original stems, he will in every case reject the hypothesis and, so much the more because such a fruitful union occurs, infer the unity of the group, as from the crossing of dogs and foxes, etc. The *unfailing inheritance* of peculiarities of both parents is thus the only true and at the same time adequate touchstone of the unity of the group from which they have sprung: namely the original seeds [*Keime*] inherent in this group developing in a succession of generations without which those hereditary variations would not have originated and would presumably not *necessarily* have *become* hereditary.[226]

The preformationist debate looms large here; the environment does not make the species, but only races or varieties.[227] What counts for a natural species is that all members share the "seeds" (*Keime*) of the original couple, in whom the species was preformed at Creation. This is a classic exemplar of the generative conception of species, based on "seeds" with a generative force unique to that kind, which featured prominently in John Ray's initial biological definition of *species*.

The *Critique of Pure Reason* was first published in 1781, the same year as Blumenbach's race work. A second edition[228] followed in 1787, and it is this version that we will follow here. In the context of discussion of whether a fundamental power exists which unites the things of understanding, Kant asks whether reason derives the unity of things by transcendental employment of understanding; in other words, if parsimony is a law of nature as well as of reason. He answers that unity is a necessity, for otherwise we would have no reason at all, and launches into a standard account of genera and species:

That the manifold respects in which individual things differ do not exclude identity of species, that the various species must be regarded merely as different determinations of a few genera, and these, in turn, of still higher genera, and so on; in short that we must seek for a certain systematic unity of all possible empirical concepts, in so far as they can be deduced from higher and more general concepts—this is a logical principle, a rule of the Schools, without which there can be no employment of reason.[229]

Kant equivocates here, it seems. On the one hand, he wants to adopt the classical process of differentiation from the *summum genera* employed by the Scholastics. On the other, he wants to derive unity from empirical data, that is, to classify from the bottom up. Parsimony is an advance, as when chemists reduce all salts to acids and alkalis

---

[226] Translated in Greene 1959, 233.
[227] For a good review see Gasking 1967.
[228] Kant 1933, B679–690. The page numbers of the second edition are conventionally preceded by the letter B, and of the first edition, by the letter A.
[229] *Op. cit.* B679f.

... but not content with this, they are unable to banish the thought that behind these varieties there is but one genus, nay, that there might even be a common principle for the earth and the salts.[230]

Parsimony is due to the need for the understanding to reduce multiplicities into unities. This is exactly what the scholastics would have sought. But he then continues:

> The logical principle of genera, which postulates identity, is balanced by another principle, namely that of *species*, which calls for manifoldness and diversity in things, notwithstanding their agreement as coming under the same genus, and which prescribes to the understanding that it attend to the diversity no less than to the identity.[231]

Observation of things must discriminate, he says, as much as the "faculty of wit" must find the appropriate universal. This differs from the scholastic account, and in some ways is more like Cusa's contraction in species. Here, observation allows us to group diversity of species under general predicates, he says, for if

> ... there were no *lower* concepts, there could not be *higher* concepts. Now the understanding can have knowledge only through concepts, and therefore, however far it carries the process of division, never through mere intuition, but always again through *lower* concepts. The knowledge of appearances in their complete determination, which is possible only through the understanding, demands an endless progress in the specification of our concepts, and an advance to yet other remaining differences, from which we have made an abstraction in the concept of the species, and still more so in that of the genus.[232]

Kant has it both ways after all: we abstract our more general categories from empirical observation, and understanding divides categories logically so that reason can deal with them.[233] But species border each other; there is a logical continuum

> admitting of no transition from one to another *per saltum*, but only through all the smaller degrees of difference between them.[234]

This he calls the logical law of the *continuum specierum*, a version of the transcendental law of *lex continui in natura*, which is Leibniz's law of continuity. While this applies to the realm of possible concepts, though, Kant rejects it in nature:

> For in the first place, the species in nature are actually divided, and must therefore constitute a *quantum discretum*. ... And further, in the second place, we could not make any determinate empirical use of this law, since it instructs us only in quite general terms that we are to seek for grades of affinity, and yields no criterion whatsoever as to how far, and in what manner, we are to prosecute the search for them.[235]

---

230 *Op. cit.* B680f.
231 *Op. cit.* B682.
232 *Op. cit.* B684.
233 *Op. cit.* B695f.
234 *Op. cit.* B687.
235 *Op. cit.* B689.

He wants species to be useful in reason and understanding—if there were no gaps in nature, then we could not make sense of it; the fact that things actually are divided by gaps is therefore fortuitous.

There is no biological discussion in this critique, but he does provide one in the *Critique of Judgement* in 1790. In section 64, he states:

> In order to see that a thing is only possible as a purpose, that is to be forced to seek the causality of its origin, not in the mechanism of nature, but in a cause whose faculty of action is determined through concepts, it is requisite that its form be not possible according to mere natural laws... The contingency of its form in all empirical natural laws in reference to reason affords a ground for regarding its causality as possible only through reason.[236]

It seems Kant is saying that contingent forms (i.e., species) can be understood not as the determinate outcome of mechanisms, but rather as the result of conceptual necessity. Eco discusses this at length and concludes that the platypus, discovered and displayed in Europe shortly after Kant's death, would have given Kant trouble unless he was able to subsume it under existing conceptual categories (such as "water mole").[237] While the platypus, or to use Kant's own example, a tree, is there and as a natural purpose produces itself as both cause and effect, generically, our ideas of it depend on our knowing the purpose or goal of such organized things (§65). Kant says in §67 of the *Judgement*,

> Hence it is only so far as matter is organized that it necessarily carries with it the concept of a natural purpose, because this its specific form is at the same time a product of nature. ...
>
> If we have once discovered in nature a faculty of bringing forth products that can only be thought by us in accordance with the concept of final causes, we go further still. We venture to judge that things belong to a system of purposes which yet do not (either in themselves or in their purposive relations) necessitate our seeking for any principle of their possibility beyond the mechanisms of causes working blindly. For the first idea, as concerns its ground, already brings us beyond the world of sense, since the unity of this supersensible principle must be regarded as valid in this way, not merely for certain species of natural beings, but for the whole of nature as a system.

Kant's teleology is too far afield from our topic,[238] but it is important to see that he saw species as the outcome of self-generative organization in nature, as he had in the lectures of 1775, as well as being things which we needed to think of as goal-directed in order to understand them. Again, here is the philosophical current of the generative notion of organic species in play. Famously, Kant was influential on Goethe, and through him, Oken, and the morphologists that followed him. In an essay in 1785, he

---

[236] Kant 1951; second edition of 1793.

[237] Eco 1999, 89–96. I very much doubt this is true, however. Eco is giving linguistic constraints undue priority here, in a Sapir-Whorfian fashion. Kant attended to the sciences because he thought that was where knowledge was gained, empirically. He therefore must think that empirical data can give us new categories.

[238] See Lenoir 1987.

discusses whether all canines are independent creations or descend from a common stock and are races of the one species. He says in "Determination of the concept of a human race"

> Species and genus [*of logic*—*JSW*] are not distinguished in natural history (which has only to do with ancestry and origin). Only in the description of nature, since it is a matter of comparing distinguishing marks, does this distinction come into play. What is species here must there often be called only race.[239]

Here Kant identified the different role the terms *genus* and *species* play in logic and natural history. When we are classifying in nature, we make use of the logic in terms of "identifying marks," but even so we are only often identifying only subspecific groups (races).

WHEN DID ESSENTIALISM BEGIN?

At this point it might be appropriate to ask when taxic essentialism actually enters the biological debate, since we have not found it so far. There are several senses in which we might call some conception of general terms "essentialist." Locke is an essentialist about terms, for example, since real essences are hidden from us, but terms are just names. Kant is an essentialist about mechanisms (such developmental preformation), particularly of generation, but in line with his phenomena–noumena distinction suggests that we can only abstract our terms from appearances.

Amundson suggested that essentialism with respect to species did not begin until the 1840s with the Strickland Code,[240] and for biological species this may be true, but only in a diagnostic sense. Essentialism, construed as the claim that a general term or concept must have necessary and sufficient inclusion criteria, or that all members of a species must share some characters, is a long standing *formal* notion, but that when it comes to applying that notion to living things, it was always understood that living species were a different category to formal species.

Let us therefore distinguish, since that is the key to this section, between several senses of "essentialism."[241] We have encountered so far *nominal essentialism* with Locke, the view that names can have essences, but only names. Is Strickland's a nominal essentialism? Not as Amundson presents it. Strickland is more correctly understood to be a *taxonomic essentialism*—that in the process of determining natural groups, one must find what actual properties (in this case biological properties) they have in common. Taxonomic essentialism is a kind of *logical essentialism*, in that it relies on the construction of formal, or logical, groups, as Aquinas posed it. But it is also a *material essentialism* in Aquinas' terms, because it relies on material properties and not just formal ones. A similar distinction was made by Buridan

---

[239] Kant 1785, translated in Greene 1959, 372n from *Gesammelte Schriften*, VIII, 100n; also translated in Kant 2007 by Holly Wilson and Günter Zöller.

[240] Amundson 2005.

[241] I have given a fuller taxonomy of essentialisms in Wilkins 2013.

between formal and material *significata*, so that he could be a "predicate essentia-list" and yet a nominalist.[242] This distinction has long standing.

A possible source of the notion that biological species before Darwin were seen to be essentialist may be John Dewey's famous essay of 1909 on "The Influence of Darwin on Philosophy."[243] Here, Dewey presents a finalistic account of Greek conception of *eidos* (anachronistically stating that the scholastics translated this as *species*, not the Latins). He says that *eidos* keeps the flux stable, makes individu-als isolated in space and time keep to a "uniform type of structure and function," and that the conception of species as a "fixed form and final cause" was the central principle of knowledge and nature.[244] But it is fixism and finalism that he opposes to Darwin's theory, not essentialism. His only mention of essence is a sneering ref-erence to "the logic that explained that the extinction of fire by water through the formal essence of aqueousness"[245]; presumably the fault is the use of a formal notion where an efficient cause is required.

Traditional essentialisms are generally nominal. From Aristotle through to the end of our period, when people discuss the essences, they are very often discussing what description or definition is essential for a universal name or term. Locke's rejection of Real Essences is a rejection not of the essences of terms, but of things. He rejects mate-rial essentialism. And it is the material essentialism of biology that is problematic—did it, as a historical fact, occur before Darwin? And is it required for taxonomy? We shall see that neither are necessarily the case, although it is likely that the issues were not so marked as I have expressed them here, and while naturalists do in fact slide from nominal to material essentialisms from time to time, it is not the identifying truth of the period that the Received View/Synthetic Historiography asserts.

ESSENTIALISM AND NATURAL SYSTEMS

Essentialism and typology are two attributes of "traditional" taxonomy that are often conflated (e.g., by Mayr). But they actually represent two distinct aspects of the old taxonomic categories. Essences are definable, and can be known by refining defini-tions, and are common to all members of a kind. No member of the kind can *not* have an essential property fully and constantly. Types, on the other hand, although the term is used in various ways throughout the modern period, are somewhat different. In biological thought, the type of a kind such as a genus can be instantiated more or less, and can be varied from. They can be abstract—not actually instantiated in any actual member of the kind, but *every* member of a kind must exhibit essences. Types are formal notions, essences are definitional, and while some types may be essential and some essences may be typical, the two concepts are not identical. Even worse, there are several types of types. Stevens cites Paul Farber's taxonomy of types. According to Farber[246] and Stevens, there are three kinds: *collection*, *classification*, and *morphological* types:

---

242 Klima 2005.
243 Reprinted in Dewey 1997.
244 Dewey 1997, 5–6.
245 Dewey 1997, 14.
246 Farber 1976.

Collection type concepts were concerned with how the name of an organism can be referred to a particular specimen or individual species. Classification type concepts were those that dealt largely with summarizing or simplifying data, whereas morphological concepts dealt with the order of nature and its laws (although Paul Farber, who I am following here, noted that these two were not always sharply distinguishable).[247]

The further discussion of types in the nineteenth century in Farber's important paper is one of the first—if not the first—challenges to the essentialism story, although he has recently been challenged on his taxonomy of types.[248] Divergences from the type of a genus were considered *terata*, or monsters. They were less than perfect, and could be individual organisms or even entire species. Hybrids were monsters too, the sense in which Linnaeus classified his hybrids. In somewhat later thought, for example, for Buffon, prior to the evolutionary period leading up to the nineteenth century, transmutation was conceived of as degradation from the type. By the time Owen proposed his notion of the *Archetype* that was to influence Darwin,[249] the transcendentalists had restored a Platonic view of types as pure forms, as ideas, and as essences,[250] but the Aristotelian account allowed only for types as actual forms. In "pure" Aristotelianism, variation from the type was an accidental difference, but as with the Linnaean fixation of the genus-species level, Enlightenment biologists were not always pure Aristotelians. Some of them weren't Aristotelians at all. Linnaeus himself had asserted that *"Natura non facit saltum,"* reiterating the principle of plenitude, at least for genera (although he crossed this apothegm out in his own copy of the *Philosophia Botanica* when he found species doing exactly that in hybrids). This principle owed most to neo-Platonic doctrines of emanation, and also relied upon the providence and benevolence of God, and it insisted upon completeness, and a grading from one form to another, as we have seen above.

At the end of the eighteenth century, classifications were commonly thought to be of three sorts: "artificial," "natural," or biological. I bracket *artificial* and *natural* in quotes because the way these terms are used in the history of systematics is at odds with the meaning of these terms in other fields (and is indeed inconsistent throughout the history of systematics itself).[251] Linnaeus felt that he was promoting an artificial classification—one based upon a *single aspect* of organisms (in plants, the sexual system). This was a dichotomous single key system based on Platonic *diairesis*—each subordinate taxon is distinguished from others in the ordinate taxon by the possession or non-possession of the key character—winged/non-winged, two-winged/four-winged, and so forth. This means that many taxa so formed are

---

[247] Stevens 1994, 134.

[248] Witteveen 2016, however, argues that the ruling notion of "type" in nineteenth-century natural history was, in fact, nomenclatural—the *name bearer* of the taxon. From this we get "type-specimen."

[249] Camardi 2001.

[250] Desmond 1984.

[251] See Winsor 2001, 2003, 2004 on the distinction between natural and artificial, and essentialist and typological taxonomies. Winsor calls typology the "method of exemplars," which is an apt term. Types applied within species, within genera and within higher taxonomic groups. See also Camardi 2001 for a discussion of the *type* concept. A useful discussion in the early nineteenth century is that of Swainson 1834, chapter 3.

privative definitions—defined by what they are *not*.[252] He did also attempt a fragmentary "natural" system—one that grouped on all available characters—but species remained those groups that shared all of some set of characters. Others, such
as Adanson,[253] based their classifications on as many characters as could be used, in
an attempt to demonstrate natural groups. Arguably, Linnaeus succeeded more than
might be expected just from it being an arbitrary system.[254] So it is perhaps better
if we adopt the distinction between *taxonomic* and other logical essentialisms and
material essentialism, and see the former as a more or less conventional and harm-
less aspect of taxonomy, and the latter as something of a rare bugaboo.

I fully concur here with Winsor's discussion[255] that the myth of essentialism in the
history of systematics is largely due to Joseph Cain's 1958 paper "Logic and Memory
in Linnaeus' System of Taxonomy,"[256] but I wonder also how much Popperian influence predisposed systematists to accept it and spread it. In any case, we have extensively supported Scott Atran's comment, quoted by Winsor:

> For my part, I have so far failed to find any natural historian of significance who ever
> adhered to the strict version of essentialism so often attributed to Aristotle. Nor is any
> weaker version of the doctrine that has indiscriminately been attributed to Cesalpino,
> Ray, Tournefort, A.-L. Jussieu and Cuvier likely to bear up under closer analysis.[257]

I am unsure what Atran means by "weaker version"—there is, at least, a diagnostic essentialism, a *taxonomic* essentialism, in play with many authors, but only
a very few people, and perhaps only Nehemiah Grew, adopted any sort of *material*
essentialism that did not end up collapsing to a causal reproductive account, prior to
the *Origin of Species*. And when it existed, before or after, it was motivated by piety
rather than science. The transcendentalists however appear to be motivated also by
a neo-Platonic philosophy.

THE ORIGINS OF SPECIES FIXISM

Species fixism, the idea that species are as they have always been, appears to have
originated with Ray, and was the standard opinion in popular botanical texts such as
James Gray's *Natural Relations of British Plants* of 1821, in which "race" is defined
as either "primitive" (*plantae primigeniae*)—"Species originally created, and not
formed by crossing with others"—or as "mule" (*hybridae*):

---

[252] Nelson and Platnick 1981.
[253] Croizat 1945, Lawrence 1963, Stafleu 1963.
[254] Cain 1995. Stresemann 1975, 52 notes that Linnaeus is *also* attempting a kind of "natural" system even in his "artificial" system, and contrasts it to the prior "classical" system—that is, the
Aristotelian system of differentiating by general features such as, in the case of birds, land or water
based lifestyles. As we saw, Bonnet retains a large amount of the classical *a priorism* of the medievals in this respect.
[255] Winsor 2001. For more recent discussions, see Khalidi 2009, Gill 2011, Müller-Wille 2011, Pedroso
2013, Clark 2015.
[256] Cain 1958.
[257] Atran 1990, 85.

Species not originally created, but formed by the pollen of one species being absorbed by the female organ of another species.[258]

Note that Gray does not assume that the hybrids are infertile.

That species were in some manner *mutable* (which is perhaps how we should contrast fixism, rather than with the later transmutationism that followed Lamarck) during the period up until the seventeenth century is held to have been the standard view.[259] Indications that species were not commonly thought to be fixed in that period can be found in the work of one of the translators of the 1611 *King James Bible*, the Calvinist George Abbot (1562–1633), Archbishop of Canterbury. In his *A briefe Description of the whole world*, (1605), he wrote

> There be other Countries in *Africke*, as *Agisimba, Libia interior, Nubia*, and others, of whom nothing is famous: but this may be said of *Africke* in generall, that it bringeth forth store of all sorts of wild Beasts, as Elephants, Lyons, Panthers, Tygers, and the like: yea, according to the Proverbe, *Africa semper aliquid apportat novi*; Often times new and strange shapes of Beasts are brought foorth there: the reason whereof is, that the Countrie being hott and full of Wildernesses, which haue in them little water, the Beastes of all sortes are enforced to meete at those few watering places that be, where often times contrary kinds haue conjunction the one with the other: so that there ariseth new kinds of species, which taketh part of both. Such alone is the Leopard, begotten of the Lion and the Beast called *Dardus*,[260] and somwhat resembling either of them.[261]

It is noteworthy, though, that according to this traditional view, novel species are formed from hybridization of extant, and presumably created, species. Thomas Huxley and others thought that fixism is due to the work of John Milton, but Milton's culpability is hard to determine.[262] In *Paradise Lost*, Book VII, the creation story of *Genesis* is repeated with little change, and the term used there is "kind," as it is in the English Bible. Nowhere in his poetical works can I find a hint of the constancy or otherwise of species.

Zirkle doesn't deal with the origin of fixism except to say

> The idea of the complete fixity of species was beginning to take shape in other quarters [*than natural history—JSW*], however, and it had become the accepted belief of the theologians. The theologians now held that species remained just as God had made them in the six days of creation—they remained just as God wanted them to be.[263]

But I would cavil at his claims that species were therefore thought to be mutable before Ray. It is not that species were not held to be fixed before the seventeenth century; they simply had no *idea* of a biological species. Remember that in

---

[258] Gray 1821, Vol. I, 41.

[259] Zirkle 1959.

[260] Possibly a panther; in Latin that would be *Pardus*. "Panther" was a generic name for big cats that were neither lions nor tigers.

[261] Nicolson 2003, 160. I am indebted to Tom Scharle for the reference. This is clearly based on Aristotle in Book VIII of the *Historia animalium*.

[262] Huxley 1893, 62.

[263] Zirkle 1959, 642.

the Latin tradition, "genus" means a general kind, while "species" means a special kind. Taking a hint from Locke and replacing genus with "kind" and species with "sort" for anyone before Ray, we can see that the problem lay in the ways things were broken down into kinds and sorts. Of *course* if you graft a branch from an apple tree onto a pear tree you have a different "sort" of fruit (examples Zirkle gives from Theophrastus). That doesn't mean it's a new *biological* kind or sort, but a new *practical* one.

Of course, a lot of the mutability of species/sorts is also a claim of biological species (in our sense) mutability. There are three kinds in the pre- and early-scientific literature that I have so far found:

1. *Mutability from hybridization*—two species can interbreed to form a novel one.
2. *Spontaneous generation* or transmutation of one form (say, a worm) into another (say, a goose). This might be seen in modern terms as a part of the single species' lifecycle.
3. *New varieties*—a species can breed untrue in some characters (even Aristotle knew this); by extension you can get new species. This is almost always credited to the influences of the local country—the soil, climate, or water—which directly changes the breed.

Only the last is relevant for us, and even here I have my doubts that many people thought there would be much change beyond the modern genus level, if that. Amundson has argued that fixism was invented in the mid-seventeenth century.[264] I think he is right, although I suspect Ray, not Linnaeus, is the culprit. But no matter who invented it, the question is why. The answer is, as Amundson shows, that fixism was an outcome of the generation debates. Spontaneous generation, a view that went back to Aristotle and earlier, did imply species mutability, since the generated form could change into a quite different form, as we saw with the Barnacle Goose case above. When the debate over generation resolved in favor of epigenesis rather than preformationism, it followed that the generation of an organism had to be controlled in some manner.[265] Even the preformationists held that something made things develop according to their kinds—the variety of preformationists known as ovists[266] called it the *emboîtement* (encasement) of generative forms, and Gasking notes:

> the seventeenth century preformationists assumed ... that all living things there were to be, had in fact been organized at God at creation... when all the created germs had reached the adult form the species would become extinct. ... it followed from such a view that there was no true generation; what appears as the formation of a new individual was simply the growth of an organised living thing which had been formed at the beginning of Time.[267]

---

[264] Amundson 2005, § 2.2.
[265] Gasking 1967, 34.
[266] Ovists held that eggs contained the preformed germ, and semen activated it. Spermists held the form was in the semen.
[267] Gasking 1967, 42.

Gasking mentions that Ray was a preformationist, although he rejected Leeuwenhoek's sperm-based version, in favor of the egg-based one that was more traditional.[268]

The kind of mutabilism that existed immediately prior to Lamarck and Erasmus Darwin, excepting that of Pierre Maupertuis, was the kind that Linnaeus allowed—species could be formed by hybridization of existing species. God, of course, formed the original stock, but new forms could arise, especially in plants, by mixing them. There was no open-ended mutabilism. However, it is equally clear that essentialism is not directly tied into the origination of fixism.

Amundson rightly argues that fixism was a precondition for the Natural System, which was itself a precondition for evolutionary theory. But that is not, of course, why it was adopted. Linnaeus did not seek to establish a natural system (not in the *Systema naturae*, at any rate), although his scheme came to be known later as *the* Natural System,[269] as a result of its title. His fixism was probably more a function of his piety than of his taxonomic concerns. So, it may be that fixism was successful simply because it was consonant with the tenor of the times, and because it came along with Linnaeus' success in establishing taxonomic nomenclature. Or it may be that Ray's institution of the natural theology movement was the more important, seeing nature as the book in which God had written. Either way, fixism comes into prominence around the end of the seventeenth century, and not in the time of Aristotle.

So we should be wary of papers like Zirkle's, and claims that mutabilism was rife before Linnaeus or Ray. In fact, lacking the basic conception of a biological kind (different from any other kind) and still hazy on the mechanisms and behaviors of generation (including spontaneous generation), the pre-Linnaeans tended to use the kind terms *genus*, *species*, *varietas*, and *formas* informally, as we would use "sort," "kind," "variety," and "form." Grand conclusions cannot be drawn from this.

BIBLIOGRAPHY

Adanson, Michel. 1763. *Familles des plantes: I. Partie. Contenant une Préface Istorike sur l'état ancien & actuel de la Botanike, & une Téorie de cete Science*. Paris: Vincent.

Allen, D. E. 2003. George Bentham's Handbook of the British flora: From controversy to cult. *Archives of Natural History* 30 (2):224–236.

Allen, Don Cameron. 1949. *The Legend of Noah. Renaissance Rationalism in Art, Science and Letters*. Urbana: University of Illinois.

Amundson, Ron. 1998. Typology reconsidered—Two doctrines on the history of evolutionary biology. *Biology and Philosophy* 13 (2):153–177.

—. 2005. *The Changing Role of the Embryo in Evolutionary Biology: Structure and Synthesis, Cambridge Studies in Philosophy and Biology*. New York: Cambridge University Press.

Anderson, Lorin. 1976. Charles Bonnet's taxonomy and the Chain of Being. *Journal of the History of Ideas* 37 (1):45–58.

Arber, Agnes. 1938. *Herbals: Their Origin and Evolution. A Chapter in the History of Botany 1470–1670*. 2nd ed. Cambridge, UK: Cambridge University Press. Reprint, 1970, Darien, CT: Hafner Publishing.

---

[268] Gasking 1967, 43, 56, see also Raven 1953.
[269] Smith 1821.

Atran, Scott. 1990. *The cognitive foundations of natural history.* New York: Cambridge University Press.

Bacon, Francis. 1913. *Advancement of Learning, and The New Atlantis, World's Classics.* London: Oxford University Press.

——. 1960. *The New Organon and Related Writings.* Translated by Fulton H. Anderson. Indianapolis: Bobbs-Merrill. Original edition, 1620.

Bonnet, Charles. 1745. *Traité d'Insectologie ou observations sur les Pucerons.* Vol. 2. Paris: Durand.

——. 1764. *Contemplation de la nature.* Amsterdam: M.-M. Rey.

——. 1769. *Contemplation de la nature.* 2nd ed. Amsterdam: M.-M. Rey.

Bovelles, Charles de. 1510. *Que hoc volumine continuentur: Liber de intellectu; Liber de sensu; Liber de Nichilo; Ars oppositorum; Liber de generatione; Liber de sapiente; Liber de duodecim numeris; Epistole complures.* Paris: Henrici Stephani.

Bowler, Peter J. 2003. *Evolution: The History of an Idea.* 3rd ed. Berkeley: University of California Press.

Breidbach, Olaf, and Michael T. Ghiselin. 2006. Athanasius Kircher (1602–1680) on Noah's Ark: Baroque "intelligent design" theory. *Proceedings of the California Academy of Science* 57 (36):991–1002.

Broberg, Gunnar. 1983. *Homo sapiens.* Linnaeus's classification of man. In *Linnaeus, The Man and His Work*, edited by Tore Frängsmyr, 156–194. Berkeley: University of California Press.

Buffon, Georges Louis Leclerc, comte de. 1749–1789. *Histoire naturelle, générale et particuliére: Avec la description du cabinet du roy.* 36 vols. Paris: De l'Imprimerie royale.

Buteo, Johannes. 1554. *Opera geometrica: De arca Noë, de sublicio ponte Caesaris, Confutatio quadraturae circuli ab Orontio Finaeo factae etc.* Lugduni: Thomas Bertellus.

——. 2008. Johannes Buteo's *The shape and capacity of Noah's Ark.* In *Issues in Creation.* Eugene, OR: Center for Origins Research.

Caesalpini, Andreae. 1583. *De plantis libri XVI.* Florentiae: Apud Georgium Marescottum.

Cain, Arthur J. 1958. Logic and memory in Linnaeus's system of taxonomy. *Proceedings of the Linnean Society of London* 169:144–163.

——. 1995. Linnaeus's natural and artificial arrangements of plants. *Botanical Journal of the Linnean Society* 117 (2):73–133.

——. 1997. John Locke on species. *Archives of Natural History* 24 (3):337–360.

——. 1999. John Ray on the species. *Archives of Natural History* 26 (2):223–238.

Camardi, Giovanni. 2001. Richard Owen, morphology and evolution. *Journal of the History of Biology* 34 (3):481–515.

Cassirer, Ernst et al., eds. 1948. *The Renaissance Philosophy of Man: Selections in Translation.* Chicago: University of Chicago Press.

Clark, Stephen R. L. 2015. Changing kinds: Aristotle and the Aristotelians. *Diametros* 45:19–34.

Classen, Albrecht. 2001. The Great Herbal of Leonhart Fuchs. *De historia stirpium commentarii insignes*, 1542. Review of The Great Herbal of Leonhart Fuchs. De historia stirpium commentarii insignes, 1542, Frederick G. Meyer, Emily Emmart Trueblood, John L. Heller. *German Studies Review* 24 (3):595–597.

Clauss, Sidonie. 1982. John Wilkins' *Essay Toward a Real Character*: Its place in seventeenth-century episteme. *Journal of the History of Ideas* 43 (4):531–553.

Coggon, Jennifer. 2002. Quinarianism after Darwin's *Origin*: The circular system of William Hincks. *Journal of the History of Biology* 35 (1):5–42.

Cohn, Norman. 1999. *Noah's Flood: The Genesis Story in Western Thought.* New Haven: Yale University Press.

Croizat, Leon. 1945. History and nomeclature of the higher units of classification. *Bulletin of the Torrey Botanical Club* 72 (1):52–75.

Dahlberg, Nicolaus E., and Carl von Linné. 1755. *Dissertatio botanica metamorphoses plantarum sistens, quam... sub praesidio... Dn. doct. Caroli Linnaei... naturae curiosorum censurae subjicit Nicolaus E. Dahlberg.* Holmiae: e Typographia regia.

Derham, William. 1718. *Philosophical letters between the late learned Mr. Ray and several of his ingenious correspondents, natives and foreigners: To which are added those of Francis Willughby esq.* London: William and John Innys.

DeLacy, Margaret. 2016. *The germ of an idea: Contagionism, religion, and society in Britain, 1660–1730, Springer English/International eBooks 2016—Full Set.* London: Springer.

Desmond, Adrian J. 1984. *Archetypes and Ancestors: Palaeontology in Victorian London, 1850–1875.* Chicago, IL: University of Chicago Press.

Dewey, John. 1997. *The Influence of Darwin on Philosophy and Other Essays, Great Books in Philosophy.* Amherst, NY: Prometheus Books.

Diderot, Denis. 1754. *Pensées sur l'interprétation de la nature.* Paris.

Dobzhansky, Theodosius. 1962. *Mankind Evolving; the Evolution of the Human Species.* New Haven, CT: Yale University Press.

Eco, Umberto. 1999. *Kant and the Platypus: Essays on Language and Cognition.* London: Vintage/Random House.

Eddy, John H., Jr. 1994. Buffon's *Histoire naturelle*: History? A critique of recent interpretations. *Isis* 85:644–661.

Fales, Evan. 1982. Natural kinds and freaks of nature. *Philosophy of Science* 49 (1):67–90.

Farber, Paul Lawrence. 1971. Buffon's concept of species. PhD thesis, Indiana University Bloomington.

—. 1976. The type-concept in zoology during the first half of the nineteenth century. *Journal of the History of Biology* 9 (1):93–119.

Farley, John. 1977. *The Spontaneous Generation Controversy from Descartes to Oparin.* Baltimore, MD: Johns Hopkins University Press.

Ferrari, Michel. 2011. Introduction to Bovelles' *Liber de Sapiente. Intellectual History Review* 21 (3):257–265.

Frängsmyr, Tore, ed. 1983. *Linnaeus, the Man and his Work.* Berkeley: University of California Press.

Freedman, Joseph S. 1993. The diffusion of the writings of Petrus Ramus in central Europe, c. 1570–c. 1630. *Renaissance Quarterly* 46 (1):98–152.

Gasking, Elizabeth B. 1967. *Investigations into Generation 1651–1828, History of Scientific Ideas.* London: Hutchinson.

Gesner, Conrad. 1551–1587. *Historia animalium.* Tiguri [Zürich]: Christ. Froschoverum.

Gill, Christopher. 2011. Essentialism in Aristotle's biology. *Critical Quarterly* 53 (4):12–20.

Glass, Bentley. 1959. The germination of the idea of biological species. In *Forerunners of Darwin, 1745–1859*, edited by Bentley Glass et al., 30–48. Baltimore, MD: Johns Hopkins Press.

Goerke, Heinz. 1973. *Linnaeus.* New York: Scribner.

Gontier, Nathalie. 2011. Depicting the tree of life: The philosophical and historical roots of evolutionary tree diagrams. *Evolution: Education and Outreach* 4 (3):515–538.

Gould, Stephen Jay. 1993. *Eight Little Piggies: Reflections in Natural History.* New York: Norton.

Gray, Samuel Frederick. 1821. *A natural arrangement of British plants: According to their relation to each other, as pointed out by Jussieu, De Candolle, Brown, & c. including those cultivated for use: with an introduction to botany in which the terms newly introduced are explained.* London: Baldwin, Cradock and Joy.

Greene, John C. 1959. *The Death of Adam: Evolution and its impact on Western Thought.* Ames: Iowa State University Press.

Gregory, Mary Efrosini. 2007. *Diderot and the Metamorphosis of Species, Studies in Philosophy*. New York: Routledge.

Grew, Nehemiah. 1682. *The anatomy of plants with an idea of a philosophical history of plants, and several other lectures, read before the Royal Society*. London: W. Rawlins, for the Author.

Hagberg, Knut. 1952. *Carl Linnaeus*. Translated by Alan Blair. London: Jonathan Cape.

Hamilton, William et al. 1874. *Lectures on metaphysics and logic*. Edinburgh: Blackwood.

Harrison, Peter. 2015. *The Territories of Science and Religion*. Chicago: University of Chicago Press.

Hermann, Johann. 1783. *Tabula affinitatum animalium olim academico specimine edita: nunc uberiore commentario illustrata cum annotationibus ad historiam naturalem animalium augendam facientibus*. Strasburg: Joh. Georgii Treuttel.

Holbach, Paul Henri Dietrich, and Jean-Baptiste de Mirabaud (nom de plume). 1770. *Système de la nature. Ou des loix du monde physique & du monde moral*. Londres.

Hopkins, Jasper. 1981. *Nicholas of Cusa on Learned Ignorance: A Translation and an Appraisal of De Docta Ignorantia*. Minneapolis, MN: The Arthur J. Benning Press.

Hotson, Howard. 2007. *Commonplace Learning: Ramism and its German Ramifications, 1543–1630*. Oxford: Oxford University Press.

Hugh of Saint Victor. 1962. *Selected Spiritual Writings*. New York: Harper & Row.

Hull, David L. 1988. *Science as a Process: An Evolutionary Account of the social and Conceptual Development of Science*. Chicago: University of Chicago Press.

Huxley, Thomas Henry. 1893. Lectures on evolution. In *Collected Essays*. London: Macmillan. Original edition, 1876.

Jones, Jan-Erik. 2007. Locke vs. Boyle: The real essence of corpuscular species. *British Journal for the History of Philosophy* 15 (4):659–684.

Jussieu, Antoine-Laurent de. 1964. *Genera plantarum*. Facsimile ed. Weinheim, Germany/Codicote, UK/New York: J. Cramer. Original edition, 1789, Paris.

Kant, Immanuel. 1775. On the different races of man. In *Lectures In Physical Geography, Summer Semester 1775*, 427–443. Königsberg: G. L. Hartung. Original edition, Von den verschiedenen Racen der Menschen zur Ankündigung der Vorlesungen der physischen Geographie im Sommerhalbenjahre.

—. 1785. Bestimmung des Begriffs einer Menschenrasse. *Berlinische Monatsschrift* (November 1785):390–417.

—. 1933. *Critique of Pure Reason*. Translated by Norman Kemp Smith. 2nd revised ed. London: Macmillan. Original edition, 1787 second edition.

—. 1951. *The Critique of Judgment*. Translation with introduction by J. H. Bernard. New York: Hafner. Original edition, 1790/1793.

—. 1969. *Kant's Werke Band II Vorkritische Schriften II 1757–1777*, edited by Königlich Preussischen Akademie der Wissenschaften. [Faks.-Ausg.], *Kant's gesammelte Schriften Erste Abtheilung Werke*. Berlin: W. de Gruyter.

—. 2007. *Anthropology, History, and Education*. (The Cambridge Edition of the Works of Immanuel Kant). Translated by Paul Guyer et al. Edited by Robert B. Louden and Günter Zöller, in translation. Cambridge: Cambridge University Press.

Kaup, Johann Jacob. 1855. Einige Worte über die systematische Stellung der Familie der Raben, Corvidae. *Journal für Ornithologie* Erinnerungsschrift zum Gedächtnisse an die VIII. Jahresversammlung der deutschen Ornithologen - Gesellschaft, abgehalten in Gotha vom 19. bis 20. Juli 1854:XLVII–LVI, foldout page.

Khalidi, Muhammad Ali. 2009. How scientific is scientific essentialism? *Journal for General Philosophy of Science* 40 (1):85–101.

Kircher, Athanasius. 1675. *Arca Noë in tres libros digesta*. Amsterdam: Joannem Janssonium à Waesberge.

Klima, Guyla. 2005. The essentialist nominalism of John Buridan. *The Review of Metaphysics* 58 (4):739–754.

Koerner, Lisbet. 1999. *Linnaeus: Nature and Nation*. Cambridge, MA: Harvard University Press.

Kuhn, Thomas S. 1959. *The Copernican Revolution: Planetary Astronomy in the Development of Western Thought*. New York: Vintage Books.

Kuntz, Marion Leathers, and Paul Grimley Kuntz, eds. 1988. *Jacob's Ladder and the Tree of Life: Concepts of Hierarchy and the Great Chain of Being*. Rev. ed. Vol. 14, American University Studies. Series V, Philosophy. New York: P. Lang.

Lankester, Edwin. 1848. *The correspondence of John Ray*. London: Ray Society.

Larson, James L. 1967. Linnaeus and the Natural Method. *Isis* 58 (3):304–320.

Lawrence, George H. M., ed. 1963. *Adanson: The Bicentennial of Michel Adanson's Familles des Plantes, The Hunt Biological Library*. Pittsburgh, PA: Carnegie Institute of Technology.

Lazenby, Elizabeth Mary. 1995. *The Historia Plantarum Generalis by John Ray. Book I. A Translation and Commentary*. Translated by Elizabeth Mary Lazenby. Newcastle, UK: University of Newcastle upon Tyne.

Lee, James. 1810. *An introduction to the science of botany: chiefly extracted from the works of Linnaeus; to which are added, several new tables and notes, and a life of the author. Corr. and enl. by James Lee, son and successor*. 4th ed. London: Printed for F.C. and J. Rivington/Wilkie and Robinson/J. Walker.

Leibniz, Gottfried Wilhelm. 1996. *New Essays on Human Understanding*. Translated by Peter Remnant and Jonathon Bennett. Cambridge, UK: Cambridge University Press. Original edition, 1765.

Lenoir, Timothy. 1980. Kant, Blumenbach, and vital materialism in German biology. *Isis* 71 (1):77–108.

—. 1987. The eternal laws of form: Morphotypes and the conditions of existence in Goethe's biological thought. In *Goethe and the Sciences: A Reappraisal*, edited by Frederick Amrine et al., 17–28. Berlin/New York: Springer Verlag.

Leroy, Jean-François. 1956. Origine de la Classification naturelle et Cartésianisme chez Tournefort. *Journal d'Agriculture Tropicale et de Botanique Appliquée* 326–327.

Liberman, Anatoly. 2013. Wrenching an etymology out of a monkey. *OUP Blog*, 23 January 2013. Available at blog.oup.com/2013/01/monkey-word-origin-etymology; accessed 1 June 2017.

Lindroth, Sten. 1983. The two faces of Linnaeus. In *Linnaeus, the Man and his Work*, edited by Tore Frängsmyr, 1–62. Berkeley: University of California Press.

Linné, Carl. 1735. *Systema naturæ, sive regna tria naturæ systematice proposita per classes, ordines, genera, & species*. Haak: Lugduni Batavorum.

Linné, Carl von. 1758. *Systema naturæ: per regna tria naturæ, secundum classes, ordines, genera, species, cum characteribus, differentiis, synonymis, locis*. Editio decima, reformata. ed. Holmiæ [Stockholm]: Impensis Direct. Laurentii Salvii.

—. 1787. *The families of plants, with their natural characters, according to the number, figure, situation, and proportion of all the parts of fructification. Tr. from the last ed. (as pub. by Dr. Reichard) of the Genera plantarum, and of the Mantissae plantarum of the elder Linneus; and from the Supplementum plantarum of the younger Linneus, with all the new families of plants, from Thunberg and the new families of plants, from Thumberg and l'Heritier. To which is prefix'd an accented catalogue*. Lichfield: Printed by J. Jackson. Sold by J. Johnson, London. T. Byrne, Dublin. And J. Balfour, Edenburgh [sic].

—. 1792. *Praelectiones in ordines naturales plantarum*. Hamburg: Hoffman.

—. 1956. *Caroli Linnaei Systema Naturae: a photographic facsimile of the first volume of the tenth edition (1758): Regnum Animale*. London: Printed by order of the Trustees, British Museum (Natural History).

Linné, Carl von. 1736. *Fundamenta botanica, quae, majorum operum prodromi instar, theoriam scientiae botanices per breves aphorismos tradunt*. Amstelodami: apud S. Schouten.

—. 1751. *Philosophia botanica, in qua explicantur Fundamenta botanica, cum definitionibus partium, exemplis terminorum, observationibus rariorum.* Stockholmiae: apud G. Kiesewetter.

Linné, Carl von. 1792. *The animal kingdom, or zoological system, of the celebrated Sir Charles Linnæus, containing a complete systematic description, arrangement, and nomenclature, of all the known species and varieties of the mammalia, or animals which give suck to their young.* Translated by Robert Kerr, Class I, Mammalia. Edinburgh: Printed for A. Strahan, and T. Cadell, London, and W. Creech, Edinburgh.

Linné, Carl von, and Stephen Freer. 2003. *Linnaeus' Philosophia Botanica.* Translated by Stephen Freer. Oxford/New York: Oxford University Press.

Llull, Ramón. 1512. Liber de ascensu et decensu intellectus (1304). In *Raymundi Lully Doctoris illuminati de nova logica de correlativis necnon de ascensu et descensu intellectus: quibus siquide tribus libellis p. brevi ad facili artificio.* Valencia: Jorge Costilla.

Lovejoy, Arthur O. 1936. *The Great Chain of Being: A Study of the History of an Idea.* Cambridge, MA: Harvard University Press. Reprint, 1964.

—. 1959. Buffon and the problem of species. In *Forerunners of Darwin 1745–1859,* edited by Bentley Glass et al., 84–113. Baltimore, MD: Johns Hopkins Press.

Lyon, John. 1976. The 'Initial Discourse' to Buffon's 'Histoire Naturelle': The first complete English translation. *Journal of the History of Biology* 9 (1):133–181.

Mackie, John L. 1976. *Problems from Locke.* Oxford: Clarendon Press.

Macleay, William Sharp. 1819. *Horae entomologicae, or, Essays on the annulose animals.* London: Printed for S. Bagster.

Mandelbrote, Scott. 2007. The uses of natural theology in seventeenth-century England. *Science in Context* 20 (3):451–480.

Mayr, Ernst. 1969. The biological meaning of species. *Biological Journal of the Linnean Society* 1 (3):311–320.

—. 1982. *The Growth of Biological Thought: Diversity, Evolution, and Inheritance.* Cambridge, MA: The Belknap Press of Harvard University Press.

Mayr, Ernst, and Peter D. Ashlock. 1991. *Principles of Systematic Zoology.* 2nd ed. New York: McGraw-Hill.

Mirbel, Charles-François Brisseau de. 1815. *Éléments de physiologie végétale et de botanique.* 3 vols. Paris: Magimel.

Morton, Alan G. 1981. *History of Botanical Science: An Account of the Development of Botany from Ancient Times to the Present Day.* London/New York: Academic Press.

Moss, Lenny. 2003. What Genes Can't Do. In *Basic Bioethics,* edited by Glenn McGee and Arthur Kaplan, Cambridge, MA: Bradford Book, MIT Press.

Müller-Wille, S., and V. Orel. 2007. From Linnaean species to Mendelian factors: Elements of hybridism, 1751–1870. *Annals of Science* 64 (2):171–215.

Müller-Wille, Staffan. 2003. Nature as a marketplace: The political economy of Linnaean botany. *History of Political Economy* 35 (Annual Supplement):154–172.

—. 2011. Making sense of essentialism. *Critical Quarterly* 53 (4):61–67.

Nelson, Gareth J. 1979. Cladistic analysis and synthesis: Principles and definitions, with a historical note on Adanson's *Familles des Plantes* (1763–1764). *Systematic Zoology* 28 (1):1–21.

Nelson, Gareth J., and Norman I. Platnick. 1981. *Systematics and Biogeography: Cladistics and Vicariance.* New York: Columbia University Press.

Nicolson, Adam. 2003. *God's Secretaries: The Making of the King James Bible.* London: HarperCollins.

Nordenskiöld, Erik. 1929. *The History of Biology: A Survey.* Translated by Leonard Bucknall Eyre. London: Kegan Paul, Trench, Trubner and Co.

Nyhart, Lynn K. 1995. *Biology Takes Form: Animal Morphology and the German Universities, 1800–1900.* Chicago: University of Chicago Press.

Ong, Walter J. 1958. *Ramus, Method, and the Decay of Dialogue. From the Art of Discourse to the Art of Reason.* Cambridge, MA: Harvard University Press.

Osborne, Richard H. 1971. *The Biological and Social Meaning of Race.* San Francisco: W. H. Freeman.

Pavord, Anna. 2005. *The Naming of Names: The Search for Order in the World of Plants.* London: Bloomsbury.

Pedroso, Makmiller. 2013. Origin essentialism in biology. *The Philosophical Quarterly* 64 (254):60–81.

Pleins, J. David. 2009. *When the Great Abyss Opened: Classic and Contemporary Readings of Noah's Flood.* Oxford: Oxford University Press.

Porter, David. 2001. *Ideographia: The Chinese Cipher in Early Modern Europe.* Stanford, CA: Stanford University Press.

Ragan, Mark A. 2009. Trees and networks before and after Darwin. *Biology Direct* 4 (1): 38. http://dx.doi.org/10.1186/1745-6150-4-43.

Raleigh, Walter. 1614. *The History of the World.* 6 vols. Vol. 1. London: William Stansby for Walter Burre.

Ramsbottom, John. 1938. Linnaeus and the species concept. *Proceedings of the Linnean Society of London* 150 (4):192–220.

Ramus, Petrus. 1756. *Professio Regia: hoc est septem artes liberales.* Edited by Johann Thomas Freige. Basil: Sebatian Henkicpitel.

Raven, Charles E. 1986. *John Ray, Naturalist: His Life and Works.* 2nd ed. Cambridge, UK/ New York: Cambridge University Press.

Raven, Charles Earle. 1953. *Natural Religion and Christian Theology: Science and Religion.* Vol. I. Cambridge: Cambridge University Press.

Ray, John. 1682. *Methodus plantarum nova brevitatis & perspicuitatis causa synoptice in tabulis exhibita; cum notis generum tum fummorum tum subalternorum characteristics, observationibus nonnullis de feminibus plantarum & indice copioso.* Londini: Impensis Henrici Faithorne & Joannis Kersey.

—. 1691. *The Wisdom of God manifested in the works of the creation. Being the substance of some common places delivered in the Chappel of Trinity College, in Cambridge.* London: Samuel Smith.

—. 2015. *Methodus Plantarum Nova.* Translated by Stephen A. Nimis et al., *Issue 176 of Ray Society.* London: Ray Society.

Richards, Robert J. 2000. Kant and Blumenbach on the Bildungstrieb: A historical misunderstanding. *Studies in History and Philosophy of Biological and Biomedical Sciences* 31 (1):11–32.

Robinet, Jean-Baptiste-René. 1768. *Considérations philosophiques de la gradation naturelle des formes de l'être, ou Les essais de la nature qui apprend à faire l'homme.* Paris: C. Saillant.

Robinet, Jean-Baptiste-René et al., eds. 1777–1783. *Dictionnaire universel des sciences morale, économique, politique et diplomatique; ou Bibliotheque de l'homme-d'état et du citoyen.* 30 vols. Vol. 18. A Londres: Chez les Libraires Associés.

Roger, Jacques. 1997. *Buffon: A Life in Natural History.* Translated by Sarah Lucille Bonnefoi. Edited by L Pearce Williams, Cornell History of Science Series. Ithaca, NY: Cornell University Press.

Roos, Anna Marie Eleanor. 2011. *Web of Nature: Martin Lister (1639–1712), the First Arachnologist.* Leiden: Brill.

Ross, David. 1949. *Aristotle.* 5th ed. London: Methuen/University Paperbacks.

Rossi, Paolo. 2000. *Logic and the Art of Memory: The Quest for a Universal Language.* London: Athlone.

Sachs, Julius V. 1890. *History of botany (1530–1860).* Translated by Henry E. F. Garnsey and Isaac B. Balfour. Oxford: Clarendon Press. Original edition, 1875.

Senn, Gustav. 1925. Die Einführung des Art—Und Gattungsbegriffs in die Biologie. *Verhandl. d. Schweizer. Naturforsh. Gesellsch.* II:183–184.

Shannon, Laurie. 2013. *The Accommodated Animal: Cosmopolity in Shakespearean Locales.* Chicago/London: The University of Chicago Press.

Singer, Charles Joseph. 1950. *A History of Biology to about the Year 1900: A General Introduction to the Study of Living Things.* 2nd ed. London: Abelard-Schuman.

Slaughter, Mary M. 1982. *Universal Languages and Scientific Taxonomy in the Seventeenth Century.* Cambridge, UK/New York: Cambridge University Press.

Sloan, Philip R. 2006. Kant on the history of nature: The ambiguous heritage of the critical philosophy for natural history. *Studies in History and Philosophy of Biological and Biomedical Sciences* 37 (4):627–648.

Sloan, Phillip R. 1979. Buffon, German biology, and the historical interpretation of biological species. *British Journal for the History of Science* 12 (41):109–153.

—. 1985. From logical universals to historical individuals: Buffon's idea of biological species. In *Histoire du Concept D'Espece dans les Sciences de la Vie,* 101–140. Paris: Fondation Singer-Polignac.

—. 2002. Preforming the categories: Eighteenth-century generation theory and the biological roots of Kant's a priori. *Journal of the History of Biology* 40 (2):229–253.

Smith, James Edward. 1821. *A grammar of botany, illustrative of artificial, as well as natural, classification, with an explanation of Jussieu's system.* London: Longman, Hurst, Rees, Orme, and Brown.

Sprague, Thomas Archibald, and Ernest Nelmes. 1928/31. The herbal of Leonhart Fuchs. *Journal of the Linnean Society. Botany* 48 (325):545–642.

Stafleu, Franz Antonie. 1963. Adanson and his *Familles des plantes.* In *Adanson: The Bicentennial of Michel Adanson's Familles des Plantes,* edited by G. H. M. Lawrence, 123–264. Pittsburgh, PA: Carnegie Institute of Technology.

—. 1971. *Linnaeus and the Linnaeans. The Spreading of Their Ideas in Systematic Botany, 1735–1789, Regnum Vegetabile.* Utrecht: Oosthoek.

Stanford, P. Kyle. 1998. Reference and natural kind terms: The real essence of Locke's view. *Pacific Philosophical Quarterly* 79 (1):78–97.

Stevens, Peter F. 1994. *The Development Of Biological Systematics: Antoine-Laurent De Jussieu, Nature, and the Natural System.* New York: Columbia University Press.

Stresemann, Erwin. 1919. Über die europäischen Baumläufer. *Verhandlungen der Ornithologischen Gesellschaft in Bayern* 14 (1):39–74.

—. 1975. *Ornithology from Aristotle to the Present.* Translated by Hans J. Epstein and Cathleen Epstein. Cambridge, MA: Harvard University Press.

Subbiondo, Joseph L., ed. 1992. *John Wilkins and 17th-Century British Linguistics.* Amsterdam/Philadelphia, PA: John Benjamins.

Swainson, William. 1834. *Preliminary discourse on the study of natural history.* London: Longman, Rees, Orme, Brown, Green and Longman.

Terrall, Mary. 2002. *The Man Who Flattened The Earth: Maupertuis and the Sciences in the Enlightenment.* Chicago: The University of Chicago Press.

Topsell, Edward. 1607. *The historie of foure-footed beastes Describing the true and liuely figure of euery beast, with a discourse of their seuerall names, conditions, kindes, vertues (both naturall and medicinall) countries of their breed, their loue and hate to mankinde, and the wonderfull worke of God in their creation, preseruation, and destruction. Necessary for all diuines and students, because the story of euery beast is amplified with narrations out of Scriptures, fathers, phylosophers, physitians, and poets: Wherein are declared diuers hyerogliphicks, emblems, epigrams, and other good histories, collected out of all the volumes of Conradus Gesner, and all other writers to this present day.* London: William Iaggard.

Tournefort, Joseph Pitton de. 1700. *Institutiones rei herbariae.* Editio altera, gallica longe auctior. Parisiis: Typographia regia. Image fixe.

Tournefort, Joseph Pitton, de. 1716–30. *The compleat herbal: Or, the botanical institutions of Mr. Tournefort. translated from the original Latin. With large additions from Ray, Gerarde, Parkinson, and others. To which are added, two alphabetical indexes. Illustrated with about five hundred copper plates. With a short account of the life and writings of the author.* Translated by John Martyn. London: Printed for R. Bonwicke, Tim Goodwin, John Walthoe, S. Wotton, Sam. Manship [and five others in London]. Original edition, Institutiones rei herbariae, 1700.

Whitehead, Alfred North, and Bertrand Russell. 1910. *Principia mathematica.* 3 vols. Cambridge: Cambridge University Press.

Wilkins, John. 1970. *The Mathematical and Philosophical Works of the Right Rev. John Wilkins.* 2nd ed. London: Frank Cass. Original edition, 1802.

—. 2002. *Essay Towards a Real Character.* London/New York: Thoemmes Continuum.

Wilkins, John, Bishop of Chester. 1668. *An essay towards a real character and a philosophical language. (An alphabetical dictionary, wherein all English words... are either referred to their places in the Philosophical tables, or explained by such words as are in those Tables.).* 2 vols. London: Sa. Gellibrand, and for John Martyn, printer to the Royal Society.

Wilkins, John S. 2013. Essentialism in biology. In *Philosophy of Biology: A Companion for Educators*, edited by Kostas Kampourakis, 395–419. Springer.

Winsor, Mary Pickard. 2001. Cain on Linnaeus: The scientist-historian as unanalysed entity. *Studies in the History and Philosophy of the Biological and Biomedical Sciences* 32 (2):239–254.

—. 2003. Non-essentialist methods in pre-Darwinian taxonomy. *Biology and Philosophy* 18 (3):387–400.

—. 2004. Setting up milestones: Sneath on Adanson and Mayr on Darwin. In *Milestones in Systematics: Essays from a symposium held within the 3rd Systematics Association Biennial Meeting, September 2001*, edited by David M. Williams and Peter L. Forey, 1–17. London: Systematics Association.

Witteveen, Joeri. 2016. Suppressing synonymy with a homonym: The emergence of the nomenclatural type concept in nineteenth century natural history. *Journal of the History of Biology* 49 (1):135–189.

Wright Henderson, Patrick Arkley. 1910. *The Life and Times of John Wilkins.* Edinburgh and London: William Blackwood.

Zirkle, Conway. 1959. Species before Darwin. *Proceedings of the American Philosophical Society* 103 (5):636–644.

4 The Nineteenth Century, a Period of Change

The ordinary naturalist is not sufficiently aware that when dogmatizing on what species are, he is grappling with the whole question of the organic world & its connection with the time past & with Man; that in involves the question of Man & his relation to the brutes, of instinct, intelligence & reason, of Creation, transmutation & progressive improvement or development. Each set of geological questions & of ethnological & zoo. & botan. are parts of the great problem which is always assuming a new aspect.

Charles Lyell[1]

NINETEENTH-CENTURY LOGIC

Early in the nineteenth century, in 1826, Archbishop Richard Whately published an influential text on logic, *The Elements of Logic*,[2] which is credited as reviving the study of logic in English-speaking countries. In this book, Whately describes *species* as *essences*, as *heads of predicables*, and as that of which genera are parts (and not species being parts of genera, since the genus partakes of the essence, or definition, of the species[3]). But he also notes that *this* sense of "species" is quite distinct from the sense in which *naturalists* use it of "organized beings," for they are real things, "unalterable and independent of our thoughts":

> ... if anyone utters such a proposition as... "Argus was a mastiff," to what head of Predicables would such a Predicate be referred? Surely our logical principles would lead us to answer, that it is the *Species*; since it could hardly be called an Accident, and is manifestly no other Predicable. And yet every Naturalist would at once pronounce that Mastiff, is no distinct Species, but is only a *variety* of the Species Dog....

---

[1] 11 February 1857 [Wilson 1970, 164]. Listed as 1851 in the printed version, but this is out of sequence, and certainly a typographical error. I am deeply indebted to Mike Dunford for drawing my attention to this comment of Lyell's (and noting the date typo), the cited note of Agassiz's, James Dana's paper, and for his conversations with me on the period covered by the "uniformitarian" and "catastrophism" debates in geology. As geology was not, at that time, held to be isolated from any other kind of natural history, Lyell felt, as did Darwin, that the issues raised in the one field (geology) had implications for issues in the other (naturalism). Mike's help has been immense here.

[2] Whately 1826, my edition being the ninth 1875 edition. Initially Whately's book was an extensive article printed as the 1823 volume of the *Encyclopaedia Metropolitana* [Whately 1823].

[3] *Elements*, Bk II, ch. 5 §3, 85.

... the solution of the difficulty is to be found in the peculiar technical sense... of the word "Species" when applied to *organized Beings*:[4] in which case it is always applied (when we are speaking strictly, as naturalists) to individuals as are supposed to be *descended from a common stock*, or which *might* have so descended; *viz.* which resemble one another (to use M. Cuvier's expression) as much as those of the same stock do.[5]

Whately expressly exempts species concepts in biology, then, from the strictures of logical notions, and that, it must be observed, includes essential characters. He notes

[The fact of two organisms being the same species] being one which can seldom be *directly* known, the consequence is, that the *marks* by which any Species of Animal or Plant is *known*, are not the very *Differentia* which *constitutes* that Species.[6]

So well prior to Darwin, and in a logical context, we find that the *species* of biology and the *species* of logic are understood to be different concepts. However, Whately expects there *will* be diagnostic "marks."

A critical review of Whately's logic entitled *An Outline of a New System of Logic* was published the next year by George Bentham, who later became a noted botanist and who was the nephew of, and was influenced by, his uncle Jeremy Bentham.[7] Bentham attacked Whately for not allowing privative classifications; oddly he thought that was required by Aristotle, rather than by the later neo-Platonists.

However, Bentham made a crucial distinction that matches Aquinas' material–formal distinction. While allowing that the differentia of a (logical) species was necessary to it, and that if the species is a universal, it is its essence,[8] that "without which the subject would not be what it is said to be,"[9] he noted also that "the peculiar sense in which naturalists make use of the word species... is very different from the logical sense of the word,"[10] and he distinguished between the *definition* of a species from its *individuation*,[11] in which "the only characteristic properties are those of *time* and *place*, which must both be exhibited," and essential definition applies only in the first case.

Further, he notes that the object of specific description is to enable a learner to recognize a species or fix the collective entity in his mind.[12] Description should thus

---

4 *Elements* Bk IV, ch. 5 §1. Terms like "organized beings," "organic beings," "natural beings," and the like refer to what we would now call "organisms." Although the term "organism" had been devised in the eighteenth century in French [Cheung 2006], the term was not introduced into English until Owen discussed the kangaroo in 1834 [Owen 1834, 359], where he said "... if the introduction of new powers into an organism necessarily requires a modification in its mode of development ...," in the context of which it is clear he means a being that has organs, or is organized.
5 *Elements*, 183. The quoted text is unchanged from the first to the ninth editions, but does not appear in the *Metropolitana* volume.
6 *Elements*, 184f.
7 Bentham 1827. See Allen 2003 for discussion of his later work in botany.
8 Bentham, *op. cit.*, 67.
9 Bentham, *op. cit.*, 68.
10 Bentham, *op. cit.*, 71.
11 Bentham, *op. cit.*, 79.
12 Bentham, *op. cit.*, 82.

be preceded by definition. However, he equivocates when discussing essential properties, and says that *"to the genus plant belong all those entities which are endowed
with the property of possessing leaves*, stalks, roots, &c." Even so, he realizes that
classification, which he calls *methodization*, is either physical or logical[13] and that
division is of several kinds, analytic when performed on individuals, logical when
performed on collective entities, and possibly also physical. There is a difference
between dividing, say, Vertebrata into mammals, birds, fishes, lizards, and serpents
in terms of obvious features if you want a "slight and general idea" of vertebrates,
but "it cannot suffice for the naturalist, who must always be assured of the all-
comprehensiveness of his classes and subclasses; he must always be enabled to ascertain
precisely to *which* of them he should refer any individual animal that comes under
his observation."[14] So, he applies a binary dichotomy of lungs/no lungs (fish), with
mammaries/not with mammaries (mammals), winged (birds) not winged (reptiles).[15]

In the rest of his work, he seems to have reinvented the Universal Language
Project. His uncle, Jeremy Bentham, in his 1817 *Chrestomathia*, revives, explicitly,[16]
Porphyry's dichotomous ("bifurcating") mode of classification, counterpointing
it to D'Alembert's classification of knowledge in the *Encyclopedia*. His division
of knowledge[17] begins with "Eudaemonics, or Ontology," which he divides into
"Coenoscopic" and "Idioscopic" ontologies (general and particular properties), and
so on, but lower down he starts introducing privative categories, such as "no-work-
producing," "not-state-producing," and so on. Moreover, some of his dichotomies,
such as Nature/Man, information/passion, seem as arbitrary as Plato's original. Still,
Jeremy Bentham's recasting of the Arbor Porphyriana as a bifurcating tree diagram
is significant.[18]

By the middle of the nineteenth century, despite the arguments naturalists were
now having over the meaning of the term *species*, the "genera plus differentia" definition remained widely accepted by logicians until, under the weight of the new set
theory and the biological pre-eminence of the use of the term, the older logic was
relegated to specialists in metaphysics and medievalists. Here, for example, is the
definition of a widely used dictionary of science and the arts in 1852:

> SPECIES. (Lat.) In Logic, a predicable which is considered as expressing the whole
> essence of the individuals of which it is affirmed. The essence of an individual is said
> to consist of two parts: 1. The material part, or genus; 2. The formal or distinctive part,
> or difference. The genus and difference together make up, in logical language, the spe-
> cies: e.g. a "biped" is compounded of the genus "animal," and the difference "having
> two legs." It is obvious that the names *species* and *genus* are merely relative; and that

---

[13] Bentham, *op. cit.*, 98.
[14] Bentham, *op. cit.*, 112.
[15] Bentham, *op. cit.*, 114.
[16] Bentham 1983, Table IV.
[17] Bentham 1983, Table V.
[18] For a more comprehensive overview, but flawed in several cases I believe, of the Benthams' logical
enterprise, see McOuat 2003. The flaws relate to identifying fixism with essentialism (McOuat agrees
in communication), and to a lesser extent not recognizing the much older tradition of the debate over
binary privative logic versus multiple species within genera. However, the paper has a much wider
agenda, and these are minor problems. Thanks to Charissa Varma for sending me this paper.

the same common terms may, in once case, be the species which is predicated of an individual, and, in another case the individual of which a species is predicated: e.g. the individual, Cæsar, belongs to the species man; but man, again, may be said to belong to the species animal, &c., as we contemplate higher and more comprehensive terms. A species, in short, when predicated of individuals, stands in the same relation to them as the genus to the species; and when predicated of other lower species, it is then, in respect of these, a genus, while it is a species in respect of a higher genus. Such a term is called a subaltern species or genus; while the highest term of all, of which nothing can be predicated, is the "summum genus;" the lowest of all, which can be predicated of nothing, the "infimæ species." The difference which, together with the genus, makes up the species, is termed the "specific difference."[19]

By the third edition in 1859, the discussion had been rewritten by Richard Owen to include the biological meaning of Cuvier, but this is as succinct a summary of the traditional conception as one will find. Moreover, we should note that it follows Aristotle in rejecting binary diairesis in favor of multiple species per genus, each of which carries its own special differentiae ("specific differences").

However, this view was not necessarily the view held by the leading philosophers of the day. Mill and Whewell, particularly, had tried to accommodate the current facts of natural history into the notion of a classification.[20] In his 1843 *System of Logic*,[21] Mill showed considerable knowledge of botanical classification conventions and awareness of variation within species. The discussion in Book I, chapter VII, especially §3–4, is well informed as to scholastic *and* biological conventions, and attempts a reconciliation of the two, without much success.

In Book I, chapter VIII, §4, he discusses the Cuvierian use of the term "Man" as the scientific definition, "Man is a mammiferous animal having two hands." This defines Man by giving "the place which the species ought to occupy in that particular [scientific] classification." He notes the Aristotelian use of *per genus et differentiam*, which he seems not to challenge. It is significant not for its resolution of the topic, but because we see here a philosopher taking pains to use as many biological examples as possible, although we also see elements and minerals appearing in the *exemplia gratia*. He defines species, at least in the sense used by naturalists, as "not, of course, the class in the sense of each individual of the class, but the individuals collectively, considered as an aggregate whole ..."[22]

Mill clearly is treating the species of the naturalist in a different sense, a "popular acceptation," more general and less logical than the sense of the logician. Nevertheless, both he and Whewell treated species as "natural classes," as Whewell stated it,[23] and in his response to Darwin's *Origin*, Whewell was dismissive of the idea that these classes could change. Hull quotes him as saying that, in botany, "... a natural class is neither more nor less than the observed steady association of certain properties, structures, and analogies, in several species and genera."[24] A Humean

---

[19] Brande and Cauvin 1853, 1137.
[20] Hull 2003.
[21] Mill 1930.
[22] Mill, *op. cit.*, Book I, chapter VII, §3.
[23] Hull 2003, 184f.
[24] Whewell 1831, 392, quoted in Hull 2003, 185.

associationist psychology is evident here, but also the Lockean idea of general terms as creations of the mind to collate past experiences. It is unclear if this was, as Hull suggests, the core of his objections to Darwin's theory of evolution, but at the least his philosophical adherence to logical essentialism certainly played a part in it.

Interestingly, *after* Darwin, a kind of essentialism regarding species was espoused by Jevons,[25] but he makes it clear that this refers to diagnostic species; that is, to classes of definitions of objects. He states that in a "natural" system of classification, all

> arrangements which serve any purpose at all must be more or less natural, because, if closely enough scrutinised, they will involve more resemblances than those whereby the class was defined[26]

and thus they are inductive groups, based, in living beings, on "inherited resemblances," such that the *"arrangement ... would display the genealogical descent of every form from the original life germ."*[27] Therefore, diagnostic essences are correlations that are causally important. Jevons follows Porphyry in treating Species as a predicable.[28] Sir William Hamilton, however, in his *Lectures on Metaphysics*, notes that we begin classification in "vague and confused" generalities, from which we refine our discriminations of things until we end, rather than commence, with the individuals, so that the genealogy of our knowledge is rather the history of how we came to know them than the history of how they came to be.[29] Species are a product of our getting to know things, giving an analogy:

> We perceive an object approaching from a distance. At first we do not know whether it be a living or an inanimate thing. By degrees we become aware that it is an animal, but of what kind—whether man or beast—we are as yet not able to determine. It continues to advance, we discover it to be a quadruped, but of what species we cannot yet say. At length, we perceive it is a horse, and again, after a season, we find that it is Bucephalus.[30]

The use of the notions of *genus* and *species* in logical discussions seems to have petered out with the introduction of set theory and formal logic by Venn, Cantor, Peirce, Frege, and others toward the end of the nineteenth century, especially around 1870–1878 in the case of Cantor. Where the inclusion of a species in a genus and of lower species in that species was the mainstay of classificatory logic prior to this development, now the talk was of sets and subsets. Moreover, the introduction of set theory itself seemed to override the older approach of *diairesis*, or top-down division. Sets could be defined from larger sets by division, or by aggregation of smaller sets. Even more radical was the distinction between intensional and extensional

---

[25] Jevons 1878, 710–713.
[26] Jevons, *op. cit.,* 680.
[27] *Loc. cit.,* italics original.
[28] Jevons, *op. cit.,* 698.
[29] Hamilton et al. 1874, Vol. 2, Lect. XXXVII.
[30] Hamilton, *op. cit.,* 334.

definitions of sets.[31] A species in the older logic had to be definable from the larger
genus. A set, however, could be described *or* defined. An extensionally defined
set is treated as isomorphic with another set, or, as Quine expresses it, "the *law
of extensionality*, which identifies sets whose members are the same"; intensions
(which Quine abhors) are specified by predicates, which "have attributes as their
'intensions' or meanings."[32] Under the Aristotelian account, all species were intensionally defined—this was the point of defining them by their essences. Now we
had aggregates that could be treated as synonymous merely by sharing all members,
irrespective of their essences. It was not immediately clear how this might apply to
the biological species problem, and it indeed took some time for it to be applied.

Similar essentialist accounts of biological species are also presented after Darwin
by several Roman Catholic authors, including a respected entomologist[33] and a
logician,[34] both Jesuits. A late example is a Dominican, Murray,[35] and a well-known
Canadian entomologist, William Thompson, another.[36] Thompson is particularly
interesting as he explicitly based his attack on Neo-Thomist philosophy. One might
conjecture that the essentialism of biological species bemoaned by Mayr and others is
in fact a *reaction* to Darwin and evolution (or at least Haeckel's Romantic version of it)
rather than something he overcame, as the Received View has it. Perhaps, like Donne,
they found all coherence gone with a temporal and gradual transmutation of species
one into another. Perhaps it was the outworking of the revival, or rather invention, of
neo-Thomism after the First Vatican Council.[37] Amundson, however, has argued that
essentialism was invented in the 1840s by Hugh Edward Strickland, but this was a
purely taxonomic essentialism with no causal or empirical consequences.[38]

Even so, some continued to use the older logical terminology and the Aristotelian
conceptions that underlay it well into the twentieth century, even if there were some
concessions to the new set theory. Husserl, for example, in 1913, writes in section 12
of his *Ideas*,

> Every essence, whether it has content or is empty (and therefore purely logical), has its
> proper place in a graded series of essences, in a graded series of *generality* and *spec-
> ificity*. The series necessarily possesses two limits that never coalesce. Moving
> downwards, we reach the *lowest specific differences* or, as we also say, the *eidetic*

[31] The terms "intension" and "extension" are medieval, according to Joseph. Mill's *Logic* introduced
the terms "connotation" and "denotation" [Joseph 1916, 146–155]. Leibniz is known to have also
contrasted the two terms.
[32] Quine 1970, 67.
[33] Wasmann 1910.
[34] Clarke 1895.
[35] Murray 1955.
[36] Thompson 1958, 1971, Thorpe 1973.
[37] Proposed in conversation by Polly Winsor. I have my doubts, though—"essence" is almost always
applied by the burgeoning Catholic intelligentsia to knowledge of the nature of God (e.g., by Cardinal
John Henry Newman), rather than to physical or material objects. The language was available, but it
appears to arise in biology much later. It is also a term in use by continental philosophers influenced
by Kant and Hegel; an obvious example is Marx; others include the existentialists such as Kierkegaard
(McOuat, *pers. comm.*). Nevertheless, I have no doubt that the Neo-Thomist revival was an influence
on this movement if only because it offered an alternative metaphysics to the Darwinian problem.
[38] Amundson 2005, 51.

singularities; and we move upwards through the essences of genus and species to a *highest genus*. Eidetic singularities are essences, which indeed have necessarily "more general" essences as their genera, but no further specifications in relation to which they themselves might be genera (proximate or mediate, higher genera). Likewise that genus is the highest which no longer has any genus above it.[39]

More interestingly, and influentially on the subsequent debate, Joseph's *Introduction to Logic* allowed that the evolutionary species of Darwin and Spencer were a different notion to that of the logical species of definitions. He goes so far as to note that species in biology cannot be defined, and that instead one must describe a type, from which individuals can diverge. Joseph continues in the tradition of Whately, separating logical species defined by essence and biological species described by types. Even at this late stage, types and essences are explicitly held to be different notions. Until Woodger introduces symbolic logic to biology,[40] such issues are discussed by a declining number of philosophers; and then of course later in the context both of cladism and the individuality thesis.

JEAN BAPTISTE DE LAMARCK: UNREAL SPECIES CHANGE

No sooner had natural history established a tradition of fixism of species than it was immediately under challenge, for example, by Pierre Maupertuis in *Vénus Physique* in 1745.[41] At the turn of the nineteenth century, there was a considerable amount of ferment over the notions of taxonomic groups or ranks. For example, Blumenbach had classified the human species into races—Caucasian, Mongolian, Ethiopian, American, and Malayan—and yet he still regarded these types as subordinate to the human species, and that all were varieties of that species,[42] although Buffon had previously denied that the notion of "race" applied to the usual human groupings.[43] Blumenbach's conception of the species was that it was formed through the action of a formative force, a *nisus formativus*, and so his is also a generative notion of species.

More influentially, Lamarck delivered a transmutationist view of species, and followed his mentor Buffon in supposing that there were no realities attaching to the term. In the *Zoological Philosophy* (Philosophie Zoologique, 1809) he writes:

> It is not a futile purpose to decide definitely what we mean by the so-called *species* among living bodies, and to enquire if it is true that species are of absolutely constancy, as old as nature, and have all existed from the beginning just as we see them to-day; or if as a result of changes in their environment, albeit extremely slow, they have not in the course of time changed their characters and shape.

...

---

[39] Husserl 1931.
[40] Woodger 1937, Woodger 1952.
[41] Maupertuis 1745. Bear in mind this is only ten years after Linnaeus' first edition of the *Systema naturae*.
[42] Nordenskiöld 1929, 306, Voegelin 1998.
[43] Roger 1997, 177f.

Let us first see what is meant by the name of species.

Any collection of like individuals which were produced by others similar to themselves is called a species.

This definition is exact: for every individual possessing life always resembles very closely those from which it sprang; but to this definition is added the allegation that the individuals composing a species never vary in their specific characters, and consequently that species have an absolute constancy in nature.

It is just this allegation that I propose to attack, since clear proofs drawn from observation show that it is ill-founded.[44]

As Gillispie puts it, "(Lamarck's) position is rather that species do not exist, than that they are mutable."[45] Lamarck reiterates in the *Zoological Philosophy* the early view of Buffon that only individual organisms exist in nature:

Thus, among living bodies, nature, as I have already said, definitely contains nothing but individuals which succeed one another by reproduction and spring from one another; but the species among them have only a relative constancy and are only invariable temporarily.[46]

It is interesting to note that here and elsewhere Lamarck explicitly restricts his comments to living bodies. His nominalism with respect to organisms is obvious, and species are not themselves in nature, but, as Locke had said, are made for communication:

Nevertheless, to facilitate the study and knowledge of so many different bodies it is useful to give the name species to any collection of like individuals perpetuated by reproduction without change, so long as their environment does not alter enough to cause variations in their habits, character and shape.[47]

The generative conception is again in play, except that Lamarck has added temporality to the mix. In the *Recherches*[48] Lamarck proposed that there was a "life-fluid" that was a variety of physical energy, a *feu éthéré*, that maintained organisms in their form, and it impelled spontaneous generation out of inanimate matter. Species was a notion that applied to the mineral kingdom as well as the biological, but mineral species differed in that they had no individuality while plants and animals did, and neither did they reproduce. In the *Zoological Philosophy* he added that *all* classifications are arbitrary products of thought, and that in nature there are only individuals. Lamarck did not accept the reality of extinction apart from human agency, but

---

[44] Lamarck 1809. English translation Lamarck 1914, 35.
[45] Gillispie 1959, 271.
[46] Lamarck 1914, 44.
[47] *Loc. cit.*
[48] Lamarck 1802.

held that fossil species merely transformed into later forms.[49] Gillispie notes that Lamarck was slightly inconsistent on species between 1797 and the statements of 1802 and 1807, but goes on to say

> … the inconsistency on species (is) trivial. … All he did between 1797 and 1800 was to assimilate the question of animal species—or rather their nonexistence—to that of species in general. For in Lamarck the word has not lost its broader connotations. It still carries the sense of all the forms into which nature casts her manifold productions in all three kingdoms (or rather in both divisions).[50]

Lamarck is still indebted to the medieval notion of species as subsidiary divisions of the *summum genus* (being) through to the infimae species of rational living things in the case of humans. In this respect, he was attempting to classify all things as the outcome of physical molecules and forces, and animal species were just an arbitrary part of that chain of being. Still, his view of these nominalistic species is that they are formed out of the generative properties of the life-fluid, and so this is a generative notion of species in that respect at least.

As to nomenclature, he accepted Linnaeus' binomial convention, and given there was no fact of the matter, held that an international agreement should be made to make names stable.[51] At this stage, Buffon's objections to the binomial nomenclature have lost the field entirely, when even his own student accepts the practice. Lamarck's view of evolution is basically a temporalization of the ladder of nature/great chain of being. He treated each species as a single lineage that had its own original spontaneous generation out of non-living material, and which ascended something like Bonnet's ladder, although the ladder could branch, as we see in the famous diagram (Figure 4.1).

The ladder that Lamarck adopted, however, was less direct than Bonnet's. He wrote:

> I do not mean that existing animals form a very simple series, regularly graded throughout; but I do mean they form a branching series, irregularly graded and free from discontinuity, or at least once free from it.[52]

Lamarck's definition was echoed by Isidore Geoffroy Saint-Hilaire (1805–1861), in 1859, in his *Histoire naturelle génerale des règnes organiques*:

> The species is a collection or a succession of individuals characterized by an ensemble of distinctive features whose transmission is natural, regular and indefinite in the current order of things.[53]

---

[49] Nordenskiöld 1929, 325.
[50] Gillispie 1959, 272.
[51] Nordenskiöld 1929, 326.
[52] Lamarck 1914; diagram from 179, quote from 37.
[53] Quoted in Lherminer and Solignac 2000, 156. The French is

> L'espèce est une collection ou une suite d'individus *caractérisés par un ensemble de traits distinctifs dont la transmission est naturelle, régulière et indéfinie dans l'ordre actuel des choses.* [Saint-Hilaire 1859, 437, italics original]

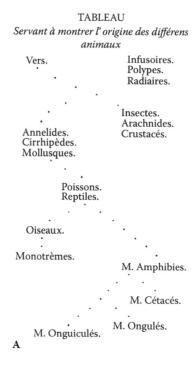

TABLEAU
*Servant à montrer l'origine des différens
animaux*

Vers.   Infusoires.
                                         Polypes.
                                         Radiaires.

                                         Insectes.
                                         Arachnides.
Annelides.   Crustacés.
Cirrhipèdes.
Mollusques.

                Poissons.
                Reptiles.

Oiseaux.

Monotrèmes.

                        M. Amphibies.

                                M. Cétacés.

                        M. Ongulés.
        M. Onguiculés.
A

FIGURE 4.1   Lamarck's "tree." **A.** Lamarck's original table. Note that he uses sparse dots to
indicate affinities.   *(Continued)*

Some of Lamarck's transformist views on species had been championed by
Etienne Geoffroy Saint-Hilaire (1772–1844), Isidore's father, although he did not
make much of the rank of species as such, preferring to focus instead upon the "unity
of plan"—relations ("analogies," or as we now call them, homologies) between different classes of organisms, in keeping with his transcendentalist views. His main
statement about species was that they were not fixed:

> The species is not fixed and does not reappear in its forms, except for the reason that
> the conditional state of its ambient medium is maintained; for, according to the bear-
> ing and under the influence of the latter, there are scarcely any changes which are not
> possible with respect to it.[54]

Geoffroy's views were popularized by Chambers' *Vestiges of the Natural History
of Creation*.[55]

---

[54] Translated in Le Guyader 2004, 228.
[55] Chambers 1844. Chambers wrote of species:

> species, the subdivision where intermarriage or breeding is usually considered as natural to ani-
> mals, and where a resemblance of offspring to parents is generally persevered in. The dog, for ins-
> tance, is a species, because all dogs can breed together, and the progeny partakes of the appearances
> of the parents. The human race is held as a species, primarily for the same reasons. [page 263]

TABLE

Showing the origin of the various animals.

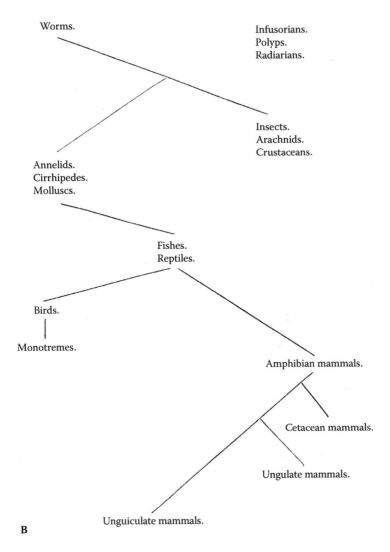

FIGURE 4.1 (CONTINUED)   Lamarck's "tree." The 1904 English version (**B**), showing lines, may be confused as a tree of descent, but Lamarck intended the lines to indicate pathways along which new monads could develop.

BARON CUVIER: FIXED FORMS AND CATASTROPHES

Lamarck's (and later Geoffroy's) nemesis, Georges Leopold Chrétien Frédéric Dagobert, Baron Cuvier (1769–1832), was by contrast a full-blooded species realist. Not only were species real, they did not transmute from one to another. Species were generated in some manner at the time of a catastrophe, and previous forms were obliterated. Nordenskiöld writes

> The immutability of species is to Cuvier's mind an absolute fact; he has not a trace of Linnaeus's hesitation, which he expressed in his old age, in face of the difficulty of drawing a line of demarcation between the species; according to Cuvier's definition, species consist of "those individuals that originate from one another or from common parents and those which resemble them as much as one another." In this definition no mention is made of the creation of the species, which, it will be remembered, Linnaeus took as his starting point, but which, on the whole, Cuvier does not discuss at all.[56]

Cuvier's definition[57] is interesting in several respects. Despite the superficial resemblance to Linnaeus' definition given above, Cuvier's more closely resembles John Ray's definition. It is a historical definition, and yet it requires resemblance, presumably to bar monsters. It is, as was Ray's, a generative and yet still a formalist definition. For Cuvier, species come into existence at the beginning of each geological epoch and never vary thereafter. He notes that if species have changed by degrees, then we ought to have found traces of these gradual modifications and intermediate forms between, say, a paleotherium and modern elephant, which we do not. Hence, species are stable. Although spare, Cuvier's definition was very influential on philosophers, as the discussion in Whately's and Mill's logics show. His fixism has the following rationale:

> These forms are neither produced nor do they change of themselves; life presupposes their existence, for it cannot change except in organisations ready prepared for it.[58]

In his *Éloge*, or funeral oration, of Lamarck, Cuvier set to demolishing Lamarck's idea of species transmuting.[59] Basing his argument upon the static nature of Egyptian mummies of various animals when compared with the modern version, and the lack of apparent progress from simple to complex forms that Lamarck's view required in the fossil record, Cuvier established the default view that species did not themselves change. He did not require, as Russell notes, that faunal epochs were an illusion—he simply claimed to have no theory of how the new faunas came into existence, as he had little time for theory without facts.[60] He tried to minimize the number of special creations, and said:

---

[56] Nordenskiöld 1929, 339.
[57] Under a section entitled "Lost species are not varieties of living species":

> Cette recherche suppose la définition de l'espèce qui sert de base à l'usage que l'on fait de ce mot, savoir que l'espèce comprend *les individus qui descendent les uns des autres ou de parens communs, et ceux qui leur ressemblent autant qu'ils se ressemblent entre eux.* [My research assumes the definition of species which serves as the basic use made of the term, understanding that the word species means *the individuals who descend from one another or from common parents and those who resemble them as much as they resemble each other.*] *Règne Animal*, i, 19. [Cuvier 1812, 74]

[58] *Loc. cit.* p20 quoted in Russell 1982.
[59] Cuvier 1835.
[60] Russell 1982, 43.

I do not pretend that a new creation was required for calling our present races of animals into existence. I only urge that they did not anciently occupy the same places, and that they must have come from some other part of the globe.[61]

JAMES PRICHARD: SPECIES ARE REAL, VARIATIONS ARE ENVIRONMENTAL

Prichard (1786–1848) published his widely read *Researches into the Physical History of Man* the year after Cuvier's "Discours" was published.[62] In it, he foreswore a priori argument, whether scriptural or not, and vowed to deal only with evidence, not speculation. That notwithstanding, Prichard declared that Linnaeus' classes were arbitrary and artificial (something we have seen Linnaeus might admit freely), but

> Not so in the case of species. Here the distinction is formed by nature, and the definition must be constant and uniform, or it is of no sort of value. It must coincide with Nature.
>
> Providence has distributed the animated world into a number of distinct species, and has ordained that each shall multiply according to its kind, and propagate the stock to perpetuity, none of them ever transgressing their own limits, or approximating in any great degree to others, or ever in any case passing into each other. Such confusion is contrary to the established order of Nature.
>
> The principle therefore of the distinction of species is constant and perpetual difference.[63]

He then used this criterion to reject Buffon's concept of species and to argue that human races were all of the same species. He denied that local variations were caused by the direct action of the soil and environment, and argued that variation occurred naturally when the environment changed and was passed on (in the context of human variation). Civilization minimized this variability in humans, and so dark skins, due to savage conditions, would turn into light skins upon civilizing. Prichard's discussion of *species* is extensive and at the time authoritative. He writes:

> The meaning attached to the term *species* in natural history is very definite and intelligible. It includes only the following conditions, namely, separate origin and distinctness of race, evinced by the constant transmission of some characteristic peculiarity of organization. A race of animals or of plants marked by any peculiar character which has always been constant and undeviating, constitutes a species; and two races are considered as specifically different, if they are distinguished from each other by some characteristic which the one cannot be supposed to have acquired, or the other to have lost through any known operation of physical causes; for we are, hence, led to

---

[61] Cuvier et al. 1818, 128 quoted in Greene 1959, 363n.
[62] Prichard 1813.
[63] Prichard 1813, 7–8, quoted in Greene 1959, 239.

conclude, that the tribes thus distinguished have not descended from the same original stock. This is the purport of the word species, as it has long been understood by writers on different departments of natural history.

Later, in his 1843 *The natural history of man*, he writes:

Species, then, are simply tribes of plants or of animals which are certainly known, or may be inferred on satisfactory grounds, to have descended from the same stocks, or from parentages precisely similar, and in no way distinguished from each other. The meaning of the term species ought always, for the reasons now explained, to have been restricted to this precise import; and when the expression is used in the following pages, it is so to be understood.[64]

Prichard's view may have influenced the well-known view of William Whewell in 1837 that *"species have a real existence in nature*, and a transition from one to the other does not exist."[65] The transition envisaged here is most probably one of plenitude than transmutation, but by 1837 Geoffroy's views on evolution were widely known, as was Cuvier's demolition job on Lamarck in the *Éloge*.

LOUIS AGASSIZ: THE LAST FIXIST AND THE LONELY PLATONIST

Louis Agassiz, Cuvier's devotee and intellectual successor, concurred with Cuvier and Prichard on the constancy of species. Notwithstanding this, Agassiz did not expect that there would be a set of characters unique to all members of a species; resemblance was not itself clear or absolute. In an early short note, he denied that characters gave the species, and instead insisted that while there was a process that underlay the forms of species, there need not be any diagnosable characters that all members of the species exhibited, bringing to mind the Lockean distinction between real and nominal essences. Like Locke, Agassiz was rejecting the nominal and accepting the real essence, and it was a generative notion of real essence at that:

... no so-termed character—that is, no observable mark—can be so striking as to indicate an absolute specific distinction; but at the same time, it should never be regarded as so trifling as to point to absolute identity; that characters do not mark off species, but that the combined relations to the external world in all circumstances of life do.[66]

Agassiz's version of the real essence here is the set of causal relations of the organism throughout its lifecycle to its environment, and he cites the sometime inclusion of the male and female of a species in separate taxa. Agassiz explicitly rejected a diagnostic notion of species, and in effect anticipated the later notion of "cryptic species" of Mayr. However, in practice, he was not so exact. He distinguished between eight

---

[64] Prichard 1843, 10.
[65] Whewell 1837, 626, Vol. 3; quoted in Hull 1973, 68.
[66] Agassiz 1842, italics original.

"species" of Man: Caucasian, Arctic, Mongol, American Indian, Negro, Hottentot, Malayan, and Australian, and claimed that these were all independent creations, not related by descent, each with its own region, flora, and fauna. The basis for this was not some generative notion, or reproductive isolation (since it was clear that human "types" could interbreed without trouble), but their clear physical differences.[67] In short, the characters indicated an absolute specific distinction. Or perhaps Man was different.

Agassiz was not a Lockean, however; he was clearly a variety of Platonist, or at least of idealist. He wrote:

> [T]here is a system in nature, to which the different systems of authors are successive approximations, more and more closely agreeing with it, in proportion as the human mind has understood nature better. This growing co-incidence between our systems and that of nature shows ... the identity of the operations of the human and the Divine intellect ...[68]

Agassiz's biographer, Lurie, calls this Agassiz's "cosmic philosophy," and notes that in his view

> [s]pecies, the individual units of identity in nature, were types of thought reflecting an ideal, immaterial inspiration. The same was true of the larger taxonomic categories— genera, families, orders, branches, and kingdoms. All such categories had no real existence in nature. Reality could be discovered only in the character of the individual animals and plants that had inhabited or were now inhabiting the material world. The individual fossil or living from represented on earth the categories of divine thought ranging from species to kingdom and ultimately symbolized a complete identity with the highest concept of being, God.[69]

Two years after the publication of his "Essay on Classification," the *Origin of Species* was released, and so the essay provides a good demarcation point between the traditional view of classification, and the revolution that was to come, even if Agassiz's views were already archaic. In that essay,[70] he argued again for the stability of species, although his primary task was to discuss ways in which naturalists could identify and name species, rather than to define them other than as the smallest division of the four great *embranchements* named by Cuvier, which Agassiz called "great types."[71] These were the ways of being, typical plans of nature. Species were the lowest group that could be differentiated out of these plans (which play the role, therefore, of Aristotle's *summum genera*). They had in themselves no identifying morphological character, because that was exhausted in the genus. Winsor notes,

> [h]aving already publicly rejected the criterion of interbreeding, during the debate on the unit of mankind, Agassiz had to ask himself what besides morphological detail

---

[67] Lurie 1960, 264ff.

[68] Agassiz 1859, 31.

[69] *Loc. cit.*

[70] Published as volume I of the *Contributions to the Natural History of the United States*, 1857 to 1862.

[71] Winsor 1979, 97. See footnote to Appendix A.

and sexual preference enables a biological species to be recognized. His answer was, its mode of reproduction and growth, its geographic distribution and fossil history, and the manifold relations that the individual organism bears to the world around it."[72]

A species is, it seems, a description of the overall biological features of organisms, for Agassiz gave the individual organism priority, not unlike Buffon. Winsor notes further, "[h]is purpose was ... to affirm the reality of all those relationships of similarity that are expressed in a natural classification." She quotes him from the *Essay*:

> Species then exist in nature in the same manner as any other groups, they are quite as ideal in the mode of existence as genera, families, etc., or quite as real. ... Now as truly as individuals, while they exist, represent their species for the time being and do not constitute them, so truly do these same individuals represent at the same time their genus, their family, their order, their class, and their type, the characters of which they bear as indelibly as those of the species.[73]

Species exist as ideas, which represent the relations actual individuals bear to the world. They are not things, in the physical sense of the term, so much as what the things represent (but do not comprise). This is very Platonic in spirit.

In his *Methods of Study in Natural History*, he further discusses classificatory categories. In this he adopts a realist view of Cuvier's *embranchements* theory as "being, so far as it is accurate, the literal interpreter of [the plan of creation],"[74] and "that classification, rightly understood, means simply the creative plan of God as expressed in organic terms,"[75] but this is at a much higher level than the species level. He notes in chapter V that the nature of Linnaean Orders is partly arbitrary, that if

> one man holds a certain kind of structural characters superior to another, he will establish the rank of the order upon that feature[76]

but overall, he says that higher taxa

> stand, as an average, relatively to each other, lower and higher.[77]

He is thus a rank realist. He argues that there is no succession between these higher types,[78] and so there is no reason to think evolution is correct.

In this post-*Origin* work Agassiz denies that Species are more real than the higher taxa, contrary to the opinions based upon de Candolle's views.[79] Species are exactly as real as these higher taxa:

---

[72] Winsor 1979, 98.
[73] Agassiz, *op. cit.*, 256f.
[74] Agassiz 1863, 41.
[75] Agassiz, *op. cit.*, 42.
[76] Agassiz, *op. cit.*, 72f.
[77] Agassiz, *op. cit.*, 86, cf. also 109 on Branches and Classes, and 127f on Genera.
[78] Agassiz, *op. cit.*, 92.
[79] Agassiz, *op. cit.*, 135.

All the more comprehensive groups, equally with Species, are based upon a positive, permanent, specific principle, maintained generation after generation with all its essential characteristics. Individuals are the transient representatives of all these organic principles, which certainly have an independent, immaterial existence, since they outlive the individuals that embody them, and are no less real after the generation that has represented them for a time has passed away, than they were before."[80]

Species are not composed of organisms, in other words; organisms at best "represent" species. And species are of the same standing as the higher taxa, which are

built upon a precise and definite plan which characterizes its Branch,—that that plan is executed in each individual in a particular way which characterizes its Class, ...[81]

and so on down to Species.

He discusses variation in domesticated animals "which has been urged with great persistency in recent discussions" (i.e., by Darwin and his followers[82]) but asserts that this is due to the "fostering care" of the breeders of freaks that are not observed in wild species,[83] and that "this in no way alters the character of the Species."[84]

They are called Breeds, and Breeds among animals are the work of man: Species were created by God.[85]

Homologies, the foundation of Darwinian argument, are "the Creative Ideas in living reality."[86] All is the work of God, and we are just making classifications to trace what God hath wrought.

Later, Agassiz further attacked "Darwinism" by means of an attack on Haeckel's genealogical classifications in Darwin's name, in a chapter added to a French edition.[87] Here he attacks the apriorism of the work of Oken and those who follow the ideal morphology school, including Haeckel in that class because he imposes his expectations on the data, which he also accuses Darwin of doing, a point he had made in an earlier review of the *Origin*. He rejects the claim of the "Darwinists and their henchmen" that organisms will not reproduce the essential characters of their ancestors. In Morris' translation, Agassiz says

---

[80] Agassiz, *op. cit.*, 136.

[81] Agassiz, *op. cit.*, 139f.

[82] Agassiz, *op. cit.*, 141.

[83] Agassiz, *op. cit.*, 145.

[84] Agassiz, *op. cit.*, 141f.

[85] Agassiz, *op. cit.*, 147. Disarmingly, Agassiz in the very next chapter refers approvingly to Darwin's work on the coral reefs as a "charming little volume" [Agassiz, *op. cit.*, 154], possibly because Darwin consciously emulated Agassiz's method from his ice age studies. Thanks again to Mike Dunford for access to his copy of this work.

[86] Agassiz, *op. cit.*, 231.

[87] Agassiz 1869, Morris 1997.

All the observations relative to domestic animals, among which there are so many and so numerous variations, again did not succeed in demonstrating a sufficiently large amplitude in these variations; never did they [the Darwinists] have as a result anything which manifests the indefinite tendency to a changeability without limit...[88]

In short, Agassiz rejected the Darwinian view of species on the grounds that the requisite variation had not been observed. Agassiz, however, has been accused of being unable to see that there could possibly be more than a certain amount of variation in his own specimens, as Lurie asserted:

When he had hundreds of fishes spread before him on a work table, these convictions (of the fixity of species) were of such force that even his keen powers of observation and his excellent ability to compare diverse types failed him. He insisted on identifying specimens that seemed even the slightest degree different from one another as separate species rather than as variants. In one analysis alone, for example, he described nine separate "species" of fishes that were in actual fact reducible to four schools of single species.[89]

If true, Agassiz, merciless on taxonomic splitters, was in practice himself a splitter because of his tendency to classify on form alone despite his stated convictions about species in theory. However, Winsor has investigated these fishes and argued that he was not so much of a splitter as Mayr had said, and when he was, he was not unusual for the time in that respect.[90]

In another, more sinister, respect though, his views did lead him to excessive splitting. He claimed that Negroes were not of the same species as whites, because, it seems, of the feelings of revulsion he had for them that led him to deny they could be conspecific to whites. This famously meant that he became the leading proponent of multigenism and hence a popular figure in the South before the Civil War.[91]

Amundson argues that Mayr's access to Agassiz's works at the Museum of Comparative Zoology at Harvard, which Agassiz had founded and Mayr was a later director, led Mayr to overgeneralize that all taxonomists before Darwin were essentialists, typologists, and fixists.[92] Agassiz was indeed a Platonist who considered species thoughts in the mind of God, and he was also undoubtedly a species fixist, particularly after Darwin. Moreover, he was both a taxonomic essentialist and a material essentialist. Possibly he was the *first* such essentialist fixist Platonist. But he was not, so to speak, typical of the time or the profession, apart from his students, and they only for a limited time.

---

[88] Agassiz 1869, 378. I have amended the translation slightly for grammar's sake. The original version was accessed on 23/9/02 at http://www.athro.com/general/atrans.html.

[89] Lurie 1960, 194f according to Mayr, in Lurie's footnote.

[90] Winsor 1979, 104ff. It would be interesting to see how those species have fared in the molecular period of systematics. Thanks to the author for pointing out my earlier mistake.

[91] Hunter Dupree 1968, 228f.

[92] Amundson 2005, 79.

JAMES DANA: A LAW OF CREATION

But Agassiz was not alone in pressing the Cuvierian view in the period leading up to the *Origin*. His very great admirer James Dana, in an essay in a journal he co-edited, reiterated the old view that "species" applies to all natural things, and that the variable characters of individuals are merely confusing. To this end, he rejected the idea that species are even groups, necessarily. Instead, he wrote

> A *species* corresponds *to a specific amount or condition of concentrated force, defined in the act or law of creation.*[93]

At least in the inorganic world: species are what they were constituted at their creation to be. In the biological world, the idea is the same, leading to the understanding that

> [t]he species is not the adult resultant of growth, nor the initial germ cell, nor its condition at any other point; it comprises the whole history of development. Each species has its own special mode of development as well as ultimate form or result, its serial unfolding, inworking and outflowing; so that the precise nature of the potentiality in each is expressed by the line that historical progress from the germ to the full expansion of its powers, and the realization of the end of its being. We comprehend the type-idea only when we understand the cycle of evolution [sensu *development—JSW*] through all its laws of progress, both as regards the living structure under development within, and its successive relations to the external world.[94]

For Dana, species are the units of the organic world as molecules are the units of the inorganic. He discusses the ranges of infertility of hybrids from the infertile mule to the continuously fertile hybrid, and says that the fully fertile hybrid is not observed in nature, at least among animals; plants are more frequently hybridizing. In a rather backhanded manner, he affirms the monogenist position—humans are one species, although non-white races are disappearing

> ... like plants beneath those of stronger root and growth, being depressed morally, intellectually and physically, contaminated by new vices, tainted variously by foreign disease, and dwindled in all their hopes and aims and means of progress, through an overshadowing race[95]

lest any of his readers get the wrong idea. At least he stood up to Agassiz on monogenism.

Species are not transmutable, for all hybridization is merely recombination of already extant variation, but there is variation within species—the unfolding of the potentiality inherent within a species according to natural law and changing circumstances.[96] Species are liable to variation as part of the law of a species, and

---

[93] Dana 1857, 306, italics original.
[94] Dana 1857, 308.
[95] Dana 1857, 311.
[96] Dana 1857, 312.

knowledge of the complete type requires knowing all these and how they relate to external circumstances. There is a higher essence, as it were, in the type. Finally, while species are real things, they are not comprehensively covered in any "material or immaterial existence"—in modern parlance, they are types, not tokens,[97] and species are both invariant and variant. In short, Dana sees species as the schematic of a developmental cycle and the ways in which it may be perturbed by the environment.

RICHARD OWEN ON THE UNITY OF TYPES

Richard Owen, who introduced many of the ideas of the ideal morphologists into British thought, did not address the question of species directly, so far as I can tell. In his Hunter Lectures of 1843, reissued in a revised edition in 1855, entitled *Lectures on the comparative anatomy and physiology of the invertebrate animals*,[98] Owen addressed the question of overall types in terms derived from Cuvier's *embranchments*, but treated species as unanalyzed units of classification.[99] He was not opposed at this time to species transmutation, although he treats it more as a formal possibility than as an actuality, and so he contributes little to the topic at hand.

Owen, as is widely known, first clearly expressed the distinction between *homologue* and *analogue* in the Glossary to his 1843 *Lectures*,[100] and further distinguished special, general, and serial homology, respectively, the correspondence of a part in one animal with a part in another, of a part in a particular animal (that is, of a species) to a fundamental or general higher type, and the repetition of a part within a particular animal or type.[101] Organic forms, he thought, were due to a mutual antagonism of two principles, one of which brought about a vegetative repetition of structure, and the other which shapes the living thing to its function, a teleological principle.[102] He allowed that species were formed through time, and that there was a naturalistic cause based on these principles, "the secret counsels of the organizing forces," as he expressed it in the *Archetype*.[103] Earlier, Owen discussed how species have changed in their meaning since the times in which Linnaeus' and Cuvier's definitions were accepted:

> I apprehend that few naturalists now-a-days, in describing and proposing a name for what they call 'a new *species*,' use that term to signify what was meant by it twenty or thirty years ago, that is, an originally distinct creation, maintaining its primitive distinction by obstructive generative peculiarities. The proposer of the new species now intends to state no more than he actually knows; as for example, that the differences in which he founds the specific character are constant in individuals of both sexes, so far

---

[97] Not *that* modern—the type–token distinction was made by C. S. Peirce [1885; see §§35–37 in Wollheim 1968 for a full discussion of this distinction] only a few decades after Dana wrote. Intriguingly, Peirce's distinction was between *icons*, *indices*, and tokens, and he referred to tokens as *replicas* of symbols [Hookway 1985, 130f].

[98] Owen 1843.

[99] Owen 1855.

[100] Owen 1843.

[101] Russell 1982, 108f.

[102] Russell 1982, 111; cf. Amundson 2005, 88-93.

[103] Owen 1848; see the discussion in Amundson 2005, chapter 4.

as observation has reached; and that they are not due to domestication or to artificially superinduced external circumstances, or to any outward influence within his cognizance; that the species is wild, or is such as it appears by nature.[104]

Owen here treats species as taxonomic objects, with the underlying implication that they are constant due to natural causal powers. The generative "peculiarities" are not ignored or forgotten when the creation part of the definition has been abandoned, and the naturalizing of species is evident. However, it is also noteworthy given Owen's later reputation that in his Presidential Address to the British Association for the Advancement of Science in 1858, Owen not only reports Darwin's and Wallace's views on speciation by natural selection, but he goes on to say:

> No doubt the type-form of any species is that which is best adapted to the conditions under which such species at the time exists; and so long as those conditions remain unchanged, so long will the type remain; all varieties departing therefrom being in the same ratio less adapted to the environing conditions of existence. But, if those conditions change, then the variety of the species at an antecedent date and state of things will become the type-form of the species at a later date, and in an altered state of things.[105]

Owen's initial reaction to Darwin and Wallace, untainted by Huxley's rhetoric, was thus fairly positive. Of special note is that Owen refers to the "type-form" of the species in a way that makes it clear that he is not referring to essences.

Ronald Amundson has argued that for the period, species were never types, as types were something that united species under larger classificatory ranks.[106] Indeed, Amundson denies that species ever were types, a point Owen, among others, shows is not correct. As many have noted, there is a distinction between Platonic forms and Aristotelian species, and the Aristotelian form *could* be varied from the standard type. Naturalists of a Platonic bent prior to Darwin, such as Agassiz, saw species as types from which *individuals* could deviate. The concept of "degeneration," which Amundson discusses, indicates that individuals were able to improperly instantiate the specific norm, as well as species as units being able to improperly or variably instantiate the generic essence.[107] So long as types are not identified with unvarying essences, it is not true that species could not be types. But if, as Amundson seems to do in his sect 4.2, types (here, morphological types as abstractions) and essences are identified, with Mayr, then there are no specific types.

That said, it is apparent both that types are almost always through this period treated as large-scale (supraspecific) types and that the ideal morphologists in the tradition of Goethe and Oken did not generally discuss species, either as biological realities or as an abstract concept, as they were more concerned with the higher types that gave a Unity of Type between species and genera.[108]

---

[104] Owen 1835, quoted in Huxley 1906, 303.
[105] Quoted in Basalla et al. 1970, 329.
[106] Amundson 2005, 81.
[107] Amundson 2005, 36, 40.
[108] Russell 1982, Nyhart 1995.

OTHER FIXIST VIEWS

The view that species were fixed as they had been created was held religiously (in the strict sense) in 1844 by the marine biologist and founder of the fashion of aquarium keeping, Philip Henry Gosse in his *An Introduction to Zoology*:

> Each order was distributed into subordinate groups, called Genera, and each genus into Species. As this last term is often somewhat vaguely used, it may not be useless to define its acceptation. It is used to signify those distinct forms which are believed to have proceeded direct from the creating hand of God, and on which was impressed a certain individuality, destined to pass down through all succeeding generations, without loss and without confusion. Thus the Horse and the Ass, the Tiger and the Leopard, the Goose and the Duck, though closely allied in form, are believed to have descended from no common parentage, however remote, but to have been primary forms of the original creation. It is often difficult in practice to determine the difference or identity of species; as we know of no fixed principle on which to found our decision, except the great law of nature, by which specific individuality is preserved—that the progeny of mixed species shall not be fertile inter se.[109]

He repeated this in his famous book, known popularly as *Omphalos*, in which he argued that the world was created as the Bible said, but was made to look old. He says

> I demand also [*as well as the creation of matter out of nothing—JSW*], in opposition to the development hypothesis [*pre-Darwinian evolutionism—JSW*], the perpetuity of specific characters, from the moment when the respective creatures were called into being, till they cease to be. I assume that each organism which the Creator educed was stamped with an indelible specific character, which made it what it was, and distinguished it from everything else, however near or like. I assume that such character has been, and is, indelible and immutable; that the characters which distinguish species from species *now*, were as definite at the first instant of their creation as now, and are as distinct now as they were then.[110]

Here we see both fixism and essentialism, and it is essentialism of the kind that Mayr and the Received View objected to—a *real* or material essentialism as well as a taxonomic one. But there is no comfort for the Received View here—Gosse was not regarded as a leading professional naturalist (though Huxley called him the "honest hodman of science"[111]), and his book was not well received; it sold so poorly that most copies were pulped, and as his son wrote, "alas! Atheists and

---

[109] Gosse 1844, xv; cf. Simpson 1925, 175.
[110] Gosse 1857, also known as *Creation (Omphalos)*. I am indebted, literally and metaphorically, to Dr Noelie Alito for purchasing an original copy of Gosse's *Creation (Omphalos)* on my behalf. A scanned copy is available on the Internet from archive.org. The title of physical copies I have seen include both *Omphalos* and the title above. Possibly the publisher reissued it with another title to increase sales by making the subject matter clearer.
[111] Numbers 1992, 141. The source of the Huxley comment is Gosse's son's book *Father and Son* [Gosse 1970, chapter V]. I cannot locate an original.

Christians alike looked at it, and laughed, and threw it away."[112] It is hard not to see Gosse's obduracy as a desperate attempt to maintain the fixity of species even in the face of Buffon's hybridization experiments, which had had a partial success, but in fact there was considerable blindness to variation, as the Agassiz example shows. Another instance involves a student of Agassiz, Stimpson, who, when finding intermediate forms of a mollusk he could not decide to place in one species or another,

> ... after he had studied it for a long time, put his heel upon it and grind[ing] it to powder, remarking, "That's the proper way to serve a damned transitional form.[113]

The religious opponents of materialism and evolution, of course, continued to assert fixism and essentialism.[114]

CHARLES LYELL: SPECIES ARE FIXED AND REAL

More significantly, for our later story, are the earlier views of Charles Lyell. It is well-known Darwin who received the second volume of Lyell's *Principles of Geology*[115] on the voyage of the *Beagle*,[116] which contained the discussion of Lamarck's views, a "book-long refutation of Lamarck," as Desmond and Moore called it.[117] Lyell presented the now-standard fixist view of species; as Kottler put it:

> Lamarck and Lyell agreed that the 'reality' of species implied their constancy. The words 'real' and 'permanent' were synonymous with respect to species in nature. Thus while Lamarck, the transmutationist, contended species were not real, Lyell, the fixist, argued they were.[118]

In this volume, Lyell concludes his discussion of Lamarck and Linnaeus, and species in general, at the end of chapter IV with this:

> For the reasons, therefore, detailed in this and the two preceding chapters, we draw the following inferences, in regard to the reality of species in nature.
>
> First, That there is a capacity in all species to accommodate themselves, to a certain extent, to a change of external circumstances, this extent varying greatly according to the species.
>
> 2dly. When the change of situation which they can endure is great, it is usually attended by some modifications of the form, colour, size, structure, or other particulars; but the mutations thus superinduced are governed by constant laws, and the capability of so varying forms part of the permanent specific character.

---

[112] Gosse 1970, 68. Gosse *fils* was fairly angry with his father, and so may have exaggerated the reaction to his father's book. See Ross 1977 for discussion.

[113] Simpson 1925, 178f, quoting Nathaniel Southgate Shaler, another student of Agassiz who ended up less disposed to fixist accounts, from Shaler and Shaler 1909, 129.

[114] For example, Goodsir 1868.

[115] Lyell 1832.

[116] Kottler 1978, 276–278.

[117] Desmond and Moore 1991, 131.

[118] Kottler 1978, 277.

3dly. Some acquired peculiarities of form, structure, and instinct, are transmissible to the offspring; but these consist of such qualities and attributes only as are intimately related to the natural wants and propensities of the species.

4thly. The entire variation from the original type, which any given kind of change can produce, may usually be effected in a brief period of time, after which no farther deviation can be obtained by continuing to alter the circumstances, though ever so gradually,—indefinite divergence, either in the way of improvement or deterioration, being prevented, and the least possible excess beyond the defined limits being fatal to the existence of the individual.

5thly. The intermixture of distinct species is guarded against by the aversion of the individuals composing them to sexual union, or by the sterility of the mule offspring. It does not appear that true hybrid races have ever been perpetuated for several generations, even by the assistance of man; for the cases usually cited relate to the crossing of mules with individuals of pure species, and not to the intermixture of hybrid with hybrid.

6thly. From the above considerations, it appears that species have a real existence in nature, and that each was endowed, at the time of its creation, with the attributes and organization by which it is now distinguished.[119]

Kottler also considers the fifth edition of the *Principles*, published in 1837, which upon his return and as he began to consider transmutation and hence the nature of species, Darwin heavily underlined and annotated. As above, Lyell relies on the infertility of hybrids. According to him, no hybrid could give rise to a new species, unless it was back bred into a pure species. In nature, an "aversion to sexual intercourse is, in general, a good test of the distinctness of original stocks, or *species*."[120]

While he (reluctantly) changed his mind after the publication of the *Origin* and many discussions with Darwin directly, his major contribution at this time is to affirm the fixity and reality of species. In dealing with the views of Lamarck and Geoffroy over transmutation, he noted:

The name of species, observes Lamarck, has been usually applied to every collection of similar individuals, produced by other individuals like themselves. This definition, he admits, is correct, because every living individual bears a very close resemblance to those from which it springs. But this is not all which is usually implied by the term species, for the majority of naturalists agree with Linnaeus in supposing that all the individuals propagated from one stock have certain distinguishing characters in common which will never vary, and which have remained the same since the creation of each species.[121]

So, he says, Lamarck must defeat this by finding no gaps as we advance in our knowledge, and in so doing show that our taxonomic characters are arbitrary. Lyell spends considerable time dealing with Lamarck's claims, resulting in the statement that the Author of Nature would foresee all conditions in which a species would exist, and so the changes Lamarck requires will not occur.[122] Mayr quotes him as saying[123]

---

[119] Lyell 1832, 64f.
[120] Quoted in Kottler 1978, 277 from Lyell 1837, 435.
[121] Lyell 1837, 363.
[122] Lyell 1837, 389f.
[123] Mayr 1982, 405.

There are fixed limits beyond which the descendants from common parents can never deviate from a common type. ... It is idle ... to dispute about the abstract possibility of the conversion of one species into another, when there are known causes, so much more active in their nature, which must always intervene and prevent the actual accomplishment of such conversions.[124]

Mayr thinks this is an expression of Lyell's essentialism: "each species had its own specific essence and thus it was impossible that it could change or evolve. This, for example, was the cornerstone of Lyell's thought." But was it really? Lyell seems to be saying not that an *essence* is causing it to remain stable, but that a species is held stable by interbreeding and "known causes" of infertility. There is typology, to be sure, but overall, Lyell's view is a *causal* one: again, we see here hints of a generative notion of species. However, Mayr correctly notes that the *Principles* was Darwin's scientific "bible"[125] and that he devoted so much time in the *Origin* to refuting special creation largely because of the challenges set by Lyell.[126]

A-P DE CANDOLLE AND ASA GRAY: THE BOTANICAL VIEW OF VARIATION

A third stream of thought in this period supposes that species are real and that so also is variation from the type. This is primarily due to the famous family of Swiss botanists, the de Candolles, in particular the elder, Augustin-Pyramus.[127] A-P de Candolle stressed the variation of living things, and defined species as

> ... the collection of all the individuals who resemble one another more than they resemble others; who are able, by reciprocal fecundation, to produce fertile individuals; and who reproduce by generation, such kind as one may by analogy suppose that all came down originally from one single individual.[128]

Elsewhere, in the *Elementary theory*, he writes

> By Species (*species*), we understand a number of plants, which agree with one another in invariable marks.[129]

So, for de Candolle species are groups of individual organisms. They are, in the tradition of Ray, to both resemble one another and to generate progeny that are fertile and resemble one another. As did Buffon and the older tradition, De Candolle treated variation as the effect of local environments and occasional hybridization. In this opinion, he was followed closely by the great American botanist, Asa Gray, who was later to become significant in the promotion of Darwinian theory in America against Agassiz. As late as 1846, in a review of the *Vestiges of Natural Creation*, Gray declared that

---

[124] From the 1835, 4th, edition, vol. II, p 433; the second sentence is not found in the first edition.

[125] Mayr 1982, 406.

[126] Mayr 1982, 407.

[127] Candolle 1819.

[128] Quoted in Hunter Dupree 1968, 54.

[129] From Candolle and Sprengel 1821, 98.

species were created as they are found, and did not transmute.[130] But by the 1850s, he had an operational view of species. While creation may have once been of importance for him, writes his biographer Hunter Dupree, what most concerned Gray was that if species transmuted as Lamarck, Geoffroy and the author of the *Vestiges* declared, then natural history would become meaningless, one presumes because we would be unable to specify the facts about the groups in biology that we encounter.[131] Hunter Dupree says, "it was this inability [*of unlike species to breed together—JSW*] which created the species border, not that he or any other could find this border easily, least of all by referring to an ideal type." Morphology was only a guide to these borders, and relied on the experience of the naturalist and the principles of classification. The Lockean character of this account is manifest. The real essence here is interfertility, not morphology. As a result, Gray worried about hybridization, and its role in speciation. This had been a concern since Linnaeus' time, and the urgency of the problem was progressively increasing among the naturalists of the period. Gray noted that hybrids would stand a good chance of being fertilized by their parents and asked

> In such cases they are said to revert to the type of the species of the impregnating parent; but would they return exactly to that type, inheriting as they do a portion of the blood of a cognate species?"

The modern problem of the introgression of genes into a species is foreshadowed here, although Gray relies on a blending inheritance model, of course, causing swamping of the variations.

Gray's other contribution to this topic is in his assertion that humans are a single species. He felt that science in general and his ideas on species and hybridization in particular pointed to the unity of the human race.[132] It is worth noting here that Gray's rather orthodox Protestantism seems to have had no particular impact on his view of species, as Cuvier's also had not. Objections to transmutation appear not to have been founded on orthodox religious doctrine.

PRE-DARWINIAN EVOLUTIONARY VIEWS OF SPECIES

> ... species, the subdivision where intermarriage or breeding is usually considered as natural to animals, and where a resemblance of offspring to parents is generally persevered in.

Vestiges of the History of Creation[133]

In German- and French-speaking countries prior to the publication of the *Origin*, there were a number of specialists propounding evolutionary views of species. Of note are Bonaparte and Unger.

---

[130] Hunter Dupree 1968, 145–147.
[131] Hunter Dupree 1968, 217.
[132] Hunter Dupree 1968, 220.
[133] Chambers 1844, 263.

Prince Charles Lucien Bonaparte (1803–1857), nephew of the famous Napoleon, was an active ornithologist, as his father Lucien had been after the British released him from detention in 1814. Exiled in Leiden by his cousin Louis Napoleon, he became friends with Hermann Schlegel, another famous ornithologist.[134] In 1851 he published the first volume of his *Conspectus generum avium*, a survey of all known species of birds worldwide. In this work, he treated extant species as the descendent forms of prior extinct forms, and in an address to an 1856 convention on "What is a species, particularly in ornithology?" he said:

> We will state with unanimous conviction that the antediluvian crocodiles, elephants and rhinoceroses were the ancestors of those living in our day, and these animals would not have been able to continue to exist without the manifold mutations that their systems produced to adapt themselves to the environment, and that became second nature to their descendants. ... If the environment remains the same, so do the species. The stabilizing influence is then by itself all-powerful. The mutating influence can succeed in opposing it only when the whole world surrounding it changes. ... But races, however different in characteristics they may be, vanish entirely or at least do not long survive as soon as the environment that produced them ceases to be the same ... The transitions between the different races and their type are the best evidence that we can supply to set aside putative species, which are to be relegated to races, with which the painstaking zoologist must nevertheless occupy himself just as earnestly.[135]

Of interest in this excerpt is the implication that species are racial groups stabilized by the influence of the environment, somewhat as stabilizing selection operates (although there is no reason to suppose Bonaparte thought selection was the reason for the stabilization). As Darwin later also argued, races are merely species in the making that are not yet made stable. Bonaparte died in 1857, leaving the *Conspectus* unfinished.

Franz Unger, an Austrian botanist at the University of Vienna, published a form of common descent with modification theory in 1852, entitled "Attempt at a History of the Vegetable Realm" (*Versuch einer Geschichte der Planzenwelt*) in which he supposed that all plant life was a single entity that had developed new forms. Unger's theory is often taken to be a forerunner of Darwin,[136] but he in fact thought that all subsequent development was an expression of the original potentiality of the *Urpflanze*:

> Nothing has been added in this regulated evolutionary process of the vegetal world that had not been previously prepared and indicated, so to speak. Neither genus, nor family, nor class of plants has manifested itself without having become necessary in time.[137]

Temkin notes that Unger held that species themselves do not change, but that some individuals metamorphosed while the old type remained in existence for some time.

---

[134] Stresemann 1975, Stroud 2000.
[135] Quoted in Stresemann 1975, 166.
[136] Cf. Temkin 1959, 339–342.
[137] Quoted in Temkin 1959, 340.

One year later, Hermann Schaaffhausen published an article "On the Constancy and Transformation of Species" rebutting Unger's ideas on the grounds they indicated man evolved from an orangutan. However, he said in his summary that

> [t]he immutability of species which most scientists regard as natural law is not proved, for there are no definite and unchangeable characteristics of the species, and the borderline between species and subspecies [*Art* and *Abart*] is wavering and uncertain.[138]

It appears then that in the post-Romantic period in Germany and German-speaking countries, naturalists were not so rigid over species as was the Swiss export to America, Agassiz. Unger's conception appears to be an entelechical view—a species was a type that was "in" the plant kingdom from the beginning, in the *Urpflanze*. However, the stasis of the species themselves is due in Unger's book to the generative powers of inner forces. Mayr quotes him saying:

> The lower as well as the higher taxa appear then not as an accidental aggregate, as an arbitrary mental construct but united with each other in a genetic manner and thus form a true intrinsic unit.[139]

Finally, mention must be made of Heinrich Georg Bronn (1800–1862), who, in his prizewinning submission to the Paris Academy of Sciences in 1857 published the *Untersuchungen über die Entwicklungs-Gesetze der organischen Welt während der Bildungs-Zeit unserer Erd-Oberfläche* (*Researches into the laws of development of the organic world during the period of development of our earth's surface*[140]), in which he presented a tree diagram for the progressive evolutionary divergence of types of animals.[141] Bronn, Nyhart tells us, was committed to the progressionism of Oken and Lamarck, and when he supervised the translation of Darwin's *Origin*, he translated "favoured races" in the subtitle as "*vervollkommneten Rassen*," or "perfect[ed] races," a subtlety that may have influenced Haeckel's later view of evolution.[142] Bronn had previously, in his 1841 *Handbuch einer Geschichte der Natur*, treated species as the result of acts of special creation.

However, he appended a critical essay to chapter 15 of the *Untersuchungen* in which he said that he doubted varieties would permanently branch off, and instead he held that species were not transformed through inherited modifications but by a law of nature, a creative force, as yet unknown.[143] Each species had its own lifespan, and then a more perfect one replaced it.

---

[138] Quoted in Temkin 1959, 342.
[139] Mayr 1982, 391.
[140] Bronn 1858.
[141] Panchen 1992, 26f, Nyhart 1995, 110–116.
[142] Junker 1991. See Gliboff 2008 for a discussion of Bronn's reputation and interpretation of Darwin's theories.
[143] Nyhart 1995, 112f. See also Junker 1991.

JOSEPH HOOKER, THOMAS WOLLASTON, AND GEORGE BENTHAM ON LOGIC AND DIVISION

Two important writers on species, Hooker and George Bentham, were collaborators, but each had a different approach. Joseph D. Hooker was one of Darwin's closest confidants.[144] He was introduced into Darwin's views on evolution as early as the 1844 manuscript, and yet he was unable to discuss these ideas with anyone but Darwin. When he finally was able to, after the 1858 reading of Wallace's and Darwin's papers, he noted to Asa Gray that he could never

> allude to his doctrine in public, & I always had in my writings to discuss the subject of variation etc & as if I had never heard of Natural Selection—which I have all along known & feel not only useful in itself as explaining many facts in variation, but as the most fatal argument about "Special Creation"...[145]

Hooker considered that much variation in plants was intraspecific, and that there were far fewer actual species than were listed and that what he called its "habit"—or general appearance and growth—was deceptive as a guide to the difference between species and varieties. He wrote to Gray in 1856,

> As to consistency with regard to species—it is a myth, a delusion ... the most consistent men are hair-splitters—they make almost every difference specific.

In the light of Whately and the logic of division in which most had been trained, at least implicitly, this tension is understandable. Every difference *was* specific, logically. The problem lay in that in natural history, specifically in botany, the line was drawn much higher than that:

> Bother variation, developement [sic] & all such subjects,! It is reasoning in a circle after all. As a Botanist I must be content to take species as *they appear to be* not as *they are,* & still less as they were or ought to be. [To Darwin, July 1845[146]]

Thomas Vernon Wollaston, in his 1856 *On the Variation of Species,*[147] defined species as a community of descent within which there was variation but between which there was no gradation. Those who Hooker had called hair-splitters he called "very hyper-accurate definers." Nevertheless, after Darwin published the *Origin,* Wollaston denied the indefinite variation that Darwin needed existed. Hooker, to the contrary, was convinced it did. In 1853, in the introduction to his *Botany of the Antarctic Voyage* volume 2, part 1, he noted that unless the systematist

---

[144] Stevens 1997 is an excellent source of material and overview for Hooker in particular. See also Turrill 1963.
[145] Quoted in Stevens 1997, 346.
[146] Stevens 1997, 349.
[147] Wollaston 1856, cf. England 1997.

act upon the idea that for practical purposes at any rate species are constant, he can never hope to give that precision to his characters of organs and functions which is necessary to render his descriptions useful to others; for in groups where the limits of species cannot be traced (or, what amounts to the same thing in the opinion of many, where they do not exist), the object of the systematist is the same as in groups where they are obvious,—to throw their forms into a natural arrangement, and to indicate them by tangible characters, whose value is approximately relative to what prevails in genera where the limitation of species is more apparent.[148]

Hooker observed that variation tended to be absent in a particular place or colony but could often be found globally. He noted, in a passage that Gray marginally wrote was "opt[issime]!" in his copy:

It is very much to be wished that the local botanist should commence his studies upon a diametrically opposite principle to that upon which he now proceeds, and that he should endeavour, by selecting good suites of specimens, to determine *how few*, not *how many* species are comprised in the flora of his district. The permanent differences will, he may depend upon it, soon force themselves upon his attention, whilst those which are non-essential will consecutively be eliminated. There is no better way of proving the validity of characters than by attempting to invalidate them.[149]

The professional botanist's role here was crucial. Hooker, George Bentham, Ferdinand Mueller, and others were better placed in virtue of having specimens from across districts and indeed the globe to mark out species than local observers or gardeners. Hooker's later collaborator Bentham early on had a fairly traditional view of species in natural history—they were created by God, of course, but he took more notice of the Cuvierian definition of descent from a common ancestor. In his early work as a botanist, a field he turned to after his logic book[150] failed to garner much contemporary attention (it got rather more later, when the issue of who invented logical quantification was being discussed), he was convinced that species could be ranked by a comprehensive survey of variation.[151] After Darwin, he was not so sure, and eventually he gave up the idea of a species rank. In 1864, he wrote to Ferdinand von Mueller that "true species are entirely limited in nature." As Stevens notes, it made little or no difference to his taxonomic work. However, in an anonymous review of de Candolle's *Geographie Botanique Raisonée* in 1856, he had stated clearly enough that a species was

... a collection of individuals which, by their resemblance to each other, or by other circumstances, we are induced to believe are descended or *may have* descended from one individual or a pair of individuals.[152]

What the inducements were, however, were unclear. Bentham almost indicates that *species* is a purely operational notion.

---

[148]   Hooker 1853, viii. Cf. Stevens 1997, *loc. cit.*
[149]   Hooker and Thomson 1855, 35. Cf. Stevens 1997, 364n362.
[150]   Bentham 1827.
[151]   Stevens 1997, 359.
[152]   Stevens 1997, 360.

A SUMMARY VIEW OF THE EARLY NINETEENTH CENTURY

It appears that while many naturalists were fixists, the leading criterion for species identification or explanation was derived from the descent of similar forms. Apart from Agassiz, nobody seems, however, to have inferred from fixism, or the pious creationism that was the usual form of words used, that species had essences or even that variation was firmly limited. In this period, variation within species was a real research difficulty. Cuvier's definition was widely disseminated and repeated, and Linnaeus' almost taken as something too obvious, and a little over-religious, to mention. This background is something shared by Darwin, and serves to highlight both his orthodoxy in this regard, and those areas in which he innovated.

BIBLIOGRAPHY

Agassiz, Louis. 1842. New views regarding the distribution of fossils in formations. *Edinburgh New Philosophical Journal* 32 (63):97–98.

—. 1859. *An essay on classification*. London: Longman, Brown, Green, Longmans and Roberts and Trubner.

—. 1863. *Methods of study in natural history*. Boston: Ticknor and Fields.

—. 1869. *De l'espece et de la classification en zoologie*. Paris: Balliere.

Allen, D. E. 2003. George Bentham's Handbook of the British flora: From controversy to cult. *Archives of Natural History* 30 (2):224–236.

Amundson, Ron. 2005. *The Changing Role of the Embryo in Evolutionary Biology: Structure and Synthesis (Cambridge Studies in Philosophy and Biology)*. New York: Cambridge University Press.

Basalla, George et al., eds. 1970. *Victorian Science: A Self-Portrait from the Presidential Addresses of the British Association for the Advancement of Science*. New York: Anchor/Doubleday.

Bentham, George. 1827. *An Outline of a New System of Logic. With a Critical Examination of Dr. Whately's "Elements of Logic."* London: Hunt and Clark.

Bentham, Jeremy. 1983. *Chrestomathia*, edited by M. J. Smith and W. H. Burston. Oxford/New York: Clarendon Press/Oxford University Press.

Brande, W. T., and Joseph Cauvin, eds. 1853. *A dictionary of science, literature, and art: Comprising the history, description, and scientific principles of every branch of human knowledge; With the derivation and definition of all the terms in general use.* 2nd ed. London: Longman.

Bronn, Heinrich G. 1858. *Untersuchungen über die Entwickelungs-Gesetze der organischen Welt während der Bildungs-Zeit unserer Erd-Oberfläche*. Stuttgart: E. Schwiezerbart'sche Verlagshandlung und Druckerei.

Candolle, Augustin-Pyramus, and Kurt Sprengel. 1821. *Elements of the philosophy of plants*. Edinburgh: W. Blackwood.

Candolle, Augustine-Pyramus de. 1819. *Théorie élementaire de la botanique, ou exposition des principes de la classification naturelle et de l'art de décrire et d'étudier les végétaux*. 2nd ed. Paris: Déterville.

Chambers, Robert. 1844. *Vestiges of the natural history of creation*. London: John Churchill.

Cheung, Tobias. 2006. From the organism of a body to the body of an organism: Occurrence and meaning of the word from the seventeenth to the nineteenth centuries. *The British Journal for the History of Science* 39 (03):319–339.

Clarke, Richard F. 1895. *Logic*. 3rd ed, Manuals of Catholic Philosophy. London: Longmans, Green.

Cuvier, Georges. 1812. Discours préliminaire. In *Recherches sur les ossemens fossiles de quadrupèdes*. Paris: Deterville.

—. 1835. Éloge de M. de Lamarck. *Mémoires de l'Académie Royale des Sciences de l'Institut de France*, 2nd series XIII:i–xxxi.

Cuvier, Georges et al. 1818. *Essay on the Theory of the Earth*. New York: Kirk & Mercein.

Dana, James D. 1857. Thoughts on species. *American Journal of Science and Arts* 24 (72):305–316.

Desmond, Adrian, and James Moore. 1991. *Darwin*. Harmondsworth, UK: Penguin.

England, Richard. 1997. Natural selection before the Origin: Public reactions of some naturalists to the Darwin-Wallace Papers (Thomas Boyd, Arthur Hussey, and Henry Baker Tristram). *Journal of the History of Biology* 30 (2):267–290.

Gillispie, Charles Coulston. 1959. Lamarck and Darwin in the history of science. In *Forerunners of Darwin 1749–1859*, edited by Bentley Glass et al., 265–291. Baltimore, MD: Johns Hopkins Press.

Gliboff, Sander. 2008. *H. G. Bronn, Ernst Haeckel, and the Origins of German Darwinism: A Study in Translation and Transformation*. Cambridge, MA/London: MIT Press.

Goodsir, Joseph Taylor. 1868. *The Bases of Life: A Discourse in Reply to Professor Huxley's Lecture on "The Physical Bases of Life."* London: Williams & Norgate.

Gosse, Edmund. 1970. *Father and Son: A Study of Two Temperaments*. London: Heinemann. Original edition, Heinemann, 1907.

Gosse, Phiil Henry. 1844. *An Introduction to Zoology*. 3 vols. Vol. 1. London: Society for Promoting Christian Knowledge.

Gosse, Philip Henry. 1857. *Creation (Omphalos): An Attempt to Untie the Geological Knot*. London: J. Van Voorst.

Greene, John C. 1959. *The Death of Adam: Evolution and Its Impact on Western Thought*. Ames: Iowa State University Press.

Hamilton, William et al. 1874. *Lectures on Metaphysics and Logic*. Edinburgh: Blackwood.

Hooker, Joseph Dalton. 1853. *The Botany of the Antarctic Voyage of H.M. Discovery Ships Erebus and Terror in the Years 1839–1843: Under the Command of Captain Sir James Clark Ross*. Vol. 2, pt.1 Flora Novae-Zelandiae. London: Reeve Brothers.

Hooker, Joseph Dalton, and Thomas Thomson. 1855. *Flora indica: Being a Systematic Account of the Plants of British India, Together with Observations on the Structure and Affinities of their Natural Orders and Genera*. Vol. 1. London: W. Pamplin.

Hookway, Christopher. 1985. *Peirce, The Arguments of the Philosophers*. London/Boston: Routledge & Kegan Paul.

Hull, David L., ed. 1973. *Darwin and His Critics; The Reception of Darwin's Theory of Evolution by the Scientific Community*. Cambridge, MA: Harvard University Press.

—. 2003. Darwin's science and Victorian philosophy of science. In *The Cambridge Companion to Darwin*, edited by Jonathon Hodge and Gregory Radick, 168–191. Cambridge, UK: Cambridge University Press.

Hunter Dupree, Anderson. 1968. *Asa Gray 1810–1888*. College ed. Vol. 132. New York: Atheneum. Original edition, The Belknap Press of Harvard University Press 1959.

Husserl, Edmund. 1931. *Ideas: General Introduction to Pure Phenomenology (Ideen au einer reinen Phänomenologie und phänomenologischen Philosophie)*. Translated by W. R. Boyce Gibson. New York: Collier Macmillan. Original edition, 1913.

Huxley, Thomas Henry. 1906. *Man's Place in Nature and Other Essays*. Everyman's Library ed. London/New York: J. M. Dent/E. P. Dutton.

Jevons, William Stanley. 1878. *The Principles of Science: A Treatise on Logic and Scientific Method*. 2nd ed. London: Macmillan. Original edition, 1873.

Joseph, Horace William Brindley. 1916. *An Introduction to Logic*. 2nd ed. Oxford: Clarendon Press. Original edition, 1906.

Junker, T. 1991. Heinrich Georg Bronn und Origin of Species. *Sudhoffs Archiv; Zeitschrift Fur Wissenschaftsgeschichte* 75 (2):180–208.

Kottler, Malcolm Jay. 1978. Charles Darwin's biological species concept and theory of geographic speciation: The transmutation notebooks. *Annals of Science* 35 (3):275–297.

Lamarck, Jean Baptiste. 1802. *Recherches sur l'organisation des corps vivants*. Paris: Dentu.

—. 1809. *Philosophie zoologique, ou, Exposition des considérations relative à l'histoire naturelle des animaux*. Paris: Dentu.

—. 1914. *Zoological Philosophy: An Exposition with Regard to the Natural History of Animals*. Translated by Hugh Elliot. London: Macmillan.

Le Guyader, Herve. 2004. *Etienne Geoffroy Saint-Hilaire, 1772–1844: A Visionary Naturalist*. Chicago/London: University of Chicago Press. Original edition, Étienne Geoffroy Saint-Hilaire, 1772–1844: Un naturaliste visionnaire, 1998.

Lherminer, Philippe, and Michel Solignac. 2000. L'espèce: Définitions d'auters. *Sciences de la vie* 153–165.

Lurie, Edward. 1960. *Louis Agassiz: A Life in Science*. Baltimore and London: Johns Hopkins University Press. Reprint, 1988.

Lyell, Charles. 1832. *Principles of Geology, Being an Attempt to Explain the Former Changes of the Earth's Surface, by Reference to Causes Now in Operation*. 2nd ed. 3 vols. London: John Murray.

Lyell, Charles Sir. 1837. *Principles of Geology, Being an Attempt to Explain the Former Changes of the Earth's Surface, by Reference to Causes Now in Operation*. 5th ed. 4 vols. London: John Murray.

Maupertuis, Pierre-Louis Moreau de. 1745. *Vénus physique*. Paris: La Haye.

Mayr, Ernst. 1982. *The Growth of Biological Thought: Diversity, Evolution, and Inheritance*. Cambridge, MA: The Belknap Press of Harvard University Press.

McOuat, Gordon R. 2003. The logical systematist: George Bentham and his *Outline of a new system of logic*. *Archives of Natural History* 30 (2):203–223.

Mill, John Stuart. 1930. *A System of Logic, Ratiocinative and Inductive: Being a Connected View of the Principles of Evidence and the Methods of Scientific Investigation*. 8th (1860) ed. London: Longmans Green. Original edition, 1843.

Morris, Paul J. 1997. Louis Agassiz's additions to the French translation of his *Essay on Classification*. *Journal of the History of Biology* 30 (1):121–134.

Murray, Desmond. 1955. *Species Revalued: A Biological Study of Species as a Unit in the Economy of Nature, Shown from Plant and Insect Life*. London: Blackfriars.

Nordenskiöld, Erik. 1929. *The History of Biology: A Survey*. Translated by Leonard Bucknall Eyre. London: Kegan Paul, Trench, Trubner and Co.

Numbers, Ronald L. 1992. *The Creationists*. New York: A. A. Knopf.

Nyhart, Lynn K. 1995. *Biology Takes Form: Animal Morphology and the German Universities, 1800–1900*. Chicago: University of Chicago Press.

Owen, Richard. 1834. On the generation of the marsupial animals, with a description of the impregnated uterus of the kangaroo. *Philosophical Transactions of the Royal Society of London* 124:333–364.

—. 1835. On the osteology of the chimpanzee and orangutan. *Transactions of the Zoological Society* 1:343–379.

—. 1843. *Lectures on the Comparative Anatomy and Physiology of the Invertebrate Animals*. Delivered at the Royal College of Surgeons, in 1843. By Richard Owen. From notes taken by William White Cooper and revised by Professor Owen. London: Longman, Brown, Green, and Longmans.

—. 1848. *The Archetype and Homologies of the Vertebrate Skeleton*. London: J. van Voorst.

—. 1855. *Lectures on the Comparative Anatomy and Physiology of the Invertebrate Animals: Delivered at the Royal College of Surgeons*. 2nd ed. London: Longman, Brown, Green and Longmans.

Panchen, Alec L. 1992. *Classification, Evolution, and the Nature of Biology*. Cambridge, UK/
   New York: Cambridge University Press.
Peirce, Charles Sanders. 1885. On the algebra of logic: A contribution to the philosophy of
   notation. *American Journal of Mathematics* 7 (2):180–202.
Prichard, James Cowles. 1813. *Researches into the Physical History of Man*. London:
   Houlston & Stoneman.
—. 1843. *The Natural History of Man; Comprising Inquiries into the Modifying Influence
   of Physical and Moral Agencies of the Different Tribes of the Human Family*. London:
   Bailliere.
Quine, Willard Van Orman. 1970. *Philosophy of Logic*. Englewood Cliffs, NJ: Prentice Hall.
Roger, Jacques. 1997. *Buffon: A Life in Natural History*. Translated by Sarah Lucille Bonnefoi.
   Edited by L. Pearce Williams, Cornell History of Science Series. Ithaca, NY: Cornell
   University Press.
Ross, Frederic R. 1977. Philip Gosse's *Omphalos*, Edmund Gosse's *Father and Son*, and
   Darwin's theory of natural selection. *Isis* 68 (1):85–96.
Russell, E. S. 1982. *Form and Function: A Contribution to the History of Animal Morphology*.
   Chicago: University of Chicago Press. Original edition, 1916, Murray.
Saint-Hilaire, Isidore Geoffroy. 1859. *Histoire naturelle générale des règnes organiques:
   principalement étudiée chez l'homme et les animaux*. Vol. 2. Paris: Victor Masson.
Shaler, Nathaniel Southgate, and Sophia Penn Page Shaler. 1909. *The Autobiography of
   Nathaniel Southgate Shaler*. Boston, New York: Houghton Mifflin.
Simpson, James Y. 1925. *Landmarks in the Struggle between Science and Religion*. London:
   Hodder and Stoughton.
Stevens, Peter F. 1997. J. D. Hooker, George Bentham, Asa Gray and Ferdinand Mueller on
   species limits in theory and practice: A mid-nineteenth century debate and its repercus-
   sions. *Historical Records of Australian Science* 11 (3):345–370.
Stresemann, Erwin. 1975. *Ornithology from Aristotle to the Present*. Translated by Hans J.
   Epstein and Cathleen Epstein. Cambridge, MA: Harvard University Press.
Stroud, Patricia Tyson. 2000. *The Emperor of Nature: Charles-Lucien Bonaparte and his
   World*. Philadelphia: University of Pennsylvania Press.
Temkin, Oswei. 1959. The idea of descent in post-Romantic German biology, 1848–1858. In
   Forerunners of Darwin 1745–1859, edited by Bentley Glass et al., 323–355. Baltimore,
   MD: Johns Hopkins University Press.
Thompson, William R. 1958. Introduction. In *The Origin of Species, Everyman Edition*.
   London: J. M. Dent. Original edition, 1928.
—. 1971. The status of species. *Studia Entomologica* 14 (Novembro):399–456.
Thorpe, William H. 1973. William Robin Thompson. 1887–1972. *Biographical Memoirs of
   Fellows of the Royal Society* 19:654–678.
Turrill, William Bertram. 1963. *Joseph Dalton Hooker. Botanist, Explorer and Administrator*.
   London: Thomas Nelson & Sons.
Voegelin, Eric. 1998. *The History of the Race Idea: From Ray to Carus*. Baton Rouge:
   Louisiana State University Press.
Wasmann, Erich. 1910. *Modern Biology and the Theory of Evolution*. Translated by A. M.
   Buchanan. 3rd ed. London: Kegan Paul, Trench, Trübner. Original edition, 1906.
Whately, Richard. 1823. *Logic*. Vol. IX, *Encyclopedia Metropolitana*. London: Applegath.
—. 1826. *Elements of Logic. Comprising the Substance of the Article in the Encyclopaedia
   Metropolitana, with Additions, & c*. London: J. Mawman.
—. 1875. *Elements of Logic*. Ninth (octavo) ed. London: Longmans, Green & Co. Original
   edition, 1826.
Whewell, William. 1831. Review of Herschel's *Preliminary Discourse* (1830). *Quarterly
   Review* 45:374–407.
—. 1837. *History of the Inductive Sciences*. 3 vols. London: Parker.

Wilson, Leonard G., ed. 1970. *Sir Charles Lyell's Scientific Journals on the Species Question, Yale Studies in the History of Science and Medicine.* New Haven: Yale University Press.

Winsor, Mary Pickard. 1979. Louis Agassiz and the species question. *Studies in History of Biology* 3:89–117.

Wollaston, Thomas Vernon. 1856. *On the Variation of Species, with Especial Reference to the Insecta: Followed by an Inquiry into the Nature of Genera.* London: J. Van Voorst.

Wollheim, Richard. 1968. *Art and Its Objects, an Introduction to Aesthetics.* New York: Harper & Row.

Woodger, Joseph Henry. 1937. *The Axiomatic Method in Biology.* Cambridge, UK: Cambridge University Press.

—. 1952. From biology to mathematics. *British Journal for the Philosophy of Science* 3 (9):1–21.

5 Darwin and the Darwinians

One of the ironies of the history of biology is that Darwin did not really explain the origin of new species in *The Origin of Species*, because he didn't know how to define a species.

Douglas Futuyma[1]

... The Origin of Species, whose title and first paragraph imply that Darwin will have much to say about speciation. Yet his magnum opus remains largely silent on the "mystery of mysteries," and the little it does say about this mystery is seen by most modern evolutionists as muddled or wrong.

Jerry Coyne and H. Allan Orr[2]

Darwin's ideas have been widely misinterpreted almost from the date of the publication of the *Origin* in November 1859. In this chapter, we shall see that he has in fact a fairly orthodox view of species as real things in nature (albeit temporary things), that he did not think interfertility was a good test of a species, and that his dismissive comments in the *Origin* have more to do with the professional nature of taxonomy and the difficulties of diagnosis and nomenclature than a claim that species did not exist at all.

DARWIN'S DEVELOPMENT ON SPECIES

It is occasionally stated that Darwin denied the reality of species, or held that "species" was an arbitrary concept (see the epigrams above). Mayr notes "one might get the impression [from the *Origin of Species*] that he considered species as something purely arbitrary and invented merely for the convenience of taxonomists."[3] Mayr goes on to note that he nevertheless treated species in a perfectly orthodox taxonomic manner and that he treated the concept purely typologically. Beatty and Ereshefsky raise similar doubts on Darwin's view of species.[4] On the contrary, it will be argued in this chapter that Darwin was a species realist, and although he developed his views over time he never ceased being a species realist, just as his mentor Lyell was.

Charles Darwin is important not so much for the novelties on the nature of the species concept that he provided—there are only really two of these, failure to breed in nature, and selection as the motive force of specific characters, as we shall see. Rather

---

[1] Futuyma 1983, 152.
[2] Coyne and Orr 2004, 9.
[3] Mayr 1982, 268.
[4] Beatty 1985, Ereshefsky 1999.

it is because his book *On the Origin of Species* changed *every* scientist's way of looking at species thereafter. He has been more closely scrutinized than anybody else, and
there is a wealth of material available. One thing that we should put out of our minds
from the beginning, though: it is *not* true that Darwin did not address the origin of
species in *On the Origin of Species*. The book is "one long argument"[5] on that very
point. Over and again, he discusses why species evolve to be distinct from parental
forms, and how they have done so. It is unclear to me how this idea gained currency.[6]

Darwin's views on species changed over time. In his earlier works he seems to
have treated species the same way as his teachers, unsurprisingly, as groups united
by some description—that which Mayr refers to as "typological, 'non-dimensional'
species of local fauna."[7] Since his views are often misrepresented[8] on the basis of
comments made in the *Origin*, this chapter will give an extensive and chronological
series of quotations from his published works, including his correspondence and the
Notebooks. While Kottler's analysis is reliable, some features of Darwin's views that
are further developments of the older notions of *species* and some of his comments
that are relevant to later debates, particularly over speciation, are to be found in comments not discussed by Kottler, Mayr, or Ghiselin.[9]

THE NOTEBOOKS

Darwin first began thinking about the nature of species in his Notebooks in 1837 and
1838. Kottler has investigated Darwin's early views on species in the Notebooks B, C,
D, and E (labeled by Kottler I–IV, which I have relabeled conventionally).[10] These are
referred to as the "transmutation notebooks" since Darwin started them after he became
convinced by Gould's investigations of the finches found on the Galápagos Islands, and
the tortoises and mockingbirds backed it up, that these were modified descendants of
South American colonists.[11] Once he started on the idea, his conjectures came thick and
fast: mammalian species were shorter lived than simpler forms because of their complexity, domestic animals were able to revert to the wild forms, or at least live like them,
perhaps species had a vital force and a fixed lifespan?, and so on. In the Notebooks, he
noted the "repugnance" of species to intercrossing (all quotations from Kottler):

---

[5] Mayr 1991.
[6] Possibly it arose from a comment made in 1866 by John Campbell, the Duke of Argyll [Campbell
1884, 240]:

> It will be seen, then, that the principle of Natural Selection has no bearing whatever on the Origin
> of Species, but only on the preservation and distribution of species when they have arisen. I have
> already pointed out that Mr. Darwin does not always keep this distinction clearly in view...
> More likely, though, it is due to the fact that the modern consensus is that species are formed
> by allopatric isolation, and Darwin held, as we shall see, that they are formed by selection on
> varieties, now called sympatric speciation. Coyne and Orr, for example, state that he "there-
> fore conflated the problem of change within a lineage with the problem of new lineages"
> [Coyne and Orr 2004, 11]. I demur: this was Darwin's *hypothesis*, rather than his confusion.

[7] Mayr 1982, 265.
[8] See Kottler 1978 for a discussion, particularly 291f.
[9] Ghiselin 1984.
[10] Kottler 1978.
[11] Desmond and Moore 1991, 224f.

... repugnance generally to marriage before domestication, ... marriage never probably excepting from strict domestication, offspring not fertile or at least most rarely and perhaps never fertile.—No offspring: physical impossibility to marriage. [B120]

Instinctive feelings against other species for sexual ends... [B161]

There is in nature a real repulsion amounting to impossibility holds good in plants between all different forms... [B189]

The dislike of two species to each other is evidently an instinct; & this prevents breeding. [B197]

The existence of wild close species of plants shows there is tendency to prevent the crossing of animals where there is much facility in crossing there comes the impediment of instinct [E143f]

Kottler observes that at this stage, Darwin agreed with Lyell that intercrossing was forced in domestication, and he made non-interbreeding a test of being a species:

... now domestication depends on perversion of instincts ... & therefore the one distinction of species would fail [B197]

Definition of species: one that remains at large with constant characters, together with beings of very near structure [B213]

My definition of species has nothing to do with hybridity, is simply, an instinctive impulse to keep separate, which no doubt be overcome, but until it is these animals are distinct species [C161]

A species as soon as once formed..., repugnance to intermarriage—settles it [B24]

Species formed... keep distinct, two species made; ... [B82]

Clearly, Darwin is more concerned with the behavior of organisms in natural conditions, not with the mere possibility of intercrossing. A species is to him at this stage an interbreeding group that is kept separate from other groups not only by the impossibility of hybridization, but also by the mating behaviors of each group. Hence, it is not a notion that can be lab-tested, although Buffon had famously, and with some success, tested his idea that Linnaean-level species were geographical variants of the *premiere souche*, or primary stock (see above). It has to be observed in the field. But there were ways to test species:

It is daily happening, that naturalists describe animals as species... There is only two ways [*sic*] of proving to them it is not; one where they can [be] proved descendant [Kottler interpolates: descent from common parents], which of course most rare, or when placed together they will breed. [B122]

The standard view of species at the time, since the original definition by Linnaeus, was that any two organisms were of the same species if they shared ancestry (in Linnaeus' pious formulation, from the pair of creatures created by God). Species are real, according to Darwin here, when they do not interbreed:

As species is real thing with respect to contemporaries—fertility must settle it [C152]

If they [*systematists—JSW*] give up infertility in largest sense as test of species— they must deny species which is absurd. [E24].

Kottler notes that Darwin is not here using Buffon's 1749 definition of species, as by the phrase "in the largest sense" Darwin is including both sterility and aversion, which Buffon had not, merely requiring sterility when crossed. Immediately before the last passage, says Kottler, Darwin had written

> ... one species may have passed through a thousand changes, keep distinct from other, & if a first & last individual were put together, they would not according to all analogy breed together.

Therefore, Darwin takes "being a species" as the *outcome of changes* that lead to a failure of the organisms to interbreed, not as the *outcome of failure to interbreed* first. He is not using a diagnostic notion of species. This is a generative notion, one to be explained by transmutation. In fact, Darwin expects that diagnosis may be nigh on impossible:

> Hence species may be good ones and differ scarcely in any external character [B213] ... we do not know what amount of difference prevents breeding... [B241]

The Notebooks were completed eight years later, in 1845. In them Darwin developed a notion of speciation as due to adaptation to local conditions, mostly due to geographical isolation, which prevented backcrossing.[12]

Darwin's Pre-*Origin* Correspondence

In his correspondence with his scientific friends, Darwin begins to ask questions about species relatively late, around 1855,[13] although Padian notes that in 1843 he did once ask the museum taxonomist G. R. Waterhouse what he meant by "relationship"[14]; Darwin's query is instructive:

> It has long appeared to me, that the root of the difficulty in settling such questions as yours,—whether the number of species &c &c should enter as an element in settling the value of existence of a group—lies in our ignorance of what we are searching after in our natural classifications.—Linnaeus confesses profound ignorance.—Most authors say it is an endeavour to discover the laws according to which the Creator has willed to produce organized beings—But what empty high-sounding sentences these are—it does not mean order in time of creation, nor propinquity to any one type, as man.—in fact it means just nothing—According to my opinion, (which I give everyone leave to hoot at, like I should have, six years since, hooted at them, for holding like views) classification consists in grouping beings according to their actual *relationship*, ie, their consanguinity, or descent from common stocks ...[15]

---

[12] Kottler 1978, 287f.
[13] Barlow 1967.
[14] Padian 1999, 353.
[15] 26 July 1843 [Burkhardt 1996, 76].

Waterhouse replied, "by relationship I mean merely resemblance." Darwin, having raised the issue of the number of species, then treats species themselves as a subordinate issue—it is the ways in which *higher* taxa are to be arranged that he is most interested in. Shortly thereafter, he mentions species in passing to Hooker:

I was so struck with the distributions of Galapagos organisms &c &c & with the character of the American fossil mammifers, &c &c that I determined to collect blindly every sort of fact, which c^d bear any way on what are species ... At last gleams of light have come, & I am almost convinced (quite contrary to the opinion I started with) that species are not (it is like confessing a murder) immutable.[16]

Some years later, he wrote to Hooker about the practical impact of the "question of species":

How painfully (to me) true is your remark, that no one has hardly a right to examine the question of species who has not minutely described many.[17]

Yet, in 1853, only four years later, Darwin noted to Hooker that his ideas on species had not made all that much difference to his classificatory work:

... in my own work, I have not felt conscious that disbelieving in the *permanence* of species has made much difference one way or the other; in some few cases (if publishing avowedly on doctrine on non-permanence) I sh^d. *not* have affixed names, & in some few cases sh^d. have affixed names to remarkable varieties. Certainly I have felt it humiliating, discussing & doubting & examining over & over again, when in my own mind, the only doubt has been, whether the forms varied *today or yesterday* (to put a fine point on it, as Snagsby would say). After describing a set of forms, as distinct species, tearing up my M.S., & then making them one again (which has happened to me) I have gnashed my teeth, cursed species, & asked what sin I had committed to be so punished: But I must confess, that perhaps the same thing w^d. have happened to me on any scheme of work—...[18]

He asked Henslow several times for an idea of the number of "close species" in botanical genera (27 June, 2 July, 7 July 1855) before he managed to make clear that he was after an impression of how many almost indistinguishable species exist in large genera. Henslow apparently succeeded, for on 21 July 1855 he replied

I thank you much for attempting to mark the list of dubious species: I was afraid it was a very difficult task, from, as you say, the want of a definition of what a species is.—I think however you were marking exactly what I wanted to know. My wish was derived as follows: I have ascertained, that APPARENTLY (I will not take up time by showing how) there is more variation, a wider geographical range, & probably more individuals, in the species of *large* genera than in the species of *small* genera. These general facts seem to me very curious, & I wanted to ascertain one point more; viz whether the closely allied and dubious forms which are generally considered as species, also belonged on average to large genera.[19]

16 11 January 1844 [Burkhardt 1996, 80].
17 Darwin to Hooker, September 1849 [Darwin 1888, 39].
18 Darwin to Hooker, 25 September 1853 [Burkhardt 1996, 128–129].
19 Barlow 1967, 182.

Note that here Darwin is also wrestling with the question of genera being real, a view he never entirely abandoned. In a letter to Asa Gray in 1857, Darwin discusses his by-now established opinion that there is no clear distinction between varieties and species, and how there seems under Darwin's evolutionary views to be no easy foundation or set of physical criteria to decide when a variety has earned a specific epithet:

> You speak of species not having any material base to rest on; but is this any greater hardship than deciding what deserves to be called a variety & be designated by a greek letter. When I was at systematic work, I know I longed to have no other difficulty (great enough) than deciding whether the form was distinct enough to deserve a name; & not to be haunted with undefined & unanswerable question whether it was a true species. What a jump it is from a well marked variety, produced by natural cause, to a species produced by the separate act of the Hand of God. But I am running on foolishly.—By the way I met the other day Phillips, the Palaeontologist, & he asked me "how do you define a species?"—I answered "I cannot" Whereupon he said "at last I have found out the only true definition,—'any form which has ever had a specific name'! ...[20]

This anecdote found its way into later mythology in Poulton's essay on the species "problem,"[21] although he dates it a week after the *Origin*. After the *Origin* was published, Darwin had sought Henslow's reaction:

> If you are *in even so slight a degree* staggered (which I hardly expect) on the immutability of species, then I am convinced with further reflection you will become more and more staggered, for this has been the process through which my mind has gone.[22]

Henslow was sufficiently staggered. He shortly afterward wondered at Owen's savage reaction to the views of the *Origin*:

> ... when his own are to a certain extent of the same character. If I understand him, he thinks the "Becoming" of species (I suppose he means the *producing* of species) a somewhat rapid and not a slow process—but he seems to think them *progressive* organised [sic] out of previously organized beings {analogous (?) to minerals (simple and compound) out of ± 60 Elements}. (5 May 1860)

And when Sedgwick attacked Darwin in an address, Henslow defended him actively and forthrightly, also saying in his lectures to students, he reported,

> ... how frequently Naturalists were at fault in regarding as *species*, forms which had (in some cases) been shown to be varieties, and how legitimately Darwin had deduced his *inferences* from positive experiment. [Letter 10 May 1860 to Hooker, which was then passed on to Darwin.]

---

[20] 29 November 1857 [Burkhardt 1996, 183].
[21] Poulton 1903, 1908.
[22] 11 November 1859. All quotations from Darwin's correspondence with Henslow are taken from Barlow 1967.

In correspondence with Huxley on September 26, 1853, and October 3, 1853, Darwin discussed the "Natural System" of classification:[23] it was merely genealogical; we did not have access to a written record, and thus we had to work it out, but the cause of analogy and homology was genealogy (i.e., descent). Huxley replied that

> Cuvier's definition of the object of Classification seems to me to embody all that is really wanted in Science—it is *to throw the facts of structure into the fewest possible general propositions.* [emphasis original]

Darwin replied

> I knew, of course, of the Cuvierian view of Classification, but I think that most naturalists look for something further, & search for 'the natural system',—'for the plan on which the Creator has worked' &c &c.—It is this further element which I believe to be simply genealogical.

In summary, his pre-*Origin* correspondence shows him to be rather cautious about showing his hand but he did seek information about what we would now, following Mayr, call "sibling species" and "cryptic species."

DARWIN'S PUBLISHED COMMENTS ON SPECIES BEFORE THE *ORIGIN*

In his *Journal of Researches*[24] Darwin makes few comments about species except to note, in the second edition of 1845, eight years after his thinking about transmutation began, that there are checks on the increases of populations.

> Every animal in a state of nature regularly breeds; yet in a species long established, any *great* increase in numbers is obviously impossible, and must be checked by some means. We are, nevertheless, seldom able with certainty to tell in any given species, at what period of life, or at what period of the year, or whether only at long intervals, the check falls; or, again, what is the precise nature of the check. Hence probably it is, that we feel so little surprise at one, of two species closely allied in habits, being rare and the other abundant in the same district; or, again, that one should be abundant in one district, and another, filling the same place in the economy of nature, should be abundant in a neighbouring district, differing very little in its conditions. If asked how this is, one immediately replies that it is determined by some slight difference, in climate, food, or the number of enemies: yet how rarely, if ever, we can point out the precise cause and manner of action of the check! We are, therefore, driven to the conclusion, that causes generally quite inappreciable by us, determine whether a given species shall be abundant or scanty in numbers.[25]

---

[23] Padian 1999, 355.
[24] Darwin 1839.
[25] Darwin 1845, chapter VIII, 167.

He also notes the difficulty of defining species in terms of morphology:

> Of the latter [rabbit, a piebald hybrid of black and gray breeds] I now possess a specimen, and it is marked about the head differently from the French specific description. This circumstance shows how cautious naturalists should be in making species; for even Cuvier, on looking at the skull of one of these rabbits, thought it was probably distinct![26]

Darwin reports that the Gauchos of South America were able to tell that these were one species because they shared the same territory and interbred, while Cuvier used only morphology. He notes in a footnote:

> The distinction of the rabbit as a species, is taken from peculiarities in the fur, from the shape of the head, and from the shortness of the ears. I may here observe that the difference between the Irish and English hare rests upon nearly similar characters, only more strongly marked. [p184n]

From this we may conclude that Darwin was no naïve morphologist, at any rate. He goes on to note that climate and ecotype are not the reason for species, nor soil (contrary to Buffon), as nearly identical regions produce different species:

> I was much struck with the marked difference between the vegetation of these eastern valleys and those on the Chilian side: yet the climate, as well as the kind of soil, is nearly the same, and the difference of longitude very trifling. The same remark holds good with the quadrupeds, and in a lesser degree with the birds and insects. I may instance the mice, of which I obtained thirteen species on the shores of the Atlantic, and five on the Pacific, and not one of them is identical. We must except all those species, which habitually or occasionally frequent elevated mountains; and certain birds, which range as far south as the Strait of Magellan. This fact is in perfect accordance with the geological history of the Andes; for these mountains have existed as a great barrier since the present races of animals have appeared; and therefore, unless we suppose the same species to have been created in two different places, we ought not to expect any closer similarity between the organic beings on the opposite sides of the Andes than on the opposite shores of the ocean. In both cases, we must leave out of the question those kinds which have been able to cross the barrier, whether of solid rock or salt-water.[5]

[5] This is merely an illustration of the admirable laws, first laid down by Mr. Lyell, on the geographical distribution of animals, as influenced by geological changes. The whole reasoning, of course, is founded on the assumption of the immutability of species; otherwise the difference in the species in the two regions might be considered as superinduced during a length of time. [chapter XV, p313]

Here Darwin is undercutting the widely held view that species are formed by climate or soil, a view that goes back to the medieval era (and which motivated Buffon's view of deviation from the *premiere souche*). Instead we have the beginnings of a

---

[26] Darwin 1845, chapter IX, 184.

biogeographic view of species as the result of geological isolation, which as we have seen in his Notebooks was a focus of his thinking at this time.

> Besides the several evident causes of destruction, there appears to be some more mysterious agency generally at work. Wherever the European has trod, death seems to pursue the aboriginal. We may look to the wide extent of the Americas, Polynesia, the Cape of Good Hope, and Australia, and we find the same result. Nor is it the white man alone that thus acts the destroyer; the Polynesian of Malay extraction has in parts of the East Indian archipelago, thus driven before him the dark-coloured native. The varieties of man seem to act on each other in the same way as different species of animals—the stronger always extirpating the weaker. [chapter XIX, p419]

Again in hindsight we can see Darwin foreshadowing the idea that better-adapted species will exclude other species in competition with them.

In *Coral Reefs*, Darwin uses the term *species* conventionally, never noting any great problem with corals.[27] In the Monograph on Cirripedia, he describes his taxonomic practice in the Preface:

> In those cases in which a genus includes only a single species, I have followed the practice of some botanists, and given only the generic character, believing it to be impossible, before a second species is discovered, to know which characters will prove of specific, in contradistinction to generic, value.

> In accordance with the Rules of the British Association, I have faithfully endeavoured to give to each species the first name attached to it, subsequently to the introduction of the binomial system, in 1758, in the tenth edition.[1] In accordance with the Rules, I have rejected all names before this date, and all MS. names. In one single instance, for reasons fully assigned in the proper place, I have broken through the great law of priority. I have given much fewer synonyms than is usual in conchological works; this partly arises from my conviction that giving references to works, in which there is not any original matter, or in which the Plates are not of a high order of excellence, is absolutely injurious to the progress of natural history, and partly, from the impossibility of feeling certain to which species the short descriptions given in most works are applicable;—thus, to take the commonest species, the *Lepas anatifera*, I have not found a single description (with the exception of the anatomical description by M. Martin St. Ange) by which this species can be certainly discriminated from the almost equally common *Lepas Hillii*. I have, however, been fortunate in having been permitted to examine a considerable number of authentically named specimens, (to which I have attached the sign (!) used by botanists,) so that several of my synonyms are certainly correct.

> [1] In the Rules published by the British Association, the 12th edition (1766) is specified, but I am informed by Mr. Strickland that this is an error, and that the binomial method was followed in the 10th edition of the 'Systema Naturæ.' [Part I, pp ix–x][28]

---

[27] Darwin 1842. Which is odd, since coral species are notoriously difficult to define or delineate [Soong and Lang 1992, Veron 2001, Bernardi et al. 2002, Carlon and Budd 2002, Pennisi 2002, Willis et al. 2006]. At this time he would still have been relying on purely morphological criteria, and this is borne out by the way in which he does describe them.

[28] Darwin, 1851.

Darwin is correct on both points. The Strickland Rules, as they came to be known, did indeed originally cite the 12th edition, but the 10th was the edition in which bionomials were first used consistently and which became the benchmark edition. Darwin was a member of the commission that determined the Strickland Rules in 1842, and which became the foundation for later strict taxonomic protocols.[29] They were published and adopted by the British Association for the Advancement of Science, forming the basis for the later Blanchard Code of 1889, itself the basis for the International Rules of 1898, adopted in 1901.[30] McOuat observes that this was part of a general professionalization of taxonomy, removing the "right" to name species from birdwatchers and gardeners to preclude confusion and synonymity.[31] It is worth noting, with Amundson, that these were only nominally essentialistic—the name had to have a definition, but there was no requirement that the species taxon had an essence.

On the *Origin of Species*, on Species

In the *Origin*[32] Darwin makes many substantive and theoretical claims about species. In the chapter entitled "Variation under domestication," he makes the following statements intended to convince those who rejected the mutability of species on logical grounds as well as practical ones. He notes that variation is an established fact within species, and that morphology is not a safe guide:

> Indefinite variability is a much more common result of changed conditions than definite variability, and has probably played a more important part in the formation of our domestic races. We see indefinite variability in the endless slight peculiarities which distinguish the individuals of the same species, and which cannot be accounted for by inheritance from either parent or from some more remote ancestor. [p16]
>
> Altogether at least a score of pigeons might be chosen, which, if shown to an ornithologist, and he were told that they were wild birds, would certainly be ranked by him as well-defined species. [p25]
>
> May not those naturalists who, knowing far less of the laws of inheritance than does the breeder, and knowing no more than he does of the intermediate links in the long lines of descent, yet admit that many of our domestic races are descended from the same parents—may they not learn a lesson of caution, when they deride the idea of species in a state of nature being lineal descendants of other species? [p29]
>
> But what concerns us is that the domestic varieties of the same species differ from each other in almost every character, which man has attended to and selected, more than do the distinct species of the same genera. [p37]

---

[29] Amundson 2005, 47.
[30] Mayr et al. 1953, 205. See Appendix A.
[31] McOuat 1996.
[32] All quotations and page numbers from the 6th edition [Darwin 1872]. Darwin's views on species as a category do not seem to have changed much between the first edition [Darwin 1859] and this edition. However, his view on gradual evolution, and the possibility of allopatric speciation, seems to have affected his expression of the nature of species. See Wilkins and Nelson 2008.

In the subsequent chapter, "Variation under nature," he points out that there are several definitions of the notion of species, including many that involve special creation:

> Before applying the principles arrived at in the last chapter to organic beings in a state of nature, we must briefly discuss whether these latter are subject to any variation. To treat this subject properly, a long catalogue of dry facts ought to be given; but these I shall reserve for a future work. Nor shall I here discuss the various definitions which have been given of the term species. No one definition has satisfied all naturalists; yet every naturalist knows vaguely what he means when he speaks of a species. Generally the term includes the unknown element of a distinct act of creation. The term "variety" is almost equally difficult to define; but here community of descent is almost universally implied, though it can rarely be proved. We have also what are called monstrosities; but they graduate into varieties. By a monstrosity I presume is meant some considerable deviation of structure, generally injurious, or not useful to the species. Some authors use the term "variation" in a technical sense, as implying a modification directly due to the physical conditions of life; and "variations" in this sense are supposed not to be inherited; but who can say that the dwarfed condition of shells in the brackish waters of the Baltic, or dwarfed plants on Alpine summits, or the thicker fur of an animal from far northwards, would not in some cases be inherited for at least a few generations? And in this case I presume that the form would be called a variety.[33]

There is, he says, a continuum of variation from occasional sports through to well-marked varieties, and some groups of organisms contain within them an enormous amount of variation:

> There is one point connected with individual differences, which is extremely perplexing: I refer to those genera which have been called "protean" or "polymorphic," in which the species present an inordinate amount of variation. With respect to many of these forms, hardly two naturalists agree whether to rank them as species or as varieties. We may instance Rubus, Rosa, and Hieracium amongst plants, several genera of insects and of Brachiopod shells. In most polymorphic genera some of the species have fixed and definite characters. Genera which are polymorphic in one country, seem to be, with a few exceptions, polymorphic in other countries, and likewise, judging from Brachiopod shells, at former periods of time. These facts are very perplexing, for they seem to show that this kind of variability is independent of the conditions of life. I am inclined to suspect that we see, at least in some of these polymorphic genera, variations which are of no service or disservice to the species, and which consequently have not been seized on and rendered definite by natural selection, as hereafter to be explained.[34]

Purely morphological notions present difficulties to all naturalists; we find intermediate forms all the time, and inductively, Darwin is suggesting that this does not stop merely within species or between extant forms within genera:

---

[33] Darwin 1875, 38.
[34] *Op. cit.*, 40.

The forms which possess in some considerable degree the character of species, but which are so closely similar to other forms, or are so closely linked to them by intermediate gradations, that naturalists do not like to rank them as distinct species, are in several respects the most important for us. We have every reason to believe that many of these doubtful and closely allied forms have permanently retained their characters for a long time; for as long, as far as we know, as have good and true species. Practically, when a naturalist can unite by means of intermediate links any two forms, he treats the one as a variety of the other; ranking the most common, but sometimes the one first described, as the species, and the other as the variety. But cases of great difficulty, which I will not here enumerate, sometimes arise in deciding whether or not to rank one form as a variety of another, even when they are closely connected by intermediate links; nor will the commonly assumed hybrid nature of the intermediate forms always remove the difficulty. In very many cases, however, one form is ranked as a variety of another, not because the intermediate links have actually been found, but because analogy leads the observer to suppose either that they do now somewhere exist, or may formerly have existed; and here a wide door for the entry of doubt and conjecture is opened.

Hence, in determining whether a form should be ranked as a species or a variety, the opinion of naturalists having sound judgment and wide experience seems the only guide to follow. We must, however, in many cases, decide by a majority of naturalists, for few well-marked and well-known varieties can be named which have not been ranked as species by at least some competent judges.[35]

In short, then, there is often no consensus, and the facts have to be worked out by the most informed majority. Even then, there is often no fact of the matter when a variety is to be distinguished from a species. Darwin is undercutting the intuitions of his professional audience here.

The geographical races or sub-species are local forms completely fixed and isolated; but as they do not differ from each other by strongly marked and important characters, "There is no possible test but individual opinion to determine which of them shall be considered as species and which as varieties." [Quoting Wallace] Lastly, representative species fill the same place in the natural economy of each island as do the local forms and sub-species; but as they are distinguished from each other by a greater amount of difference than that between the local forms and sub-species, they are almost universally ranked by naturalists as true species. Nevertheless, no certain criterion can possibly be given by which variable forms, local forms, sub-species, and representative species can be recognised.[36]

In fact, he says, sometimes the distinctness of species is due to the systematist classifying every variant as a distinct species (the splitters of modern taxonomy).

Some few naturalists maintain that animals never present varieties; but then these same naturalists rank the slightest difference as of specific value; and when the same identical form is met with in two distinct countries, or in two geological formations, they believe that two distinct species are hidden under the same dress. The term species thus comes to be a mere useless abstraction, implying and assuming a separate act

[35] *Op. cit.*, 41.
[36] *Op. cit.*, 42.

of creation. It is certain that many forms, considered by highly competent judges to be varieties, resemble species so completely in character, that they have been thus ranked by other highly competent judges. But to discuss whether they ought to be called species or varieties, before any definition of these terms has been generally accepted, is vainly to beat the air.[37]

I have been struck with the fact, that if any animal or plant in a state of nature be highly useful to man, or from any cause closely attracts his attention, varieties of it will almost universally be found recorded. These varieties, moreover, will often be ranked by some authors as species. Look at the common oak, how closely it has been studied; yet a German author makes more than a dozen species out of forms, which are almost universally considered by other botanists to be varieties; and in this country the highest botanical authorities and practical men can be quoted to show that the sessile and pedunculated oaks are either good and distinct species or mere varieties.

The ubiquitous variation of organisms, supposed by some to have been Darwin's major contribution to "population thinking,"[38] is something he derives from the work of Alphonse de Candolle:

I may here allude to a remarkable memoir lately published by A. [Alphonse] de Candolle, on the oaks of the whole world. No one ever had more ample materials for the discrimination of the species, or could have worked on them with more zeal and sagacity. ... De Candolle then goes on to say that he gives the rank of species to the forms that differ by characters never varying on the same tree, and never found connected by intermediate states. After this discussion, the result of so much labour, he emphatically remarks: "They are mistaken, who repeat that the greater part of our species are clearly limited, and that the doubtful species are in a feeble minority. This seemed to be true, so long as a genus was imperfectly known, and its species were founded upon a few specimens, that is to say, were provisional. Just as we come to know them better, intermediate forms flow in, and doubts as to specific limits augment."[39]

Intermediates are common, then, and claims of distinctness seem to rely on an a priori notion of descent from created parents, rather than be evidence in favor of the idea.

When a young naturalist commences the study of a group of organisms quite unknown to him, he is at first much perplexed in determining what differences to consider as specific, and what as varietal; for he knows nothing of the amount and kind of variation to which the group is subject; and this shows, at least, how very generally there is some variation. But if he confine his attention to one class within one country, he will soon make up his mind how to rank most of the doubtful forms. His general tendency will be to make many species, for he will become impressed, just like the pigeon or poultry fancier before alluded to, with the amount of difference in the forms which he is continually studying; and he has little general knowledge of analogical variation in other groups and in other countries, by which to correct his first impressions. As he extends the range of his observations, he will meet with more cases of difficulty; for he will encounter a greater number of closely allied forms. But if his observations be

---

[37] *Op. cit.*, 43.
[38] Sober 1980; but see Levit and Meister 2006.
[39] Darwin 1872, 43f.

widely extended, he will in the end generally be able to make up his own mind; but he will succeed in this at the expense of admitting much variation, and the truth of this admission will often be disputed by other naturalists. When he comes to study allied forms brought from countries not now continuous, in which case he cannot hope to find intermediate links, he will be compelled to trust almost entirely to analogy, and his difficulties will rise to a climax.

Certainly no clear line of demarcation has as yet been drawn between species and sub-species—that is, the forms which in the opinion of some naturalists come very near to, but do not quite arrive at, the rank of species: or, again, between sub-species and well-marked varieties, or between lesser varieties and individual differences. These differences blend into each other by an insensible series; and a series impresses the mind with the idea of an actual passage.[40]

This passage is most critical—Darwin has moved from the formal variation of groups to the idea that these forms are the result of a temporal sequence. Moreover, the only difference between slight variants, marked variations, and species is a matter of time passed. The difference of *rank* is arbitrary:

From these remarks it will be seen that I look at the term species as one arbitrarily given, for the sake of convenience, to a set of individuals closely resembling each other, and that it does not essentially differ from the term variety, which is given to less distinct and more fluctuating forms. The term variety, again, in comparison with mere individual differences, is also applied arbitrarily, for convenience' sake.[41]

We should be cautious here. It does not seem to me that Darwin is saying that the *groupings* are arbitrary, and the Notebook comment that species are real to their contemporaries backs up the claim that he thinks the groups are natural. What he thinks is arbitrary is where the distinction between the ranks of *species* and *variety* is to be drawn. And genera are the result of the age since common parenthood; some are larger than others and have more variation because the conditions that cause species to form have remained favorable for a long time:

From looking at species as only strongly marked and well-defined varieties, I was led to anticipate that the species of the larger genera in each country would oftener present varieties, than the species of the smaller genera; for wherever many closely related species (i.e., species of the same genus) have been formed, many varieties or incipient species ought, as a general rule, to be now forming. Where many large trees grow, we expect to find saplings. Where many species of a genus have been formed through variation, circumstances have been favourable for variation; and hence we might expect that the circumstances would generally be still favourable to variation. On the other hand, if we look at each species as a special act of creation, there is no apparent reason why more varieties should occur in a group having many species, than in one having few.[42]

Moreover, the species of the larger genera are related to each other in the same manner as the varieties of any one species are related to each other. No naturalist pretends

---

[40] *Op. cit.*, 44f.
[41] *Op. cit.*, 46.
[42] *Op. cit.*, 47.

that all the species of a genus are equally distinct from each other; they may generally be divided into sub-genera, or sections, or lesser groups. As Fries has well remarked, little groups of species are generally clustered like satellites around other species. And what are varieties but groups of forms, unequally related to each other, and clustered round certain forms—that is, round their parent-species? Undoubtedly there is one most important point of difference between varieties and species; namely, that the amount of difference between varieties, when compared with each other or with their parent-species, is much less than that between the species of the same genus.[43]

So Darwin presents species themselves as real, but not as a formal and fixed rank. Genera are like species and varieties in that they are the result of groupings of variation. They too do not seem to be a fixed rank, commensurate across all genera. He summarizes the argument in this chapter thus:

Finally, varieties cannot be distinguished from species,—except, first, by the discovery of intermediate linking forms; and, secondly, by a certain indefinite amount of difference between them; for two forms, if differing very little, are generally ranked as varieties, notwithstanding that they cannot be closely connected; but the amount of difference considered necessary to give to any two forms the rank of species cannot be defined. In genera having more than the average number of species in any country, the species of these genera have more than the average number of varieties. In large genera the species are apt to be closely, but unequally, allied together, forming little clusters round other species. Species very closely allied to other species apparently have restricted ranges. In all these respects the species of large genera present a strong analogy with varieties. And we can clearly understand these analogies, if species once existed as varieties, and thus originated; whereas, these analogies are utterly inexplicable if species are independent creations.[44]

Darwin had no doubt that species were formed through selection on varietal forms, and this provides the missing mechanism for how conditions of life can give rise to varieties:

Again, it may be asked, how is it that varieties, which I have called incipient species, become ultimately converted into good and distinct species, which in most cases obviously differ from each other far more than do the varieties of the same species? How do those groups of species, which constitute what are called distinct genera, and which differ from each other more than do the species of the same genus, arise? All these results, as we shall more fully see in the next chapter, follow from the struggle for life. Owing to this struggle, variations, however slight and from whatever cause proceeding, if they be in any degree profitable to the individuals of a species, in their infinitely complex relations to other organic beings and to their physical conditions of life, will tend to the preservation of such individuals, and will generally be inherited by the offspring. The offspring, also, will thus have a better chance of surviving, for, of the many individuals of any species which are periodically born, but a small number can survive. I have called this principle, by which each slight variation, if useful, is preserved, by the term Natural Selection, in order to mark its relation to man's power of selection.[45]

---

[43] *Op. cit.*, 49.

[44] *Op. cit.*, 49f.

[45] *Op. cit.*, chapter III, p51f.

Selection is for Darwin not restricted to the intraspecific, but can be interspecific or even merely a matter of simple survival:

> Hence, as more individuals are produced than can possibly survive, there must in every case be a struggle for existence, either one individual with another of the same species, or with the individuals of distinct species, or with the physical conditions of life.[46]
>
> But the struggle will almost invariably be most severe between the individuals of the same species, for they frequent the same districts, require the same food, and are exposed to the same dangers. In the case of varieties of the same species, the struggle will generally be almost equally severe, and we sometimes see the contest soon decided...[47]

And in chapter IV, he notes

> In order that any great amount of modification should be effected in a species, a variety, when once formed, must again, perhaps after a long interval of time, vary or present individual differences of the same favourable nature as before; and these must be again preserved, and so onward, step by step.[48]

Darwin had no trouble finding links between asexuals and self-fertilizing species, and it is clear that he did not exclude asexuals from *being* species, as we see:

> It must have struck most naturalists as a strange anomaly that, both with animals and plants, some species of the same family and even of the same genus, though agreeing closely with each other in their whole organisation, are hermaphrodites, and some unisexual. But if, in fact, all hermaphrodites do occasionally intercross, the difference between them and unisexual species is, as far as function is concerned, very small.[49]

Darwin defined the processes that keep species distinct in even occasionally sexual organisms as the result of intercrossing. The benefits of sexual reproduction include "vigour and fertility," both germane to selection (although he doesn't explain exactly why, which resulted in extensive and ongoing debates in the following century on the evolutionary benefits of sex). But equally interesting here is that, contrary to many twentieth-century Darwinians (for example, Fisher and Simpson[50]), Darwin himself has no problem explaining asexual species, and even explains them, as Manfred Eigen does today,[51] as the result of natural selection entirely.

> Intercrossing plays a very important part in nature by keeping the individuals of the same species, or of the same variety, true and uniform in character. It will obviously thus act far more efficiently with those animals which unite for each birth; but, as already stated, we have reason to believe that occasional intercrosses take place with all animals and plants. Even if these take place only at long intervals of time, the

---

[46] *Op. cit.*, 53.
[47] *Op. cit.*, 60.
[48] *Op. cit.*, 66.
[49] *Op. cit.*, 77f.
[50] See pages 246 *et seq.*
[51] Eigen 1993. See also Ereshefsky 2010, Caro-Quintero and Konstantinidis 2012.

young thus produced will gain so much in vigour and fertility over the offspring from long-continued self-fertilisation, that they will have a better chance of surviving and propagating their kind; and thus in the long run the influence of crosses, even at rare intervals, will be great. With respect to organic beings extremely low in the scale, which do not propagate sexually, nor conjugate, and which cannot possibly intercross, uniformity of character can be retained by them under the same conditions of life, only through the principle of inheritance, and through natural selection which will destroy any individuals departing from the proper type. If the conditions of life change and the form undergoes modification, uniformity of character can be given to the modified offspring, solely by natural selection preserving similar favourable variations.[52]

Darwin at the time of the sixth edition thought that the formation of species did not rely on isolation and here he takes Moritz Wagner's view to task.[53] Isolation does, he thought, make species formation easier, but he cannot agree it is required, as Wagner thought. And if the isolated population is *too* small, then isolation can in fact *prevent* speciation from occurring due to a lack of variation. The founder effect or drift through biased sampling has not occurred to him, as it later did to Weismann.[54]

Isolation, also, is an important element in the modification of species through natural selection. In a confined or isolated area, if not very large, the organic and inorganic conditions of life will generally be almost uniform; so that natural selection will tend to modify all the varying individuals of the same species in the same manner. Intercrossing with the inhabitants of the surrounding districts will, also, be thus prevented. Moritz Wagner has lately published an interesting essay on this subject, and has shown that the service rendered by isolation in preventing crosses between newly-formed varieties is probably greater even than I supposed. But from reasons already assigned I can by no means agree with this naturalist, that migration and isolation are necessary elements for the formation of new species. The importance of isolation is likewise great in preventing, after any physical change in the conditions such as of climate, elevation of the land, &c., the immigration of better adapted organisms; and thus new places in the natural economy of the district will be left open to be filled up by the modification of the old inhabitants. Lastly, isolation will give time for a new variety to be improved at a slow rate; and this may sometimes be of much importance. If, however, an isolated area be very small, either from being surrounded by barriers, or from having very peculiar physical conditions, the total number of the inhabitants will be small; and this will retard the production of new species through natural selection, by decreasing the chances of favourable variations arising.[55]

Although isolation is of great importance in the production of new species, on the whole I am inclined to believe that largeness of area is still more important, especially for the production of species which shall prove capable of enduring for a long period, and

---

[52] Darwin 1872, 79.

[53] Wagner 1889.

[54] Weismann 1904, 286:

> ... there are very variable species and very constant species, and it is obvious that colonies which are founded by a very variable species can hardly ever remain exactly identical with the ancestral species; and that several of them will turn out differently, even granting that the conditions of life be exactly the same, for no colony will contain all the variants of the species in the same proportion, but at most only a few of them, and the result of mingling these must ultimately result in the development of a somewhat different form in each colonial area.

[55] Darwin 1872, 79f.

of spreading widely. Throughout a great and open area, not only will there be a better chance of favourable variations, arising from the large number of individuals of the same species there supported, but the conditions of life are much more complex from the large number of already existing species; and if some of these many species become modified and improved, others will have to be improved in a corresponding degree, or they will be exterminated. Each new form, also, as soon as it has been much improved, will be able to spread over the open and continuous area, and will thus come into competition with many other forms. Moreover, great areas, though now continuous, will often, owing to former oscillations of level, have existed in a broken condition; so that the good effects of isolation will generally, to a certain extent, have concurred. Finally, I conclude that, although small isolated areas have been in some respects highly favourable for the production of new species, yet that the course of modification will generally have been more rapid on large areas; and what is more important, that the new forms produced on large areas, which already have been victorious over many competitors, will be those that will spread most widely, and will give rise to the greatest number of new varieties and species. They will thus play a more important part in the changing history of the organic world.[56]

Ironically, as Kottler noted, Darwin in the early Notebooks believed that isolation *was* a *sine qua non* for speciation, in part following the views of Leopold von Buch.[57] But by this later stage, Darwin appears to have made Natural Selection the primary cause of species, requiring that variation needs to occur *in situ* as it were, and so needing larger populations to give it opportunity to do so.

He also notes that the structures that are useful in classifying species are often not of any adaptive value:

> Hence modifications of structure, viewed by systematists as of high value, may be wholly due to the laws of variation and correlation, without being, as far as we can judge, of the slightest service to the species.[58]

Considering the traditional logical notions of generic and specific characters from which the concept of *species* was drawn by natural historians in the seventeenth century, it is interesting to note that in contrast Darwin expects the specific characters to vary more than the generic, and moreover that if a character is variable in the genus between closely allied species, it is more variable in the individual species as well:

> ... on the view that species are only strongly marked and fixed varieties, we might expect often to find them still continuing to vary in those parts of their structure which have varied within a moderately recent period, and which have thus come to differ. Or to state the case in another manner: the points in which all the species of a genus resemble each other, and in which they differ from allied genera, are called generic characters; and these characters may be attributed to inheritance from a common progenitor, for it can rarely have happened that natural selection will have modified several distinct species, fitted to more or less widely different habits, in exactly the same manner:—and as these so-called generic characters have been inherited from before the period when the several species first branched off from their common progenitor,

---

[56] *Op. cit.,* 80f.
[57] Kottler 1978, 285–288.
[58] Darwin 1872, 110.

and subsequently have not varied or come to differ in any degree, or only in a slight degree, it is not probable that they should vary at the present day. On the other hand, the points in which species differ from other species of the same genus are called specific characters; and as these specific characters have varied and come to differ since the period when the species branched off from a common progenitor, it is probable that they should still often be in some degree variable,—at least more variable than those parts of the organisation which have for a very long period remained constant.[59]

Asking in effect why the Great Chain of Being principle of *lex completio* does not result in no species being seen at all but instead one variable mass, Darwin answers,

I believe that species come to be tolerably well-defined objects, and do not at any one period present an inextricable chaos of varying and intermediate links"[60]

due to

1. The fact that variation and selection take time, and may not have yet occurred
2. When smaller isolated populations spread out, selection exterminates the older intermediate forms (in modern terms, when in sympatry, allopatric variations exclude the less fit older forms)
3. Intermediates are subject to accidental extinction because they are less widely spread
4. "Numberless intermediate varieties, linking closely together all the species of the same group, must assuredly have existed; but the very process of natural selection constantly tends, as has been so often remarked, to exterminate the parent-forms and the intermediate links"[61]

When reading the *Origin* one is struck by Darwin's repeated locutions "to the benefit of the species" or "of advantage to the species." Darwin seems to be making the (now) classical blunder of group selectionism. However, as one reads these examples it becomes clear that Darwin is using it as a circumlocution for "of advantage to the members of the species that carry this trait." It is clear that he thinks that selection occurs, in the main, through competition between varieties (and of course species are just well-marked varieties as he has said above), increasing or decreasing in relative numbers as they carry beneficial or non-beneficial traits. In this usage, *species* is just a way of marking the variety that is subjected to selection, or has been so subjected and gone to fixation; it is roughly equivalent, therefore, in Darwin's mind, to the sum total of what G. C. Williams later called "evolutionary genes"[62]— those hereditable variations that are selectable.

---

[59] *Op. cit.*, 115.
[60] *Op. cit.*, 127.
[61] *Op. cit.*, 128.
[62] Williams 1966.

But Darwin did not think that selection *caused* sterility and hence species. Rather, he held that selection incidentally resulted in forms that were unable to breed together:

> The view commonly entertained by naturalists is that species, when intercrossed, have been specially endowed with sterility, in order to prevent their confusion. This view certainly seems at first highly probable, for species living together could hardly have been kept distinct had they been capable of freely crossing. The subject is in many ways important for us, more especially as the sterility of species when first crossed, and that of their hybrid offspring, cannot have been acquired, as I shall show, by the preservation of successive profitable degrees of sterility. It is an incidental result of differences in the reproductive systems of the parent-species.[63]
>
> The fertility of varieties, that is of the forms known or believed to be descended from common parents, when crossed, and likewise the fertility of their mongrel offspring, is, with reference to my theory, of equal importance with the sterility of species; for it seems to make a broad and clear distinction between varieties and species.[64]

However, on examination, he does not think this holds generally true:

> It is certain, on the one hand, that the sterility of various species when crossed is so different in degree and graduates away so insensibly, and, on the other hand, that the fertility of pure species is so easily affected by various circumstances, that for all practical purposes it is most difficult to say where perfect fertility ends and sterility begins. I think no better evidence of this can be required than that the two most experienced observers who have ever lived, namely Kölreuter and Gärtner, arrived at diametrically opposite conclusions in regard to some of the very same forms. It is also most instructive to compare—but I have not space here to enter into details—the evidence advanced by our best botanists on the question whether certain doubtful forms should be ranked as species or varieties, with the evidence from fertility adduced by different hybridisers, or by the same observer from experiments made during different years. It can thus be shown that neither sterility nor fertility affords any certain distinction between species and varieties. The evidence from this source graduates away, and is doubtful in the same degree as is the evidence derived from other constitutional and structural differences.[65]

And also

> By the term systematic affinity is meant, the general resemblance between species in structure and constitution. Now the fertility of first crosses, and of the hybrids produced from them, is largely governed by their systematic affinity. This is clearly shown by hybrids never having been raised between species ranked by systematists in distinct families; and on the other hand, by very closely allied species generally uniting with facility. But the correspondence between systematic affinity and the facility of crossing is by no means strict. A multitude of cases could be given of very closely allied species which will not unite, or only with extreme difficulty; and on the other hand of very distinct species which unite with the utmost facility. ...

---

[63] Darwin 1872, 209.
[64] *Op. cit.*, 209f.
[65] *Op. cit.*, 210f.

And, reiterating the comment in the Notebook:

> No one has been able to point out what kind or what amount of difference, in any recognisable character, is sufficient to prevent two species crossing.[66]

Darwin notes that species are almost always intersterile, but that this is often due to the fact that as soon as this intersterility is noticed, taxonomists will rank these varieties as species:

> It may be urged, as an overwhelming argument, that there must be some essential distinction between species and varieties, inasmuch as the latter, however much they may differ from each other in external appearance, cross with perfect facility, and yield perfectly fertile offspring. With some exceptions, presently to be given, I fully admit that this is the rule. But the subject is surrounded by difficulties, for, looking to varieties produced under nature, if two forms hitherto reputed to be varieties be found in any degree sterile together, they are at once ranked by most naturalists as species.[67]

Anyway, some obvious varieties within species are sterile together, even though they are able to breed with mutual races:

> From these facts it can no longer be maintained that varieties when crossed are invariably quite fertile. From the great difficulty of ascertaining the infertility of varieties in a state of nature, for a supposed variety, if proved to be infertile in any degree, would almost universally be ranked as a species;—from man attending only to external characters in his domestic varieties, and from such varieties not having been exposed for very long periods to uniform conditions of life;—from these several considerations we may conclude that fertility does not constitute a fundamental distinction between varieties and species when crossed. The general sterility of crossed species may safely be looked at, not as a special acquirement or endowment, but as incidental on changes of an unknown nature in their sexual elements.[68]

So in the end, Darwin refuses to make sterility a test of species, or even to expect that sterility will correlate with systematic affinity, summarizing the arguments in that chapter thus:

> First crosses between forms, sufficiently distinct to be ranked as species, and their hybrids, are very generally, but not universally, sterile. The sterility is of all degrees, and is often so slight that the most careful experimentalists have arrived at diametrically opposite conclusions in ranking forms by this test. The sterility is innately variable in individuals of the same species, and is eminently susceptible to the action of favourable and unfavourable conditions. The degree of sterility does not strictly follow systematic affinity, but is governed by several curious and complex laws. It is generally different, and sometimes widely different in reciprocal crosses between the same two species. It is not always equal in degree in a first cross and in the hybrids produced from this cross.

---

[66] *Op. cit.*, 215f.
[67] *Op. cit.*, 226.
[68] *Op. cit.*, 229.

In the same manner as in grafting trees, the capacity in one species or variety to take on another, is incidental on differences, generally of an unknown nature, in their vegetative systems, so in crossing, the greater or less facility of one species to unite with another is incidental on unknown differences in their reproductive systems. There is no more reason to think that species have been specially endowed with various degrees of sterility to prevent their crossing and blending in nature, than to think that trees have been specially endowed with various and somewhat analogous degrees of difficulty in being grafted together in order to prevent their inarching in our forests.[69]

In the chapter on classification, Darwin attended to the problems that this view brings with it for working naturalists, but also the problems it solves, especially in understanding the reason for the systematic affinities:

Naturalists, as we have seen, try to arrange the species, genera, and families in each class, on what is called the Natural System. But what is meant by this system? Some authors look at it merely as a scheme for arranging together those living objects which are most alike, and for separating those which are most unlike; or as an artificial method of enunciating, as briefly as possible, general propositions,—that is, by one sentence to give the characters common, for instance, to all mammals, by another those common to all carnivora, by another those common to the dog-genus, and then, by adding a single sentence, a full description is given of each kind of dog. The ingenuity and utility of this system are indisputable. But many naturalists think that something more is meant by the Natural System; they believe that it reveals the plan of the Creator; that unless it be specified whether order in time or space, or both, or what else is meant by the plan of the Creator, it seems to me that nothing is thus added to our knowledge. Expressions such as that famous one by Linnæus, which we often meet with in a more or less concealed form, namely, that the characters do not make the genus, but that the genus gives the characters, seem to imply that some deeper bond is included in our classifications than mere resemblance. I believe that this is the case, and that community of descent—the one known cause of close similarity in organic beings—is the bond, which though observed by various degrees of modification, is partially revealed to us by our classifications.[70]

The importance, for classification, of trifling characters, mainly depends on their being correlated with many other characters of more or less importance. The value indeed of an aggregate of characters is very evident in natural history. Hence, as has often been remarked, a species may depart from its allies in several characters, both of high physiological importance, and of almost universal prevalence, and yet leave us in no doubt where it should be ranked. Hence, also, it has been found that a classification founded on any single character, however important that may be, has always failed; for no part of the organisation is invariably constant. The importance of an aggregate of characters, even when none are important, alone explains the aphorism enunciated by Linnæus, namely, that the characters do not give the genus, but the genus gives the characters; for this seems founded on the appreciation of many trifling points of resemblance, to slight to be defined.[71]

---

[69] *Op. cit.*, 233.
[70] *Op. cit.*, 319f.
[71] *Op. cit.*, 321.

All the foregoing rules and aids and difficulties in classification may be explained, if I do not greatly deceive myself, on the view that the Natural System is founded on descent with modification;—that the characters which naturalists consider as showing true affinity between any two or more species, are those which have been inherited from a common parent, all true classification being genealogical;—that community of descent is the hidden bond which naturalists have been unconsciously seeking, and not some unknown plan of creation, or the enunciation of general propositions, and the mere putting together and separating objects more or less alike.

But I must explain my meaning more fully. I believe that the *arrangement* of the groups within each class, in due subordination and relation to each other, must be strictly genealogical in order to be natural; but that the *amount* of difference in the several branches or groups, though allied in the same degree in blood to their common progenitor, may differ greatly, being due to the different degrees of modification which they have undergone; and this is expressed by the forms being ranked under different genera, families, sections, or orders.[72]

Given that there is some continuing dispute over whether or not Darwin was a cladist,[73] it is worth stating here my view that Padian is right—for Darwin in this last passage classification is necessarily natural only if it matches genealogy,[74] but it may also be represented as well in terms of grade of organization. If it is, though, such grades must not trim away the genealogical relationships. Darwin was not a cladist,[75] but he was pretty close to it. However, he recognizes the practical necessity of representing overall differences in a classification scheme. It's just not the same as saying these differences are part of a *natural* classification; this point shall become significant in the argument presented in the final chapters. But even if Darwin were an "eclectic" in his approach to classification as Mayr suggests, this would be due to the fact that he had not yet completely worked out the implications of his view of species related by common descent.

Darwin then summarizes the difference between the process by which things have evolved and the patterns that diagnose them—his argument is remarkably similar to the so-called "father of cladistics," Willi Hennig's, view on "reciprocal illumination," in which he argues that knowledge of groups illuminates the uncovering of the knowledge of other groups, which then help refine the initial groups:[76]

With species in a state of nature, every naturalist has in fact brought descent into his classification; for he includes in his lowest grade, that of species, the two sexes; and how enormously these sometimes differ in the most important characters, is known to every naturalist: scarcely a single fact can be predicated in common of the adult males and hermaphrodites of certain cirripedes, and yet no one dreams of separating them. ... The naturalist includes as one species the various larval stages of the same

---

[72] *Op. cit.*, 323.

[73] Mayr 1982, 209–213, Mayr 1994.

[74] Padian 1999.

[75] *Cladism* is a term that denotes a taxonomic methodology of classifying in terms of shared ancestry or characters that derive from shared ancestry, properly known as "phylogenetic systematics," developed in the 1950s. It is strictly anachronistic to apply to Darwin terms that are only meaningful to describe a later school of thought.

[76] Hennig 1966, 21f, 148, 206, 222.

individual, however much they may differ from each other and from the adult, as well as the so-called alternate generations of Steenstrup, which can only in a technical sense be considered as the same individual. He includes monsters and varieties, not from their partial resemblance to the parent-form, but because they are descended from it.

As descent has universally been used in classing together the individuals of the same species, though the males and females and larvæ are sometimes extremely different; and as it has been used in classing varieties which have undergone a certain, and sometimes a considerable, amount of modification, may not this same element of descent have been unconsciously used in grouping species under genera, and genera under higher groups, all under the so-called natural system? I believe it has been unconsciously used; and thus only can I understand the several rules and guides which have been followed by our best systematists. As we have no written pedigrees, we are forced to trace community of descent by resemblances of any kind. Therefore we choose those characters which are the least likely to have been modified, in relation to the conditions of life to which each species has been recently exposed. Rudimentary structures on this view are as good as, or even sometimes better than, other parts of the organisation. We care not how trifling a character may be—let it be the mere inflection of the angle of the jaw, the manner in which an insect's wing is folded, whether the skin be covered by hair or feathers—if it prevail throughout many and different species, especially those having very different habits of life, it assumes high value; for we can account for its presence in so many forms with such different habits, only by inheritance from a common parent. We may err in this respect in regard to single points of structure, but when several characters, let them be ever so trifling, concur throughout a large group of beings having different habits, we may feel almost sure, on the theory of descent, that these characters have been inherited from a common ancestor; and we know that such aggregated characters have especial value in classification.

We can understand why a species or a group of species may depart from its allies, in several of its most important characteristics, and yet be safely classed with them. This may be safely done, and is often done, as along as a sufficient number of characters, let them be ever so unimportant, betray the hidden bond of community of descent. Let two forms have not a single character in common, yet, if these extreme forms are connected together by a chain of intermediate groups, we may at once infer their community of descent, and we put them all into the same class. As we find organs of high physiological importance—those which serve to preserve life under the most diverse conditions of existence—are generally the most constant, we attach especial value to them; but if these same organs, in another group or section of a group, are found to differ much, we at once value them less in our classification. We shall presently see why embryological characters are of such high classificatory importance. Geographical distribution may sometimes be brought usefully into play in classing large genera, because all the species of the same genus, inhabiting any distinct and isolated region, are in all probability descended from the same parents.[77]

In the final chapter, the "Recapitulation and Conclusion," Darwin makes the now-famous comment about the reality of species which led so many to conclude he was a mere nominalist:

When the views advanced by me in this volume, and by Mr. Wallace, or when analogous views on the origin of species are generally admitted, we can dimly foresee that

---

[77] Darwin 1872, 325f.

there will be a considerable revolution in natural history. Systematists will be able to pursue their labours as at present; but they will not be incessantly haunted by the shadowy doubt whether this or that form be a true species. This, I feel sure and I speak after experience, will be no slight relief. The endless disputes whether or not some fifty species of British brambles are good species will cease. Systematists will have only to decide (not that this will be easy) whether any form be sufficiently constant and distinct from other forms, to be capable of definition; and if definable, whether the differences be sufficiently important to deserve a specific name. This latter point will become a far more essential consideration than it is at present; for differences, however slight, between any two forms, if not blended by intermediate gradations, are looked at by most naturalists as sufficient to raise both forms to the rank of species.

Hereafter we shall be compelled to acknowledge that the only distinction between species and well-marked varieties is, that the latter are known, or believed, to be connected at the present day by intermediate gradations whereas species were formerly thus connected. Hence, without rejecting the consideration of the present existence of intermediate gradations between any two forms, we shall be led to weigh more carefully and to value higher the actual amount of difference between them. It is quite possible that forms now generally acknowledged to be merely varieties may hereafter be thought worthy of specific names; and in this case scientific and common language will come into accordance. In short, we shall have to treat species in the same manner as those naturalists treat genera, who admit that genera are merely artificial combinations made for convenience. This may not be a cheering prospect; but we shall at least be freed from the vain search for the undiscovered and undiscoverable essence of the term species.[78]

It will be worthwhile to look at this matter closely, given the excerpts above as cross bearings on the subtlety of Darwin's views on species, and ask, has Darwin said that species are "merely artificial combinations made for convenience?" The answer is, no. He has said rather that naturalists shall be forced to treat them that way. We have seen repeatedly that Darwin did not insist *species* were unreal, merely that the *rank* was arbitrarily assigned and that we could not see if they were real on the basis of characters. The reason the term *species* has no discoverable essence (but does that imply it has a Real Essence in the Lockean sense?) is that each case is different in the biological particulars. But they are separated, he says, in that the "intergradations" between them are extinct. *That* is real enough. Darwin's definition of species is simply that they do not interbreed, or, in the case of "unisexual" organisms, that Natural Selection keeps them isolated in the "proper type" suited to the conditions of life in which they live. In this I am concurring with Kottler's and Ghiselin's conclusions; where Darwin seems to be a nominalist, he is in fact describing the problems of current taxonomic criteria, founded on creationist views of species, and so describing what species are *not*.[79]

The glossary to the later editions of the *Origin* does not give us a definition for species, so we lack from Darwin the sort of epigrammatic slogan for species that so many other writers have given us, and even if it had, it would be the definition of the

---

78 *Op. cit.*, 371.
79 Kottler 1978, 293f, Ghiselin 1984.

compiler W. S. Dallas, not of Darwin. David Williams notes that Dallas' definition of terms like *homology* differed from Darwin's use of the term.[80]

AFTER THE *ORIGIN*

After the *Origin*, Darwin is able to publish some generalizations about species. He does so in chapter 2 of the *Descent of Man*:

> If we consider all the races of man as forming a single species, his range is enormous; but some separate races, as the Americans and Polynesians, have very wide ranges. It is a well-known law that widely-ranging species are much more variable than species with restricted ranges; and the variability of man may with more truth be compared with that of widely-ranging species, than with that of domesticated animals.[81]

The "well-known law" is in part of Darwin's own construction, as we have seen. In the *Variation of Plants and Animals under Domestication*, published in 1868 and revised in 1875,[82] Darwin discusses how species arise in nature, and proposes intersterility as a criterion of species-status.

> ... how, it may be asked, have species arisen in a state of nature? The differences between natural varieties are slight; whereas the differences are considerable between the species of the same genus, and great between the species of distinct genera. How do these lesser differences become augmented into the greater difference? How do varieties, or as I have called them, incipient species, become converted into true and well-defined species?[83]

Using domestic varieties as a guide to variation in nature, as he had in the *Origin*, he notes

> ... that the sterility of distinct species when crossed, and of their hybrid progeny, depends exclusively on the nature of their sexual elements, and not on any differences in their structure or general constitution. ... That excellent observer, Gärtner, likewise concluded that species when crossed are sterile owing to differences confined to their reproductive systems[84]

and in answer to the question of how it is that extremely different domesticated varieties are "perfectly fertile" while closely allied species are not, he responds:

> [p]assing over the fact that the amount of external difference between two species is no sure guide to the degree of their mutual sterility, so that similar differences in the case of varieties [*of domestic animals and plants—JSW*] would be no sure guide, we know that with species the cause lies exclusively in differences in their sexual constitution.[85]

---

[80] *Pers. comm.*
[81] Second edition, Darwin 1871, 112.
[82] Darwin 1868, 1875; modern edition 1998.
[83] Darwin 1868, Vol. I, 45.
[84] Darwin 1868, Vol. II, 168f.
[85] Darwin 1868, Vol. II, 172.

In a short letter to *Nature*, while discussing whether variations in brain structure could enable the evolution of instincts, Darwin noted in his own defense:

[t]he writer of the article in referring to my words "the preservation of useful variations of pre-existing instincts" adds "the question is, whence these variations?"
Nothing is more to be desired in natural history than that some one would be able
to answer such a query. But as far as our present subject is concerned, the writer
probably will admit that a multitude of variations have arisen, for instance in colour
and in the character of the hair, feathers, horns, &c., which are quite independent of
habit and of use in previous generations. It seems far from wonderful, considering the
complex conditions to which the whole organisation is exposed during the successive
stages of its development from the germ, that every part should be liable to occasional
modifications: the wonder indeed is that any two individuals of the same species are
at all closely alike.[86]

So, in the end, Darwin proposed a "snowflake" theory of species—all members are alike in some ways, but they are also unique individuals. Variation occurs naturally, and it is weeded out according to how well suited it is to the conditions of life. Species are held distinct incidentally to their being adapted to those conditions; they are real at the time, although no rank seems to be absolute, and there is no particular amount or kind of difference between them that marks out, or correlates with being, distinct species. For sexual organisms all that can be said is that they do not, in nature, interbreed (no matter whether they can be made to in captivity). He is not a nominalist, as he notes in his response to Agassiz's criticism,

[I]f species do not exist at all, as the supporters of the transmutation theory maintain,
how can they vary? And if individuals alone exist, how can differences which may be
observed among them prove the variability of species?[87]

Darwin's reply, to Asa Gray, was

I am surprised that Agassiz did not succeed in writing something better. How absurd
that logical quibble "if species do not exist how can they vary"? As if anyone doubted
their temporary existence?[88]

We need to note with Darwin, as with others, that there is a difference between denying that the *rank* of species has a definition, and denying that the *term species* has one. Darwin denies the former, but not the latter. He is not a nominalist but a pluralist with regard to what makes species distinct. Nevertheless, all these causes resolve down to an aversion to interbreeding in sexual organisms and differences in their sexual structures and constitutions, and selection maintaining the appropriate forms and organs for living in the conditions in which they find themselves, for asexuals. Of course, he also allowed that conditions of life may directly affect both

---

[86] Darwin 1873.
[87] Quoted in Lurie 1960, 297.
[88] Quoted in Gayon 1996, 229; cf. also Ghiselin 1984, chapter 4.

the sexual organs and the structures of the organisms, although he rejected it in a particular case.[89] Nevertheless, I think we can dispose of Futuyma's and others' mischaracterization. Darwin knew very well how to define species, given the fact of evolution.

In summary, Darwin was a species realist, but denied the absolute rank of Linnaean classification, although he used it in practice and was a contributor to the Strickland Rules. He allowed for asexual species, and his views focused primarily on the process of speciation, which he initially entertained might be due to geographical isolation (and possibly the effects of the local climate, as in the traditional view) but later held was due to the fixing of subspecific varieties due to natural selection. Speciation is a side-effect of selection between varieties that causes changes in the sexual reproduction system of sexual species.[90]

INTERPRETATIONS OF DARWIN'S IDEA OF SPECIES

Both historically and recently, Darwin has been variously interpreted on species. We will deal with some of the historical cases below, but it is worth reviewing the recent literature, as Darwin is something of a palimpsest in this, as other topics, on which thinkers project their own views. We will see that around the end of the nineteenth century, it became the consensus that Darwin denied that species were real, or that they had any special status in biology. However, the most influential writer on Darwin's supposed "species nominalism" has been Ernst Mayr, who published numerous pieces claiming that Darwin had the "wrong view of species."[91]

Ernst Mayr held that Darwin had the "right" understanding of species, early in his Notebooks, basing his claim on Kottler's review of them. However, Mayr thought Darwin "failed" because of his insistence upon the primacy of natural selection:

> Darwin succeeded in convincing the world of the occurrence of evolution and ... he found (in natural selection) the mechanism that is responsible for evolutionary change and adaptation. It is not nearly so widely recognized that Darwin failed to solve the problem indicated by the title of his work. Although he demonstrated the modification of species in the time dimension, he never seriously attempted a rigorous analysis of the problem of the multiplication of species, the splitting of one species into two.
>
> ...
>
> In retrospect, it is apparent that Darwin's failure ... resulted to a large extent from a misunderstanding of the true nature of species.[92]

As Mayr was an advocate of allopatric speciation as the major, if not sole, cause of speciation, natural selection, which required sympatry, was not the cause of new species; hence Darwin was wrong. However, Mayr held, Darwin did get right

---

[89] Darwin 1875, vol. II, 413.
[90] A slightly different interpretation is given by Mallet 2008.
[91] Mayr 1942, Burma and Mayr 1949, Mayr 1957, 1959, 1982.
[92] Mayr 1963, 12, 14. Quoted in Mallet 2013.

that species were variable populations, and so he adopted "population thinking." However, according to Mayr, Darwin treated the division into species as arbitrary, and approached the matter "purely typologically, as characterized by 'degree of difference.'"[93]

Recent authors have argued that Darwin was, as Poulton and his generation had assumed, a species denier. Mallet, for example, says this:

> Throughout the history of the species debate, starting with Darwin, there have been some who argue that species are not individual real objects, but should instead be considered merely as man-made constructs, merely useful in understanding biodiversity and its evolution. These people are not necessarily nihilists, who deny that species exist: they simply argue that actual morphological and genetic gaps between populations would be more useful for delimiting species than inferred processes underlying evolution or maintenance of these gaps. By their refusal to unite these ideas under a single named concept, this biologist "silent majority" has rarely found a common voice.[94]

This view, which has come to be known as "species nominalism," has been challenged about Darwin.[95] Stamos has dealt with these interpretations in the first chapter of his 2003, but then argues that Darwin was not merely a species *taxa* realist, as interpreters like Beatty[96] held, but a species *category* (or rank) realist. I must disagree with Stamos here. To my mind, the category that Darwin accepted was, as Beatty suggests, a taxonomic category in practice, rather than an entity-class. For a start, taxonomic discussions on variation in species was of long standing when the *Origin* was published, and was a particularly hot issue when Darwin began his Notebooks. Second, Darwin's practice was standard well before he announced his evolutionary theory. Beatty and Stamos think this is because he had a strategy to "sell" his views against the prevailing fixism, but this is unlikely if fixism was already in the wind when he published, for example, his *Cirripedia* studies, as I think it was. Darwin was most definitely careful in his rhetorical strategies, but not about the concept of species as such.

Others have attempted to argue that Darwin held an individualistic view of species, that is, as spatiotemporally located and bordered singular objects.[97] Again, this seems to be reading issues back into the past based on the concerns of the future, but it is not, *ipso facto*, false. However, I think that the class/individual dichotomy assumed here is somewhat overstated. If the Essentialism Story is wrong, historically, then there is no conflict between Darwin doing traditional classification, morphologically, and the metaphysical notion, to which he never alludes, despite Stamos' linking of Locke's conventionalism and Darwin's Notebooks, that species are particulars. I return to this in the third section of the book. For the moment, I think it worth noting that Darwin was doing traditional taxonomy with "classes" and

[93] Mayr 1982, 269.
[94] Mallet 2007, 685. See also Mallet 2010.
[95] For example, by Stamos 1996, 1999, 2003.
[96] Beatty 1985.
[97] See Gayon 1996.

"types," and we should take him at his word that his theories had no effect upon his classificatory activities.

MORITZ WAGNER, PIERRE TRÉMAUX, AND GEOGRAPHIC SPECIATION

Moritz Wagner (1813–1887) was a celebrated explorer and geographer, and his writings were influential. He had proposed, in opposition to Darwin's notion that species are formed from racial types through selection, that species must be isolated geographically.[98] Mayr discusses the reaction of Darwin and his contemporaries to Wagner's isolation model.[99] Dismissed by Weismann and Wallace, as we shall see, the geographical isolation thesis was nevertheless adopted by the Rev. J. T. Gulick, whose work on Hawaiian landsnails (gen. *Achatinella*) led him also to claim that much evolutionary variation was due to chance.[100] Thereafter, there seemed to be two camps: those who thought that isolation was the *sine qua non* of speciation and that chance was the cause of variation between populations, and those who thought that speciation occurred equally if not entirely through the action of natural and perhaps sexual selection. After the later work of Sewall Wright had been accepted and promoted in the work of Dobzhansky during the mid-twentieth century, and following Poulton's and Mayr's coinages, this view came to be known as the *allopatric* theory of speciation, and Darwin's published view as the *sympatric* theory.

A possible unrecognized source for the notion of geographic isolation as a mechanism for speciation is the work of the amateur anthropologist and architecture scholar, Pierre Trémaux (1818–1895), whose work *Origin et transformations de l'homme et des autres êtres* (*The origin and evolution of man and other beings*), published in 1865, proposed that the effects of local climate and conditions (*sol*) would fully determine racial and species characters by adaptation, which would then be maintained by interbreeding, or as he calls it, "crossing" (*croisement*).[101] Such changes would happen rapidly, evolutionarily speaking, and then reach an *equilibrium*. Charitably interpreted, for the language of *Origin et transformations* is florid and non-technical, Trémaux proposed both an allopatric theory and a punctuated equilibrium theory. However, his amateur standing in the French scientific community meant that his self-published works were not taken seriously, although Darwin had two copies in his library, and had read at least one of them,[102] and as a result may have revised his expression in the 1866 fourth edition of the *Origin of Species* regarding the rates of change to include more or less rapid changes.[103] Trémaux has been unjustly tarred with being a crank because Marx wrote to Engels to recommend him over Darwin and Engels replied that Trémaux was simply silly and ignorant of the facts. Nearly every commentator since has relied on Engels'

---

98 Wagner 1868, 1873, 1889.
99 Mayr 1982, 562–566.
100 Gulick 1873, 1888, 1890, 1908.
101 Trémaux 1865.
102 Wilkins and Nelson 2008.
103 Darwin 1866, 409f.

interpretation,[104] which is to my mind clearly a strawman interpretation based on Engels' own ignorance of the use of the French term *sol* by Buffon and others to mean "habitat." In his 1865 work there is an extended discussion of the definitions of "species," the first such discussion I know of, mostly focusing on French definitions from Cuvier onward. It is possible, although there is no direct evidence as yet, that Wagner also had read Trémaux's ideas when he published three years afterward. Trémaux later published a more "cosmic" form of his ideas,[105] which are reminiscent of Spencer's cosmic evolution.

WALLACE AND WEISMANN'S ADAPTATIONIST DEFINITION

Alfred Russel Wallace (1823–1913), co-discoverer of natural selection as an agent of evolution with Darwin, never really admitted the action of anything else in evolution. It followed, therefore, that he would insist that natural selection was the agent of speciation, and hence that species are to be identified with their special adaptations.

Before he had gone public with his own evolutionism, Wallace asked if there was only an indefinable amount of difference that separated permanent varieties from species:

> If there is no other character, that fact is one of the strongest arguments against the independent creation of species, for why should a special act of creation be required to call into existence an organism differing only in degree from another which has been produced by existing laws? If an amount of permanent difference, represented by any number up to 10, may be produced by the ordinary course of nature, it is surely most illogical to suppose, and very hard to believe, that an amount of difference represented by 11 required a special act to call it into existence.[106]

Kottler describes how Wallace's idea of species in this period involved lack of interbreeding, like Darwin's: "contact without intermixture being a good test of specific difference."[107] In his *Darwinism* he defined species as

> An assemblage of individuals which have become somewhat modified in structure, form, and constitution, so as to adapt them to slightly different conditions of life; which

---

[104] Two exceptions: Diane Paul [1981], and Stephen Jay Gould [1999]. Gould, however, seems to have wanted to distance himself from the Engels' misinterpretation of Trémaux in order to shield punctuated equilibrium theory from another charge of being unsupportable, and so he says:

> I had long been curious about Trémaux and sought a copy of his book for many years. I finally purchased one a few years ago—and I must say that I have never read a more absurd or more poorly documented thesis. Basically, Trémaux argues that the nature of the soil determines national characteristics and that higher civilizations tend to arise on more complex soils formed in later geological periods. If Marx really believed that such unsupported nonsense could exceed the Origin of Species in importance, then he could not have properly understood or appreciated the power of Darwin's facts and ideas. [page 90]

In fact, we think that Marx had the right of it. Trémaux actually is putting forward a mechanism for speciation that explains species (at least, of animals) and human geographical differentiation [Wilkins and Nelson 2008]. By failing to read *sol* (soil) as a general term for habitat rather than, as Engels did, a term for "geological formations," Gould unfairly dismissed Trémaux.

[105] Trémaux 1874.
[106] Wallace 1858; quoted in Kottler 1978, 294.
[107] Kottler 1978, 295.

can be differentiated from allied assemblages; which reproduce their like; which usually breed together; and, perhaps, when crossed with their near allies, always produce offspring which are more or less sterile *inter se*.[108]

Earlier, in his *Contributions to the Theory of Natural Selection*,[109] while discussing the numbers of species in the Malayan Archipelago of Papilionidae (a group of butterflies) he discussed the principle by which he should give them the specific rank. He noted,

One of the best and most orthodox definitions is that of Pritchard [sic],[110] the great ethnologist, who says, that "separate origin and distinctness of race, evinced by a constant transmission of some characteristic peculiarity of organization" constitutes a species. Now leaving out the question of "origin" which we cannot determine, and taking only the proof of separate origin, "the constant transmission of some characteristic peculiarity of organization," we have a definition which will compel us to neglect altogether the amount of difference between any two forms, and to consider only whether the differences that present themselves are permanent. The rule, therefore, I have endeavoured to adopt is, that when the difference between two forms inhabiting separate areas seems quite constant, when it can be defined in words, and when it is not confined to a single peculiarity only, I have considered such forms to be species.

This is a generative conception again. He then discusses variation, polymorphisms, local varieties, co-existing varieties, and sub-species, and says,

Species are merely those strongly marked races or local forms which when in contact do not intermix, and when inhabiting distinct areas are generally believed to have had a separate origin, and to be incapable of producing a fertile hybrid offspring. ... it will be evident that we have no means whatever of distinguishing so-called "true species" from the several modes of variation here pointed out, and into which they so often pass by an insensible gradation.[111]

Wallace is taking Darwin's approach to its natural conclusion, as it were. If varieties are incipient species, and species merely well-marked and more or less permanent varieties, then there is nothing *sui generis* about *species* as a categorical rank. In this he was followed also by August Weismann (1834–1914),[112] who treated species entirely as complexes of adaptations. Discussing a number of examples of series of forms that can only be arbitrarily delimited, Weismann notes,

All the individual members of these series are connected by intermediate forms in such a manner that a long period of constancy of forms seems to be succeeded by a shorter period of transformation, from which again a relatively constant form arises.

[108] Wallace 1889, 167.
[109] Wallace 1870, 142.
[110] Wallace is referring to James Prichard [1813, Book II, chapter 1; 3rd edition quoted, 1836]. See above.
[111] Wallace 1870, 161.
[112] Weismann 1904.

We see, therefore, that the idea of species is fully justified in a certain sense; we find indeed at certain times a breaking up of the fixed specific type, the species becomes variable, but soon the medley of forms clears up again, and a new constant form arises— *a new species*, which remains the same for a long series of generations, until ultimately it too begins to waver, and is transformed once more. But if we were to place side by side the cross-sections of this genealogical tree at different levels, we should only see several well-defined species between which no intermediate forms could be recognized; these would only be found in the intermediate strata.[113]

... the species is essentially a complex of adaptations, of modern adaptations which have been recently acquired, and of inherited adaptations handed down from long ago— a complex which might well have been other than it is, and indeed must have been different if it had originated under the influence of other conditions of life.[114]

In contradiction to Karl von Nägeli (Swiss, 1817–1891), who thinks of species as "a vital crystallization" (in Weismann's words[115]), Weismann denies that there is an evolutionary force that impels species to evolve, and defends natural selection as the entirety of evolutionary mechanism. Since the major features of evolutionary groups are adaptive in their origin, Weismann says[116]

... if the step from one species to the next succeeding one does not depend on adaptation, then the greater steps to genera, families, and orders cannot be referred to it either, since these can only be thought of as depending upon a long-continued splitting up of species.

Weismann appears to think that entire species transmute and are changed into new species after a period of fragmentation of forms. However, he realizes that species are variable:

But of course species are not exclusively complicated systems of adaptations, for they are at the same time 'variation complexes,' the individual components of which are not all adaptive, since they do not all reach the limits of the useful or the injurious.[117]

He recognizes that selection applies to sub-organismic "vital units," and that there are "indifferent characters," or non-adaptive characters, which result as by-products, as it were, of selection for "a harmonious whole." Selection in the "germ plasm" may

... give rise to correlative variations in determinants next to them or related to them in any way, and that these may possess the same stability as the primary variation. This seems to me sufficient reason why biologically unimportant characters may become constant characters of the species.[118]

---

[113] *Op. cit.*, vol. II, 305.

[114] *Op. cit.*, vol. II, 307.

[115] Mazumdar 1995 gives details of Nägeli and Schleiden's views on species. Nägeli's [1898] account of species relied upon there being a mechanical "idioplasm" or reproductive plasm.

[116] *Op. cit.*, vol. II, 306.

[117] *Op. cit.*, vol. II, 307.

[118] *Op. cit.*, vol. II, 308.

An example of this is the vestigial hind limbs in the Greenland whale.[119] So Weismann was not exactly the panadaptationist he is sometimes made out to be,[120] and he also allowed for the existence of neutral characters. However, he rejected outright the views of those who thought that isolation was a precondition to new species and that the characters that formed them were in any way neutral. In discussing variation, he notes that

> ... there are very variable species and very constant species, and it is obvious that colonies which are founded by a very variable species can hardly ever remain exactly identical with the ancestral species; and that several of them will turn out differently, even granting that the conditions of life be exactly the same, for no colony will contain all the variants of the species in the same proportion, but at most only a few of them, and the result of mingling these must ultimately result in the development of a somewhat different form in each colonial area.[121]

This is in some ways an early forerunner of the "founder effect" conception of the origin of new species proposed by Mayr and developed further by Hampton Carson.[122] What is most striking about this is that Weismann is effectively ascribing speciation in this case to stochastic sampling. This is something that, as the strict selectionist Romanes held him to be, he should not have adopted. Weismann opposed, though, an exclusivist position such as that of Wagner's that *all* species had to be formed in this way.

BIBLIOGRAPHY

Amundson, Ron. 2005. *The Changing Role of the Embryo in Evolutionary Biology: Structure And Synthesis, Cambridge Studies in Philosophy and Biology*. New York: Cambridge University Press.

Barlow, Nora, ed. 1967. *Darwin and Henslow: The Growth of an Idea; Letters, 1831–1860*. London: Murray for Bentham-Moxon Trust.

Beatty, John. 1985. Speaking of species: Darwin's strategy. In *The Darwinian Heritage*, edited by David Kohn. Princeton PA: Princeton, University Press.

Bernardi, G. et al. 2002. Species boundaries, populations and colour morphs in the coral reef three-spot damselfish (Dascyllus trimaculatus) species complex. *Proceedings of the National Academy of Sciences of the United States of America* 269 (1491):599–605.

Burkhardt, Frederick, ed. 1996. *Charles Darwin's Letters: A Selection, 1825–1859*. Cambridge/New York: University of Cambridge.

Burma, Benjamin H., and Ernst Mayr. 1949. The species concept: A discussion. *Evolution* 3 (4):369–373.

Campbell, George J. D., Duke of Argyll. 1884. *The reign of law*. 18th ed. London: Alexander Strahan. Original edition, 1866.

Carlon, D. B., and A. F. Budd. 2002. Incipient speciation across a depth gradient in a scleractinian coral? *Evolution; International Journal of Organic Evolution* 56 (11):2227–2242.

---

[119] *Op. cit.*, vol. II, 313.
[120] For example, by Gould 2002, 198ff.
[121] Weismann 1904, 286.
[122] Mayr 1954, Carson 1957, 1971, Coyne 1994.

Caro-Quintero, Alejandro, and Konstantinos T. Konstantinidis. 2012. Bacterial species may exist, metagenomics reveal. *Environmental Microbiology* 14 (2):347–355.

Carson, Hampton L. 1957. The species as a field for gene recombination. In *The Species Problem: A Symposium Presented at the Atlanta Meeting of the American Association for the Advancement of Science, December 28–29, 1955*, Publication No 50, edited by Ernst Mayr, 23–38. Washington, DC: American Association for the Advancement of Science.

—. 1971. Speciation and the founder principle. *Stadler Genetics Symposium* 3:51–70.

Coyne, Jerry A. 1994. Ernst Mayr and the origin of the species. *Evolution* 48 (1):19–30.

Coyne, Jerry A., and H. Allen Orr. 2004. *Speciation*. Sunderland, MA: Sinauer Associates.

Darwin, Charles Robert. 1839. *Journal of researches into the geology and natural history of the various countries visited by H.M.S. Beagle under the command of Captain Fitzroy, R.N. from 1832 to 1836*. London: Henry Colburn.

—. 1842. *The structure and distribution of coral reefs. Being the first part of the Geology of the voyage of the Beagle, under the command of Capt. Fitzroy, R.N., during the years 1832 to 1836*. London: Smith Elder and Co.

—. 1845. *Journal of researches into the natural history and geology of the countries visited during the voyage of H.M.S. Beagle round the world: Under the command of Capt. Fitz Roy*. 2nd ed. London: John Murray.

Darwin, Charles Robert. 1851. *A monograph on the sub-class Cirripedia: with figures of all the species*. London: Printed for the Ray Society.

—. 1859. *On the origin of species by means of natural selection, Or The preservation of favoured races in the struggle for life*. London: John Murray.

—. 1866. *On the origin of species by means of natural selection: or the preservation of favoured races in the struggle for life*. 4th ed. London: John Murray.

—. 1868. *The variation of animals and plants under domestication*. 2 vols. London: John Murray.

—. 1871. *The descent of man and selection in relation to sex*. London: John Murray.

—. 1872. *The origin of species by means of natural selection, or, The preservation of favoured races in the struggle for life*. 6th ed. London: John Murray.

—. 1873. Origin of certain instincts. *Nature* vii:417.

—. 1875. *The variation of animals and plants under domestication*. 2nd revised ed. London: John Murray. Original edition, 1868.

—. 1998. *The variation of animals and plants under domestication*. 2nd ed. 2 vols. Baltimore: Johns Hopkins University Press. Original edition, 1888 Appleton American edition.

Darwin, Francis, ed. 1888. *The life and letters of Charles Darwin: Including an autobiographical chapter*. 3 vols. London: John Murray.

Desmond, Adrian, and James Moore. 1991. *Darwin*. Harmondsworth, UK: Penguin.

Eigen, Manfred. 1993. The origin of genetic information: Viruses as models. *Gene* 135 (1–2):37–47.

Ereshefsky, Marc. 1999. Species and the Linnean hierarchy. In *Species, New Interdisciplinary Essays*, edited by R. A. Wilson, 285–305. Cambridge, MA: Bradford/MIT Press.

—. 2010. Microbiology and the species problem. *Biology and Philosophy* 25 (4):553–568.

Futuyma, Douglas J. 1983. *Science on Trial: The Case for Evolution*. New York: Pantheon.

Gayon, Jean. 1996. The individuality of the species: A Darwinian theory?—From Buffon to Ghiselin, and back to Darwin. *Biology and Philosophy* 11 (2):215–244.

Ghiselin, Michael T. 1984. *The Triumph of the Darwinian Method, with a New Preface*. rev. ed. Chicago: University of Chicago Press. Original edition, 1969.

Gould, Stephen Jay. 1999. A Darwinian gentleman at Marx's funeral. *Natural History* 108 (7):32–41.

—. 2002. *The Structure of Evolutionary Theory*. Cambridge, MA: The Belknap Press of Harvard University Press.

Gulick, John Thomas. 1873. On diversity of evolution under one set of external conditions. *Zoological Journal of the Linnean Society* 11 (56):496–505.

—. 1888. Divergent evolution through cumulative segregation. *Zoological Journal of the Linnean Society* 20 (120):189–274.

—. 1890. Intensive segregation, or divergence through independent transformation. *Zoological Journal of the Linnean Society* 23 (145):312–380.

—. 1908. Isolation and selection in the evolution of species. The need of clear definitions. *The American Naturalist* 42 (493):48–57.

Hennig, Willi. 1966. *Phylogenetic Systematics*. Translated by D. Dwight Davis and Rainer Zangerl. Urbana, IL: University of Illinois Press.

Kottler, Malcolm Jay. 1978. Charles Darwin's biological species concept and theory of geographic speciation: The transmutation notebooks. *Annals of Science* 35 (3):275–297.

Levit, Georgy S., and Kay Meister. 2006. The history of essentialism vs. Ernst Mayr's "Essentialism Story": A case study of German idealistic morphology. *Theory in Biosciences* 124:281–307.

Lurie, Edward. 1960. *Louis Agassiz: A Life in Science*. Baltimore and London: Johns Hopkins University Press. Reprint, 1988.

Mallet, James. 2007. Species, concepts of. In *Encyclopedia of Biodiversity*, edited by Simon A. Levin, 679–691. New York: Academic Press.

—. 2008. Mayr's view of Darwin: Was Darwin wrong about speciation? *Biological Journal of the Linnean Society* 95 (1):3–16.

—. 2010. Why was Darwin's view of species rejected by twentieth century biologists? *Biology and Philosophy* 25 (4):497–527.

—. 2013. Darwin and species. In *The Cambridge Encyclopedia of Darwin and Evolutionary Thought*, edited by Michael Ruse, 109–115. Cambridge, UK: Cambridge University Press.

Mayr, Ernst. 1942. *Systematics and the Origin of Species from the Viewpoint of a Zoologist*. New York: Columbia University Press.

—. 1954. Change of genetic environment and evolution. In *Evolution as a Process*, edited by Julian Huxley et al., 157–180. London: Allen and Unwin.

—. 1957. Species concepts and definitions. In *The Species Problem: A Symposium Presented at the Atlanta Meeting of the American Association for the Advancement of Science, December 28–29, 1955*, Publication No 50, edited by Ernst Mayr, 1–22. Washington, DC: American Association for the Advancement of Science.

—. 1959. Darwin and the evolutionary theory in biology. In *Evolution and Anthropology: A Centennial Appraisal*, edited by B. J. Meggers, 1–10. Washington, DC: Anthropological Society of Washington.

—. 1963. *Animal Species and Evolution*. Cambridge, MA: The Belknap Press of Harvard University Press.

—. 1982. *The Growth of Biological Thought: Diversity, Evolution, and Inheritance*. Cambridge, MA: The Belknap Press of Harvard University Press.

—. 1991. *One Long Argument: Charles Darwin and the Genesis of Modern Evolutionary Thought*. Cambridge, MA: Harvard University Press.

—. 1994. Ordering systems. *Science* 266 (5186):715–716.

Mayr, Ernst et al. 1953. *Methods and Principles of Systematic Zoology*. New York: McGraw-Hill.

Mazumdar, Pauline M. H. 1995. *Species and Specificity: An Interpretation of the History of Immunology*. Cambridge/New York: Cambridge University Press.

McOuat, Gordon R. 1996. Species, rules and meaning: The politics of language and the ends of definitions in 19th century natural history. *Studies in History and Philosophy of Science Part A* 27 (4):473–519.

Nägeli, Karl Wilhelm von. 1898. *A Mechanico-Physiological Theory of Organic Evolution.* Chicago: Open Court.

Padian, Kevin. 1999. Charles Darwin's view of classification in theory and practice. *Systematic Biology* 48 (2):352–364.

Paul, Diane. 1981. "In the interests of civilization": Marxist views of race and culture in the nineteenth century. *Journal of the History of Ideas* 42 (1):115–138.

Pennisi, Elizabeth. 2002. Ecology. A coral by any other name. *Science* 296 (5575):1949–1950.

Poulton, Edward Bagnall. 1903. What is a species? *Proceedings of the Entomological Society of London* lxxvii–cxvi.

—. 1908. *Essays on Evolution.* Oxford: Clarendon.

Prichard, James Cowles. 1813. *Researches into the Physical History of Man.* London: Houlston & Stoneman.

—. 1836. *Researches into the Physical History of Mankind.* 3rd ed. 5 vols. Vol. 1. London: Sherwood, Gilbert, and Piper.

Sober, Elliott. 1980. Evolution, population thinking, and essentialism. *Philosophy of Science* 47 (3):350–383.

Soong, K., and J. C. Lang. 1992. Reproductive integration in reef corals. *Biological Bulletin* 183 (3):418–431.

Stamos, David N. 1996. Was Darwin really a species nominalist? *Journal of the History of Biology* 29 (1):127–144.

—. 1999. Darwin's species category realism. *History and Philosophy of the Life Sciences* 21 (2):137–186.

—. 2003. *The Species Problem: Biological Species, Ontology, and the Metaphysics of Biology.* Lanham, MD: Lexington Books.

Trémaux, Pierre. 1865. *Origin et transformations de l'homme et des autres êtres.* Paris: L. Hachette.

—. 1874. *Origine des Espèces et de l'Homme, avec les causes de Fixité et de Transformation, et Principe Universel du Mouvement et de la Vie ou loi des Transmissions de Force.* 4th 1878 ed. Bale, Switzerland/Paris: Librairie Française/Sagnier.

Veron, John Edward Norwood. 2001. The state of coral reef science. *Science* 293 (5537):1996–1997.

Wagner, Moritz. 1868. *Die Darwin'sche Theorie und das Migrationsgesetz der Organismen.* Leipzig: Duncker & Humbolt.

—. 1873. *The Darwinian Theory and the Law of the Migration of Organisms.* Translated by James L. Laird. London: Edward Stanford.

—. 1889. *Die Entstehung der Arten durch raumliche Sonderung.* Basel: Benno Schwalbe.

Wallace, Alfred Russel. 1858. Note on the theory of permanent and geographical varieties. *The Zoologist* 16:5887–5888.

—. 1870. *Contributions to the Theory of Natural Selection: A Series of Essays.* London: Macmillan.

—. 1889. *Darwinism: An Exposition of the Theory of Natural Selection, with Some of Its Applications.* London: Macmillan.

Weismann, August. 1904. *The Evolution Theory.* Translated by J. Arthur Thompson and Margaret R. Thompson. 2 vols. London: Edward Arnold.

Wilkins, John S., and Gareth J. Nelson. 2008. Trémaux on species: A theory of allopatric speciation (and punctuated equilibrium) before Wagner. *History and Philosophy of the Life Sciences* 30 (2):179–206.

Williams, George C. 1966. *Adaptation And Natural Selection: A Critique of Some Current Evolutionary Thought.* Princeton, NJ: Princeton University Press.

Willis, Bette L. et al. 2006. The role of hybridization in the evolution of reef corals. *Annual Review of Ecology, Evolution, and Systematics* 37 (1):489–517.

6 The Species Problem Arises

OTHER DARWINIANS: LANKESTER, ROMANES, HUXLEY, POULTON, KARL JORDAN

At least one of Darwin's most prominent followers, E. Ray Lankester, exceeded Darwin's published suggestion that the term "species" was arbitrary. Poulton said that Lankester was

> ... inclined to think that we should discard the word species not merely momentarily but altogether.[1]

Ernst Haeckel concurred, stating in his *The Evolution of Man* that

> Endless disputes arose among the "pure systematizers" on the empty question, whether the form called a species was "a good or bad species, a species or a variety, a sub-species or a group," without the question being even put as to what these terms really contained and comprised. If they had earnestly endeavoured to gain a clear conception of the terms, they would long ago have perceived that they have no absolute meaning, but are merely stages in the classification, or systematic categories, and of relative importance only.[2]

T. H. Huxley, who had disagreed in correspondence with Darwin over saltative evolution, likewise wrote of species that

> [a]nimals and plants are divided into groups, which become gradually smaller, beginning with a KINGDOM, which is divided into SUB-KINGDOMS; then come the smaller divisions called PROVINCES; and so on from a PROVINCE to a CLASS, from a CLASS to an ORDER, from ORDERS to FAMILIES, and from these to GENERA, until at length we come to the smallest groups of animals which can be defined one from the other by constant characters, which are not sexual; and these are what naturalists call SPECIES in practice, whatever they may do in theory.

> If in a state of nature you find any two groups of living beings, which are separated one from the other by some constantly-recurring characteristic, I don't care how slight or trivial, so long as it is defined and constant, and does not depend on sexual

---

[1] Poulton 1908, 62. Lankester writes of Agassiz's objectivity of species, that he wrote at the moment Darwin published the Origin

> ... by which universal opinion has been brought to the position that species, as well as genera, orders and classes, are the subjective expressions of a vast ramifying pedigree in which the only objective existence are individuals ... [Lankester 1890, 319]

[2] 1874, third edition cited [Haeckel 1896, vol. 1, 115].

peculiarities, then all naturalists agree in calling them two species; that is what is meant by the word species—that is to say, it is, for the practical naturalist, a mere question of structural differences.[1]

. . .

[1] I lay stress here on the *practical* signification of "Species." Whether a physiological test between species exist or not, it is hardly ever applicable by the practical naturalist.[3]

He repeats the requirement for the use of non-sexual characters in a later essay.[4] Huxley treats species in this limited taxonomic context[5] as merely conventional definitional entities; that is, as a nominal essence. The note implies that he expects there may be some real essence in the form of physiological (i.e., reproductive?) differences, but that they are not useable by practicing taxonomists. So long as there is a constant unvarying character and a smallest diagnosable group, there is a species. In his 1859 review of the *Origin*, when discussing Darwin's views on sterility of hybrids, Huxley asks what is known of the "essential properties of species." He summarizes:

Living beings, whether animals or plants, are divisible into multitudes of distinctly defineable kinds, which are morphological species. They are also divisible into groups of individuals, which breed freely together, tending to reproduce their like, and are physiological species.[6]

However, Huxley, following and in concert with Darwin, denied that reproductive isolation was sufficient to make a species, and noted that the degree of infertility between species varies from absolute to minor. Elsewhere, in his 1863 critical review of the *Origin*,[7] he distinguishes between internal ("physiological") and external ("anatomical" or "morphological") characterizations of species,[8] and in a paper "Darwin on The Origin of Species" published in the *Westminster Review* in 1860 he notes that naturalists employ the term *species* in a double sense to denote "two very different orders of relations":

When we call a group of animals, or of plants, a species, we may imply thereby either, that all these animals and plants have some common peculiarity of form or structure; or, we may mean that they possess some common functional character. That part of biological science which deals with form and structure is called Morphology—that which concerns itself with function, Physiology—so that we may conveniently speak of these two senses or aspects of "species"—the one as morphological, the other as physiological.[9]

---

[3] Huxley 1906, 226f.
[4] Huxley 1906, 301ff.
[5] This series of lectures was collected in 1864 and published in Huxley 1895, but the essay was first delivered in 1863.
[6] Huxley 1893, 50.
[7] Reprinted in Huxley 1906 as chapter IX.
[8] Forsdyke 2001, 31.
[9] Huxley 1906, 302.

George Romanes, who coined the term "neo-Darwinism" to sneer at the extreme selectionist views of Wallace and Weismann,[10] in his *Darwin and after Darwin* gave five definitions of species:[11]

1. "A group of individuals descended by way of natural generation from an original and specially created type."
   He calls this "virtually obsolete."
2. "A group of individuals which, while fully fertile inter se, are sterile with all other individuals—or, at any rate, do not generate fully fertile hybrids."
   Romanes calls this the "physiological definition" and claims that it is not entertained by any naturalist at that time, as it is incomplete.
3. "A group of individuals which, however many characters they share with other individuals, agree in presenting one or more characters of a peculiar kind, with some certain degree of distinctness."
   Romanes claims this is practically followed by all naturalists. But it is insufficient to enable a uniform standard of specific distinction, and so he adduces two more definitions, "which will yield to evolutionists the steady and uniform criterion required."
4. "A group of individuals which, however many characters they share with other individuals, agree in presenting one or more characters of a peculiar and hereditary kind, with some certain degree of distinctness."
   This merely adds hereditary characters to the previous definition.
   And finally, he adds one for the "ultra-Darwinians" who insist that species are formed through natural selection.
5. "A group of individuals which, however many characters they share with other individuals, agree in presenting one or more characters of a peculiar, hereditary, and adaptive kind, with some certain degree of distinctness."

These are the "logically possible" definitions of species, meaning by this that they include all the differentiae of the thing defined. Romanes presents something rather similar to some versions of the modern phylogenetic species concepts, but the "degree of distinctness" is still relatively vague, and relies on the judgment of the specialists. He rejects absolutely the idea that the characters that mark species from each other must or even can be adaptive, and a discussion he instigated in the Biological Section of the British Association ended in

> … as complete a destruction as was possible of the doctrine that all the distinctive characters of every species must necessarily be useful, vestigial or correlated. For it became unquestionable that the same generalization admitted of being made, with the same degree of effect, touching all the distinctive characters of every "snark."[12]

---

[10] Although Gould 2002, 198 ascribes this to Spencer in 1893, Romanes claimed the honor in his 1895 as an earlier coinage of his own.

[11] Romanes 1895, vol. II, 229–234.

[12] Romanes 1895, 235.

Lewis Carroll must have been pleased. Romanes then takes Wallace to task for his definition, and Weismann for his rejection of the inheritance of acquired characters,[13] which Romanes as a "proper" Darwinian had followed Darwin in defending; in fact this topic was the occasion for the "more Darwinian than Darwin" comment when he coined the terms "neo-Darwinian" and "ultra-Darwinism." He distinguished between species formed through non-hereditary influence of the environment as *somatogenetic* species, and those in which the environment makes changes to heritable material—the germ plasm in Weismann's terms—as *blastogenetic species*. Neither term survived the rejection of neo-Lamarckism with the rise of Mendelian genetics at the turn of the century.[14]

British entomologist Edward Bagnall Poulton wrote an extensive and influential essay on species in which he coined the term *syngamy*.[15] He noted that the fixity of species was not required by Augustine or Aquinas, and cites Aubrey Moore who fingers Milton as the guilty party for this doctrine, although he thinks it was due to the spirit of the age (which, as we have seen, is pretty well right; the received view of the Middle Ages was non-historical).[16] He also cites Sir William Thistleton-Dyer as claiming that fixity was traceable to Caspar Bauhin and Joachim Jung, and discusses Darwin's own views in some detail.[17] He coined several terms to set up the discussion about species, some of which have entered into common usage.[18] First, he defines groups formed by Linnaean diagnosis of forms as *Syndiagnostic*. Then, he names groups that freely interbreed as *Syngamic*, and in a privative fashion terms those that do not as *Asyngamic* (with the substantive noun forms *Syngamy* and *Asyngamy*). Next, he coins the term *Epigony* to mean breeding from a common parent. Finally, he provides a term for the organic forms that live in the same region: *Sympatric*. Again, he uses a privative term—*Asympatric*—for those that do not. *Allopatry* had to wait for Mayr.

With the technical apparatus in hand, Poulton moves from diagnosis to the underlying reality of species. He says:

> Diagnosis ... is founded upon the conception that there is an unbroken transition in characters of the component individuals of a species. Underlying this idea are the more fundamental conceptions of species as groups of individuals related by Syngamy and Epigony.

He argues, contrary to Darwin's views, that sterility is an effect of *Asyngamy*, not vice versa; reiterating Max Muller's and Moritz Wagner's views on *speciation*, as the topic came to be known later. Thistleton-Dyer had said that older writers employed

> ... the word species as a designation for the totality of all individuals differing from all others by marks or characters which experience showed to be reasonably constant or trustworthy, as is the practice of modern naturalists.[19]

[13] Butler 1884, 241.
[14] Bowler 2003.
[15] Poulton 1903.
[16] Moore 1889, 179f. Moore's short passage is testament to his scholarship; he also rightly identifies Ray as the originator of both a natural definition of *species* and also of fixity of species.
[17] Thistleton-Dyer 1902, 370. Poulton is mistaken. Thistleton-Dyer had ascribed binomials to Bauhin and Jung, not fixity of species, which he also ascribes to Ray.
[18] Poulton 1908, 60–62.
[19] Thistleton-Dyer 1902, 370.

He (inaccurately, see above) cites Darwin a week after the publication of the *Origin*:

> … I met Phillips, the palaeontologist, and he asked me, "How do you define a species?" I answered, "I cannot." Whereupon he said, "At last I have found out the only true definition,—any form which has ever had a species name!"[20]

Similar views were held in the 1920s by British ichthyologist C. Tate Regan,[21] and were named by Blackwelder the "taxonomic species concept," and by Kitcher the "cynical species concept."[22]

According to Poulton, though, species are interbreeding communities, syngamic communities. This came later to be termed by Johannes Lotsy; briefly, a *Syngameon* as a neutral term for such communities, irrespective of their rank,[23] although as Dobzhansky noted, Lotsy seemed to equate the syngameon with "species" anyway.[24] It underlay the gradual variation in forms from one end of a range of organisms to another, which he referred to as a transition. But while when clear this made diagnosis possible, there were cases in which it would fail: polymorphisms, seasonal dimorphisms, individual developmental adaptation (he calls it *individual modification*, and cites Baldwin), geographical races and sub-species, and artificial selection. Moreover, says Poulton, interspecific sterility is not an infallible test of specificity, because, as Darwin knew, related forms often can interbreed. Instead, sterility of hybrids is "an incidental consequence of asyngamy," of separation for a long period.[25] In addition, asyngamy itself is usually the byproduct of asympatry,[26] although he allowed that Karl Jordan was correct when he said that it could be due to *mechanical incompatibility*, or the lack of fit between sexual organs (in the 1896 report using the bird genus *Papilio* as the test case).[27] Poulton also accepts Henry Bates' 1862 claim of preferential mating.[28] All this notwithstanding, sterility is not, in his opinion, due to the action of selection. This seminal paper, republished in 1905, was greatly influential in setting up the terms, both literally and metaphorically, of the twentieth century debate over species and speciation. It subsequently inspired the writing of a text that summarized the issues as understood at the end of the period in which neo-Lamarckian mechanisms were still viable hypotheses, shortly before Dobzhansky's paper and book.[29] We may usefully date the *Species Problem* from Poulton's paper.

---

[20] Poulton cites *More Letters of Charles Darwin* 1903 [Darwin 1972, vol. I, 127]. As we see above, this anecdote occurs in Darwin's correspondence to Gray in 1857.

[21] Regan 1926.

[22] Blackwelder 1967, Kitcher 1984.

[23] Lotsy 1931.

[24] Dobzhansky 1941, 311.

[25] Poulton 1908, 80.

[26] Poulton 1908, 84.

[27] Jordan 1896; cf. later, Jordan 1905.

[28] Bates 1862, 501:

> The process of the creation of a new species I believe to be accelerated in the Ithomiae and allied genera by the strong tendency of the insects, when pairing, to select none but their exact counterparts: this also enables a number of very closely allied ones to exist together, or the representative forms to live side by side on the confines of their areas, without amalgamating.

[29] Robson 1928.

Karl Jordan's papers[30] seem to have had some impact on the way people discussed variation within species, and it contains the core of the biospecies concept later propounded by Mayr, who mentions Jordan as a "forerunner,"[31] and quotes Jordan's statement

> Individuals connected by blood relationship form a single faunistic unit in an area ... the units, of which the fauna of an area is composed, are separated from each other by gaps which at this point are not bridged by anything.[32]

Jordan gave the criteria that he believed defined two organisms as being in the same or different species in the context of a discussion of the variation of physical traits in organisms that we use to establish the "blood relationships of individuals and thus the physical gaps between species."[33] He said that individuals of two species announce themselves by having physiological differences that they always reproduce in their progeny, and by being able to live in the same region without blending into each other. He listed three criteria:

> The criteria of the *species* [*he uses the Latin word "species" here—JSW*] (= *Art*) concept are thus threefold, and each individual point is an applicable test: a species requires known traits, it does not beget descendants equally well with individuals of other species, and it does not coalesce into other species.[34]

Jordan's view seems to require not only geographical isolation, but also constancy of characters, which Mayr's later definition does not. He goes on to say that the non-coalescence (non-hybridization) of species explains the "enormous number" of extant species, and that this is due to the internal organization of the species (by which I take him to mean of the typical genetic structure of the species). He makes the comment that species act as if there were "*no* relationship between them, but as if they were doing business for themselves."[35]

---

[30] Especially Jordan 1905.

[31] Mayr 1982, 285.

[32] Jordan 1905, 157:

> *Solche blutsverwandte Individuen bilden eine faunistische Einheit in einem Gebeit ...*

which Jordan follows with the elided comment

> *zu welcher Einheit wir erführungsgemäß alle andern Individuen des Gebeits rechnen müssen, welche ihnen gleichen*

meaning roughly that we assign all individuals to these units when we empirically classify them as identical. Again, there is an epistemological element here that is overlooked. The subsequent statement Mayr quotes comes after a number of examples of this classification.

[33] Jordan 1905, 159.

[34] Thanks to Ian Musgrave for help with the translation:

> Das Kriterium des Begriffs Species (=Art) ist daher ein dreifaches, und jeder einzelne Punkt ist der Prüfung zugänglich: Eine Art bat gewisse Körpermerkmale, erzeugt keine den Individuen anderer Arten gleich Nachkommen und verschmilzt nicht mit andern Arten.

[35] *Op. cit.*, 160: "... als ob *nie* ein Zusammenhang zwischen ihnen gewesen, als ob jede Art für sich geschaffen ware." Italics represent emphatic spacing.

NON-DARWINIAN IDEAS AFTER DARWIN

The variable ensemble of Darwinian ideas was not universally adopted.[36] During the so-called "eclipse of Darwinism" period, in which neo-Lamarckian ideas overtook Darwin's mechanism of natural selection, species were often thought to be types again. Such American neo-Lamarckians as Cope and Hyatt adopted an "orthogenetic Lamarckism" in which species underwent a series of developmental changes in a kind of embryological analogy between individuals and species.[37] In the period from Edward Drinker Cope's 1868 essay "On the origin of genera"[38] through to the period immediately before World War I, species were thought by this school to be the result of internal forces rather than selection or geographical isolation. Cope called this growth force "bathmism," while others, such as Alpheus Hyatt and Alpheus Packard, and Henry Fairfield Osborn, had additional mechanisms. Osborn, in particular, adopted the Baldwin Effect as (he thought) a non-Darwinian mechanism which enabled organisms to direct their own evolution through individual adaptation,[39] which marked the Baldwin Effect as anti-Darwinian, or Lamarckian, for a long time to come.[40] Osborn, under attack from Baldwin for saying that variation is non-random, then made the claim that there were linear variational trends for which Darwinism could not account.[41]

William Bateson produced a lengthy book describing the sorts of variations that occurred from the type, in which he treated species as morphological classes, with no continuity of form between species, and hence no reason to think they varied sufficiently for Darwinian evolution to occur. Species are groups of organisms united by a common form, but there is no definite difference that divides them. He writes of what he calls "the Problem of Species":

> No definition of a Specific Difference has been found, perhaps because these Differences are indefinite and hence not capable of definition. But the forms of living things, taken at a moment, do nevertheless most certainly form a discontinuous series and not a continuous series. ...
> The existence, then, of Specific Differences is one of the characteristics of the forms of living things. This is no merely subjective conception, but an objective, tangible fact. This is the first part of the problem.
> In the next place, not only do Specific forms exist in Nature, but they exist in such a way as to fit the place in Nature in which they are placed; that is to say, the Specific form which an organism has, is *adapted* to the position which it fills. This again is a relative truth, for the adaptation is not absolute.[42]

For Bateson, though, form is more than a way of describing or diagnosing species, it *is* what species are—they are classes of forms. Adaptation is not relevant to

---

[36] Hull 1973.
[37] Bowler 1983, 121ff.
[38] Cope 1868.
[39] Bowler 1983, 131.
[40] Turney et al. 1996.
[41] Bowler 1983, 132f.
[42] Bateson 1894, 2.

the origin of these discontinuous forms, but variation is. Bateson stops short of saying that form is a causal factor in evolution, but he does say that symmetry of form and the repetition of forms (*merism* is his term for this) are causes of speciation.[43] Indeed, variation from the type (the Specific Differences) is due to a "pathological accident."[44] Later, as one of the Mendelian geneticists, Bateson opposed the idea that there was genetic variation of the kind Darwinian selection required to form species. In 1913, he wrote:

> All constructive theories of evolution have been built upon the understanding that we know if the relation of varieties to species justifies the assumption that the one phenomenon is a *phase* of the other, and that each species arises ... from another species either by one, or several, genetic steps. ... [However,] complete fertility of the results of intercrossing [between members of different "species"] is, and I think must rightly be regarded, as *inconsistent* with actual specific difference.[45]

Another Mendelian, Hugo de Vries, shortly after proposed a concept of "elementary species" as pure genetic lines in his 1904 lectures "Species and Varieties"[46] and in the earlier *Die Mutationstheorie*.[47] According to the Mendelian view that de Vries adopted, species in the Linnaean sense actually comprised a number of smaller lines of pure genetic stock:

> Species is a word, which has always had a double meaning. One is the systematic species, which is the unit of our system. But these units are by no means indivisible. ... These minor entities are called varieties in systematic works. ... Some of these varieties are in reality just as good as species, and have been "elevated," as it is called, by some writers, to this rank. This conception of the elementary species would be quite justifiable, and would get rid of all difficulties, were it not for one practical obstacle. The number of species in all genera would be doubled and tripled, and as these numbers are already cumbersome in many cases, the distinction of the native species of any given country would lose most of its charm and interest.
>
> In order to meet this difficulty we must recognize two sorts of species. The systematic species are the practical units of the systematists and florists, and all friends of wild nature should do their utmost to preserve them as Linnaeus has proposed them. These units, however, are not really existing entities; they have as little claim to be regarded as such as genera and families. The real units are the elementary species; their limits often apparently overlap and can only in rare cases be determined on the sole ground of field-observations. Pedigree-culture is the method required and any form which remains constant and distinct from its allies in the garden is to be considered as an elementary species.[48]

De Vries came to his views through the observation of what we now know to be alloploid forms in *Oenothera lamarckiana*, the evening primrose, which de Vries

---

[43] Bateson 1894, 19ff.
[44] As he approvingly quotes Virchow [Bateson 1894, 95].
[45] Quoted in Forsdyke 2001, 31.
[46] de Vries 1912.
[47] de Vries 1901, de Vries 1911.
[48] de Vries 1912, 11.

had cultivated and maintained pedigrees.[49] He was of the view that these were straight mutations forming a single new elementary species at once. In a way, given the way alloploids occur, he was right, but his idea led fairly directly to the later views of Goldschmidt, who, like de Vries, felt that species *always*, or almost always, arose in sudden saltative leaps.

In the lecture discussing the evening primrose, de Vries further defines the marks of an elementary species:

> Elementary species differ from their nearest allies by progressive changes, that is by the acquisition of some new character. The derivative species has one unit more than the parent.[50]

This meant that if the new elementary species were crossed with its parental elementary species, the progeny would be incomplete for that character (he clearly means Mendelian factor) and would therefore be unnatural.[51] Hence backcrossing would not occur in the wild or under cultivation.[52] Elementary species thus do not exhibit subvarieties, for they are the "real type."[53] De Vries' concept was fundamentally anti-Darwinian, in the sense that he rejected the idea of there being continuous variation within species on which selection could act in such a way as to form new species. In fact, he argued for his theory on the grounds that the length of time required for evolution would be noticeably shorter on his account. At that time, Lord Kelvin's arguments against Darwinian evolution—that reasoning from the rate of cooling of the earth, evolution would need to have happened in tens, not hundreds or thousands of millions of years—were still current. Rayleigh's discovery of radioactivity as a source of planetary heat was not announced until about this time (1906), and it did not immediately filter through to the wider scientific community.[54]

Prior to Bateson's and Poulton's essays, there was no species *problem* as such, but only a species *question*. The latter is concerned primarily with the origins of species, how they come to be. The *problem* arises when we have accounts of species formation, whether by selection or something else, that do not involve immediate saltation from one to another or *creatio de novo*. It is the problem of defining what rank it might be that species achieve when they become species, and this sets the agenda for Dobzhansky's 1935 paper, and the remainder of the twentieth-century debates.

There were a few Darwinians in the period before the Synthesis who discussed the concept of species. For example, in 1934, J. Arthur Thompson defined species extensively in the classical terms but with an emphasis on the role played by selection.[55] Thompson defines it in the context of the human species and races of man, and gives four criteria: non-trivial difference, true breeding and constancy of characters,

---

[49] de Vries 1912, 17, lecture IX.
[50] de Vries 1912, 253.
[51] *Op. cit.*, 254.
[52] *Op. cit.*, 527.
[53] *Op. cit.*, 127.
[54] Bowler 1989, 207.
[55] Thompson 1934, vol. 2, 1333f.

interfertility and production of fertile offspring, and the constancy of specific characters in different environments. He says,

> To sum up: *A species is a group of similar individuals differing from other groups in a number of more or less true-breeding characters, greater than those which often occur within the limits of a family, and not the direct result of environmental or other nurtural influences. The members of a species are fertile with one another, but not readily with other species.*[56]

Races share some of these features, i.e., they breed true, but they are less marked than specific characters, and they may be maintained by natural and sexual selection. However, Darwinian notions of species were the exception rather than the rule for some time into the new century. Of more significance was the influence of Mendelian genetics, and importantly hybridization.

LOTSY AND THE EVOLUTION OF SPECIES BY HYBRIDIZATION

In a work published in 1916 in English, Dutch botanist Johannes Paulus Lotsy (1867–1931) proposed both a definition of species and a conception of evolution. He begins by noting that the concept of species is vague:

> All theories of evolution have, until quite recently, been guided by a *vague* knowledge of what a species is, and consequently have been vague themselves.[57]

Lotsy therefore proposes a definition based on "identity of constitution," and, citing Ray, discusses Alexis Jordan's discovery of variety within all Linnaean species. He says

> Jordan *consequently discarded morphological comparison as a criterium for specific purity* and, falling back to Ray (whom he may or may not have known) *substituted for it: nulla certior.... quam distincta propagatio ex semine.*[58]

From this, he says, Jordan drew the "well founded conclusion": "*The Linnaean species is no species.*" Lotsy therefore proposed a term, the *Linneon*, for the product of Linnaean classification defined as "*the total of individuals which resemble one another more than they do any other individuals*" (italics original). The types contained *within* a Linneon Jordan called species, and Lotsy calls *Jordanons*, since "*[b] eeding true to type is ... by itself no reliable test for specific purity.*"[59] He then gives his own, proper, definition of a species:

> *A species consists of the total of individuals of identical constitution unable to form more than one kind of gametes.*

---

[56] Thompson 1934, vol. 2, 1334, italics original.
[57] Lotsy 1916, 14.
[58] Lotsy 1916, 21f, italics original. The Jordan here is Claude Thomas Alexis Jordan (1814–1897), a French botanist who named every variety as a species.
[59] Lotsy 1916, 23.

Moreover, Lotsy proposes a genetic test:

> *Specific purity is indicated by the uniformity and identity of the F1 genera-*
> *tions obtained by crossing the individuals to be tested,* RECIPROCALLY.[60]

Thus, a species is for Lotsy an operationally applicable concept. The result is three definitions of terms:

> LINNEON: *to replace the term species in the Linnaean sense, and to desig-*
> *nate a group of individuals which resemble one another more than they do*
> *any other individuals.*

To establish a Linneon consequently requires careful morphological comparison only.

> JORDANON: *to replace the term species in the Jordanian sense, viz: mikro-*
> *species [sic], elementary species etc. and to designate a group of externally*
> *alike individuals which all propagate their kind faithfully, under conditions*
> *excluding contamination by crossing with individuals belonging to other*
> *groups, as far as these external characters are concerned, with the only*
> *exception of noninheritable modifications of these characters, caused by the*
> *influences of the surroundings in the widest sense, to which these individu-*
> *als or those composing the progeny may be exposed.*

To establish a Jordanon, morphological comparison alone consequently does not suffice; the transmittability of the characters by which the form was distinguished, must be experimental breeders.

> SPECIES: *to designate a group of individuals of identical constitution, unable*
> *to form more than one kind of gametes; all monogametic individuals of iden-*
> *tical constitution consequently belonging to one species.*[61]

Lotsy rejects the idea of intraspecific variation in Darwin's sense, and is of the view that every homozygotic form is itself a species. It follows that every mixing of these "pure forms" is the origination of a new species if that novelty persists. Moreover, he thinks that species can be polyphyletic—they can arise more than once, because a species is formed in virtue of its "constitution," not in terms of its history.[62] Moreover, "nature primarily can make nothing but individuals,"[63] and it can secondarily group those individuals in various ways. Linneons are not natural, though—they are groupings formed by the human mind. He gives an example using human races supposing that even if there were four "pure races" arranged in army battalions we could divide them up in various ways, according to tattoo marks given from parent to child so that each child has two marks.[64] If there has been no intercrossing, then although the individuals have changed, he says, the constitution would remain the same and we could arrange the progeny into those groups, but if crossings occurred, we could not assign them to the right armies.

---

[60] Lotsy 1916, 24.
[61] Lotsy 1916, 27.
[62] Lotsy 1916, 45.
[63] Lotsy 1916, 46.
[64] Lotsy 1916, 47–49.

In the forefront of his definitions, Lotsy has the Mendelian genetics then being first investigated in detail. He thinks that classification by the genetic constitution forms a kind of Lockean "real essence," which is the point of the race example. If our groupings match nature's real essence (i.e., genetic constitution), then they are natural groups. Otherwise they are not, and given the typological nature of his definition, ordinary species (i.e., what came to be called "biological species") were not natural entities. The remainder of his argument relies on Mendelian assortment forming novel varieties, as we would call them, or "allogamous forms" as he calls them, from mutations.[65]

GÖTE TURESSON ON ECOSPECIES AND AGAMOSPECIES

The Swedish botanist Göte Turesson undertook a series of experiments in the 1920s and early 1930s,[66] in which he transplanted the "ground stock" of widely distributed Swedish plants into various different habitats—dunes, seacliffs, woodlands, high altitudes and so on,[67] and discovered that differing forms arose as a response to climate. He made a number of influential distinctions on taxa which later formed the basis for ecological species concepts. He proposed *ecospecies* "to cover the Linnean species or genotype compounds as they are realized in nature" and related *coenospecies* (the Linnaean taxon) to ecospecies in a diagram (Figure 6.1).[68]

Coenospecies are "the total sum of possible combinations in a genotype compound," and include one or more ecospecies, which include one or more *ecotypes*, the forms that develop in different *ecosystems* or habitat types. These comprise all the "reaction-types" of ecotypes that are elicited by extreme habitats, called *ecophenes*. In a similar inclusive hierarchy, he listed a genetical array of concepts—*genospecies* (the genetical construction of ecospecies), *genotypes* (Johannsen's 1909 term) and the "reaction-types" of genotypes, *genophenes*. This dual hierarchy between the ecological and genealogical was repeated later by Eldredge and Salthe.[69] Turesson's "reaction-types" are in modern terms the reaction norms of genes, although the idea of a reaction norm for an ecotype appears to have been abandoned.

In one paper, he gave succinct definitions of the major terms:

a) *Ecospecies*: An amphimict-population the constituents of which in nature produce vital and fertile descendants with each other giving rise to less vital or more or less sterile descendants in nature, however, when crossed with constituents of any other population. ...

b) *Agamospecies*: An apomict-population the constituents of which, for morphological, cytological or other reasons, are to be considered as having a common origin. ...

[65] Lotsy 1916, 159.
[66] Turesson 1922a, 1922b, 1925, 1927, 1929, 1930.
[67] Turrill 1940, 52.
[68] Turesson 1922a, 344.
[69] See Salthe 1985, Eldredge 1989.

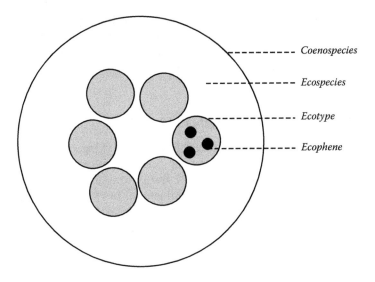

Coenospecies

Ecospecies

Ecotype

Ecophene

FIGURE 6.1   Ecospecies and coenospecies.

c) *Coenospecies*: A population-complex the constituents of which group themselves in nature in species units of lower magnitude on account of vitality and sterility limits having all, however, a common origin so far as morphological, cytological or experimental facts indicate such and origin. ...[70]

Turesson criticizes the Linnaean conception of species for not helping us to determine the *natural* limitations of species. He says that the Linnaean notion covers several senses, defined above, and that the

... existence in nature of units of these different orders also makes it a logical impossibility to reach one standard definition of the "species."

This is, incidentally, the first use of the term "agamospecies" I have encountered, so he may very well have coined it. As far as I can tell, Turesson does not require that agamospecies cannot also be ecospecies or coenospecies.

Mayr accuses Turesson of being typological, and claims that he gives the impression of plant species comprised "a mosaic of ecotypes rather than as an aggregate of variable populations,"[71] but this is not how contributors to the 1940 *New Systematics* volume read him. Turrill notes:

He [Turesson] has clearly shown not only that the species, as usually accepted by the taxonomist, is a complex assemblage of biotypes, but also that the species population varies in its biotype composition with habitat conditions.[72]

---

[70] Turesson 1929, 332–333.
[71] Mayr 1982, 277.
[72] Turrill 1940, 52.

Another contributor also regards Turesson's approach as a populational one.[73] In any event, it is clear that Turesson's conception of species taxa included *both* a "biological" (i.e., a genetic) and an ecological aspect. It may be that Turesson's scheme is itself "typological," but its implementation and later influence is not, in itself, necessarily so.

GERMAN THINKERS: ISOLATION IS THE KEY

Species concepts played an increasing role in the thinking of several German-speaking biologists in the early part of the century. In particular, the views of Erwin Stresemann, curator of birds at the Berlin Museum, and mentor to both Ernst Mayr and Bernhard Rensch, were influential. In 1919, he had written that morphology as a criterion of species had been abandoned by the late 1890s among ornithologists in favor of physiological divergence, as evidenced by reproductive isolation. He said:

> forms of the rank of species have physiologically diverged from each other to such an extent, that they can come together again without mixing with each other. ... Morphological divergence is thus independent of physiological divergence.[74]

Bernhard Rensch was strongly influenced by this approach, and he defined several terms to deal with the reproductive isolation of species and within species in the case of races that do not interbreed although overall the species has a shared gene pool.[75] He called these *Rassenkreise*, or "race circles," and complexes of incipient species that replaced each other geographically, he called *Artenkreise*, or "species circles."[76] Although he doesn't define species in the 1959 work, he does talk frequently about "good species" being those that are isolated by sexual or genetic differences when in contact.

A review of the state of play at the end of the period before the Synthesis began was published by British zoologist Guy C. Robson,[77] and much of the modern debate was prefigured there—polymorphisms, reproductive isolation, allopatry (under another name, of course), and genetic variance are all discussed. At that time, however, it was unclear whether or not the neo-Lamarckian view of the inheritance of acquired characters was a viable view or not. Robson tended to think not, but he allowed that later research may show otherwise. For him, species are not the necessary outcome of evolution, and are recognized by correlations of differentiated characters that "hang together."[78] He notes that sampling of specimens fails to reproduce the wider diversity of characters of the larger assemblage of the taxon. Races are localized more or less homogeneous groups in what is a continuous distribution of forms. He rejects inability or disinclination to hybridize

[73] Salisbury 1940, 332.
[74] Stresemann 1919, 64, 66, quoted in Mayr and Provine 1980, 415.
[75] Rensch 1980, 294f.
[76] Rensch 1928, Rensch 1929, Rensch 1947, Rensch 1959.
[77] Robson 1928.
[78] Robson 1928, 223–224.

as a good test of species, and also the proposal to give up the notion of species. Although there are no absolute criteria for judging species, Robson does think that well-defined groups do occur in nature, and he rejects the "morphological," "genetic," "physiological," and "ecological" criteria as being either incomplete or ill-defined tests of species rank.[79]

THE MENDELIANS: MORGAN AND STURTEVANT

The famous Chicago geneticist Thomas Hunt Morgan was initially of the view that species were nonexistent, arbitrary units devised for the convenience of taxonomists (and hence is a conventionalist species denier at this stage). He held that, with Buffon, only individual organisms existed:

> We should always keep in mind the fact that the individual is the only reality with which we have to deal, and that the arrangement of these into species, genera, families, etc. is only a scheme invented by man for purposes of classification. Thus, there is no such thing in nature as a species, except as a concept of forms more or less alike.[80]

His view of species as types determined by convention never changed thereafter, although he did later indicate that we think of species in terms of their adaptations.

He did allow that there was an enormous amount of variability in genes, though. His students, in particular Alfred H. Sturtevant, had a stronger interest in species, and Sturtevant held that there was a "wild-type" of each species from which variants were insignificant and that species actually differed in few genes.[81] H. J. Muller, however, later adopted in full the neo-Darwinian account of the Modern Synthesis, treating species as formed through hybrid infertility due to genetic variations such as chromosomal rearrangements. Even in his *Drosophila*, the nature of species was plastic and variable, and

> [i]t becomes, then a matter of definition and of convenience, in any given series of cases, just where we decide to draw the line above which two groups will be distinguished as separate species, and below which they are denoted subspecies or races, since in nature there is no abrupt transition here.[82]

Despite this, he said, a "well-knit" species is qualitatively different from individuals of two closely related species, and the word "species" denotes

> at least to a rough approximation, how these groups stand in relation to that general level at which 'speciation' takes place.[83]

---

[79] Robson 1928, 21f.
[80] Morgan 1903, 33 quoted in Allen 1980, 359f.
[81] Dobzhansky 1980.
[82] Muller 1940, 252f.
[83] Muller 1940, 254.

BIBLIOGRAPHY

Allen, Garland E. 1980. The evolutionary synthesis: Morgan and natural selection revisited. In *The Evolutionary Synthesis*, edited by Ernst Mayr and William B. Provine, 356–382. New York: Columbia University Press.

Bates, Henry Walter. 1862. XXXII. Contributions to an Insect fauna of the Amazon Valley. Lepidoptera: Heliconidæ. *Transactions of the Linnean Society of London* 23 (3):495–566.

Bateson, William. 1894. *Material for the study of variation treated with especial regard to discontinuity in the origin of species*. London: Macmillan. Reprint, 1992, Johns Hopkins edition, with introduction by Peter J. Bowler and an essay by Gerry Webster.

Blackwelder, Richard Eliot. 1967. *Taxonomy: A Text and Reference Book*. New York: John Wiley & Sons.

Bowler, Peter J. 1983. *The Eclipse of Darwinism: Anti-Darwinian Evolution Theories in the Decades around 1900*. Baltimore and London: John Hopkins University Press.

—. 1989. *Evolution: The History of an Idea*. Rev. ed. Berkeley: University of California Press. Original edition, 1984.

—. 2003. *Evolution: The History of an Idea*. 3rd ed., completely rev. and expanded ed. Berkeley: University of California Press.

Butler, Samuel. 1884. *Selections from previous works, with remarks on Mr. G.J. Romanes' "Mental evolution in animals", and, A psalm of Montreal*. London: Trübner.

Cope, Edward Drinker. 1868. On the origin of genera. *Proceedings of the Academy of Natural Sciences of Philadelphia* 20:242–300.

Darwin, Charles Robert. 1972. *More Letters of Charles Darwin; a Record of His Work in a Series of Hitherto Unpublished Letters*, edited by Appleton Francis Darwin and A. C. Seward. New York, 1903. New York: Johnson Reprint Corp.

de Vries, Hugo. 1901. *Die Mutationstheorie: Versuche und Beobachtungen über die Entstehung von Arten im Pflanzenreich*. 2 vols. Leipzig: Veit.

—. 1911. *The Mutation Theory: Experiments and Oobservations on the Origin of Species in the Vegetable Kingdom*. London: Kegan Paul, Trench Trubner.

—. 1912. *Species and Varieties: Their Origin by Mutation. Lectures Delivered at the University of California*. 3rd ed. Chicago, IL: Open Court. Original edition, 1904.

Dobzhansky, Theodosius. 1941. *Genetics and the Origin of Species*. 2nd ed. New York: Columbia University Press.

—. 1980. Morgan and his school in the 1930s. In *The Evolutionary Synthesis*, edited by Ernst Mayr and William B. Provine, 445–452. New York: Columbia University Press.

Eldredge, Niles. 1989. *Macroevolutionary Dynamics: Species, Niches, and Adaptive Peaks*. New York: McGraw-Hill.

Forsdyke, Donald R. 2001. *The Origin of Species Revisited: A Victorian Who Anticipated Modern Developments in Darwin's Theory*. Kingston, ON: McGill-Queen's University Press.

Gould, Stephen Jay. 2002. *The Structure of Evolutionary Theory*. Cambridge, MA: The Belknap Press of Harvard University Press.

Haeckel, Ernst Heinrich Philipp August. 1896. *The evolution of man: A popular exposition of the principal points of human ontogeny and phylogeny*. 3rd ed. New York: Appleton.

Hull, David L., ed. 1973. *Darwin and His Critics; The Reception of Darwin's Theory of Evolution by the Scientific Community*. Cambridge, MA: Harvard University Press.

Huxley, T. H. 1895. *Man's place in nature and other anthropological essays, Collected Essays by T. H. Huxley*; vol. 7. London: Macmillan and Co.

Huxley, Thomas Henry. 1893. *Darwiniana: Essays*. London: Macmillan.

—. 1906. *Man's Place in Nature and Other Essays*. Everyman's Library ed. London/New York: J. M. Dent/E. P. Dutton.

Jordan, Karl. 1896. On mechanical selection and other problems. *Novitates Zoologicae* 3:426–525.

—. 1905. Der Gegensatz zwischen geographischer und nichtgeographischer Variation. *Zeitschrift für Wissenschaftliche Zoologie* 83:151–210.

Kitcher, Philip. 1984. Species. *Philosophy of Science* 51 (2):308–333.

Lankester, Edwin Ray. 1890. *The Advancement of Science*. London/New York: Macmillan.

Lotsy, Johannes Paulus. 1916. *Evolution by Means of Hybridization*. The Hague: Martinus Nijhoff.

—. 1931. On the species of the taxonomist in its relation to evolution. *Genetica* 13 (1–2):1–16.

Mayr, Ernst. 1982. *The Growth of Biological Thought: Diversity, Evolution, and Inheritance*. Cambridge, MA: The Belknap Press of Harvard University Press.

Mayr, Ernst, and William B. Provine. 1980. *The Evolutionary Synthesis: Perspectives on the Unification of Biology*. Cambridge, MA: Harvard University Press.

Moore, Aubrey L. 1889. *Science and the Faith. Essays on apologetic subjects, with an introduction*. London: Kegan Paul & Co.

Morgan, Thomas Hunt. 1903. *Evolution and Adaptation*. New York: Macmillan.

Muller, Hermann Joseph. 1940. Bearings of the "Drosophila" work on systematics. In *The New Systematics*, edited by Julian Huxley, 185–268. London: Oxford University Press.

Poulton, Edward Bagnall. 1903. What is a species? *Proceedings of the Entomological Society of London* lxxvii–cxvi.

—. 1908. *Essays on Evolution*. Oxford: Clarendon.

Regan, C. Tate. 1926. Organic evolution. *Report of the British Association for the Advancement of Science* 1925:75–86.

Rensch, Bernhard. 1928. Grenzfälle von Rasse und Art. *Journal für Ornithologie* 76:222–231.

—. 1929. *Das Prinzip geographischer Rassenkreise und das Problem der Artbildung*. Berlin: Gebrüder Borntraeger.

—. 1947. *Neuere Probleme der Abstammungslehre*. Stuttgart: Enke.

—. 1959. *Evolution above the Species Level*. New York: Columbia University Press. Original edition, 2nd ed. of *Neuere Probleme der Abstammungslehre*, Ferdinand Enke, Stuttgart, 1954.

—. 1980. Historical development of the present Synthetic Neo-Darwinism in Germany. In *The Evolutionary Synthesis*, edited by Ernst Mayr and William B. Provine, 284–303. Cambridge, MA/London: Harvard University Press.

Robson, Guy Coburn. 1928. *The Species Problem: An Introduction to the Study of Evolutionary Divergence in Natural Populations, Biological Monographs and Manuals*. Edinburgh: Oliver and Boyd.

Romanes, George John. 1895. *Darwin, and after Darwin: An Exposition of the Darwinian Theory and a Discussion of the Post-Darwinian Questions*. 3 vols. Vol. II. *Post Darwinian Questions: Heredity and Utility*. London: Longmans, Green and Co.

Salisbury, E. J. 1940. Ecological aspects of plant taxonomy. In *The New Systematics*, edited by Julian Huxley, 329–340. London: Oxford University Press.

Salthe, Stanley N. 1985. *Evolving Heirarchical Systems: Their Structure and Representation*. New York: Columbia University Press.

Stresemann, Erwin. 1919. Über die europäischen Baumläufer. *Verhandlungen der Ornithologischen Gesellschaft in Bayern* 14 (1):39–74.

Thistleton-Dyer, Sir William. 1902. The rise and influence of Darwinism. *The Edinburgh Review* 216 (October):366–407.

Thompson, J. Arthur. 1934. *Biology for Everyman*. 2 vols. London: J. M. Dent.

Turesson, Göte. 1922a. The genotypical response of the plant species to the habitat. *Hereditas* 3 (3):211–350.

—. 1922b. The species and variety as ecological units. *Hereditas* 3 (1):10–113.

—. 1925. The plant species in relation to habitat and climate. *Hereditas* 6 (2):147–236.

—. 1927. Contributions to the genecology of glacial relics. *Hereditas* 9 (1–3):81–101.

—. 1929. Zur natur und begrenzung der artenheiten. *Hereditas* 12 (3):323–334.

—. 1930. The selective effect of climate upon plant species. *Hereditas* 14 (2):99–152.

Turney, Peter et al. 1996. Evolution, learning, and instinct: 100 years of the Baldwin Effect. *Evolutionary Computation* 4 (3):iv–viii.

Turrill, Walter B. 1940. Experimental and synthetic plant taxonomy. In *The New Systematics*, edited by Julian Huxley, 47–72. London: Oxford University Press.

7 The Synthesis and Species

First, then, we have the problem involved in the origin of species. As a pre-liminary to that, logic demands that we should define the term. It may be that logic is wrong, and that it would be better to leave it undefined, accepting the fact that all biologists have a pragmatic idea at the back of their heads. It may even be that the word is undefinable. However, an attempt at a definition will be of service in throwing light on the difficulties of the biological as well of the logical problems involved.

Ronald Aylmer Fisher[1]

RONALD FISHER AND WILD-TYPE SPECIES

Ronald A. Fisher is famous as the founder of the modern synthesis between Mendelian genetics and Darwinian natural selection. *The Genetical Theory of Natural Selection* is a seminal work that introduced mathematical models to genetics and selection, and while often cited is rarely quoted.[2] But Fisher addressed a number of questions in that book, including a rarely mentioned discussion about eugenics (Fisher was in favor of a form of eugenics, and chapters 8 through 12 are an argument for it), and one of these questions, almost parenthetically, is about species in the context of sexual and asexual reproduction.

The tradition in British evolutionary biology since Fisher has, on the whole, tended to treat species as names of convenience for communication, in the style of Darwin of the *Origin*, and of Locke. It is therefore somewhat surprising to note that Fisher had a realist approach to species, and one that predated the 1935 article by Dobzhansky that is widely seen as kicking off the species debate of the modern era. Fisher says of species that

> [t]he genetical identity in the majority of loci, which underlies the genetic variability presented by most species, seems to supply the systematist with the true basis of his concepts of specific identity or diversity.[3]

Sexual species are, in fact, the "wild type"—the sum of all the genes in any species the great majority of which are uniform:

---

[1] Huxley 1940, 154.
[2] Fisher 1930.
[3] Fisher 1930, 138.

... we have some reason to suppose that they [*allelomorphic loci, or alleles in modern parlance—JSW*] form a very small minority of all the loci, and that the great majority exhibit, within the species, substantially that complete uniformity, which has been shown to be necessary, if full advantage is to be taken of the chances of favourable mutations.[4]

Fisher is here dealing with the existence of *asexual* species, as they present a problem for him, or rather, they would have if he had been (at that time) certain that any organisms existed without any sexual reproduction (the claim was not revised in the 1958 edition). In this case, he says:

In such an asexual group, systematic classification would not be impossible, for groups of related forms would exist which had arisen by divergence from a common ancestor. Species, properly speaking, we could scarcely be expected to find, for each individual genotype would have an equal right to be regarded as specifically distinct. And no natural groups would exist bound together by constant interchange of their germ-plasm.[5]

Clearly, this exchange of genes in germ-plasm is the *sine qua non* of a species for Fisher. But, he goes on to say, there *would* be an analogue of species in asexuals:

The groups most nearly corresponding to species would be those adapted to fill so similar a place in nature that any one individual could replace another, or more explicitly that an evolutionary improvement in any one individual threatens the existence of all the others.

So while Fisher is a realist about asexual groups, they are ecological groups adapted to the environment in which they find themselves, and kept distinct in virtue of selection against less-fit variants. This resolves also the problem of favorable mutations—if a novelty of value arises in an asexual lineage then it will not spread throughout the population, but instead it will replace the population. In sexual organisms that have proper species, selection maintains the identity of populations rather than of lineages, and favorable mutations can be spread by recombination of genes. But Fisher does not think that there will be many of these groups, and that if they did exist they would be those groups "of so simple a character that their genetic constitution consisted of a single gene."[6]

Of sexual species proper, Fisher presents the view that apart from geographical isolation, in which

the two separated moieties thereafter evolv[e] as separate species, in almost complete independence, in somewhat different habitats, until such time as the morphological differences between them entitle them to 'specific rank'" [p. 139].

Species are also caused to fission by what we now call sympatric speciation, because in "many cases it may safely be asserted that no geographic isolation at all

[4] *Loc. cit.*
[5] Fisher 1930, 135.
[6] Fisher 1930, 137.

can be postulated." A species subject to different conditions at the extremes of its range will adapt at those extremes, and hybrid forms will be disadvantageous if the migration rate is less than the rate of increase of the favorable forms in the environment to which they are adapted. Fisher champions as a mode of speciation the selectionist account of Darwin and Wallace, and yet still allows for the Wagner-style mode of allopatric isolation.

At this point we have reached the beginnings of the Modern Synthesis, and hence the modern debate. We are now equipped to put the modern debate into context, especially claims of conceptual novelty. To that we now turn, beginning with Theodosius Dobzhansky's discussion shortly after Fisher's book, which defined the modern species debate.

THEODOSIUS DOBZHANSKY'S DEFINITION

We may arbitrarily mark the beginnings of the modern debate with Dobzhansky's classic 1935 essay, "A critique of the species concept in biology,"[7] although Poulton's essay is also crucial. It's not really so arbitrary—from Dobzhansky's essay and the book that followed it[8] flowed both the present debate over what species are, and the birth of the Modern Synthesis. Dobzhansky's work was an attempt to take Darwin seriously about species by a working systematist not imbued with the British deflationary tradition. And he introduced two of the major innovations in the debate— the idea of species as evolutionary players, and the idea that genetic exchange marked out these players. It seems that when typostrophic views of evolution were abandoned, the issue from the Great Chain of what divisions nature forced upon us and what divisions were of our own convenience came back to the fore. Later, Mayr described the Synthesis as being a "shared species problem."[9] It remains a shared problem.

Theodosius Dobzhansky was perhaps the most significant of all the synthesists[10] through the middle of the twentieth century. He introduced Sewall Wright's ideas on drift into the synthetic orthodoxy (often to considerable opposition), and his work on the laboratory and field genetics of *Drosophila* spp. revolutionized the field.[11]

Dobzhansky published his paper, later substantially included in the chapter on species in his *Genetics and the Origin of Species*, which discussed Lotsy's revision of Poulton's notion of syngamy as the foundation for a genetic population: "an habitually interbreeding community of individuals."[12] Dobzhansky says of the syngameon approach to species that it applies only to panmictic populations of organisms, and which, although attractive in its simplicity, is therefore inapplicable in the case of many species that are divided into reproductively separated populations.

---

[7] Dobzhansky 1935.
[8] Dobzhansky 1937.
[9] Mayr and Provine 1980, 1.
[10] In order to identify those active in the so-called "Modern Synthesis," which is not looking so modern any more, we need a term. I trust I can be excused this one.
[11] Depew and Weber 1995, 291–297, 300–302.
[12] Dobzhansky 1941, 311.

He proposes instead to base specific rank on the existence of reproductive isolating mechanisms, and provides a revision to Lotsy's definition:

> ... a species is a group of individuals fully fertile inter se, but barred from interbreeding with other similar groups by its physiological properties (producing either incompatibility of parents, or sterility of the hybrids, or both).[13]

The *rank* of *species* is something that arises in evolution when continuity of reproduction becomes discontinuous:

> Considered dynamically, the species represents that stage of evolutionary divergence, at which the once actually or potentially interbreeding array of forms becomes segregated into two or more separate arrays which are physiologically incapable of interbreeding.[14]

By "array of forms," Dobzhansky can be interpreted as meaning either diagnostic morphs, as Mayr does, or as the types within a population that affect reproductive isolation. His discussion makes it clear that he does not intend "form" in an essentialistic sense, I believe, but in a causal sense. In the 1935 paper, he treats reproductive isolation in terms of physiological differences; this is the sense I interpret him to mean by "form," although he may have equivocated on the distinction between diagnostic and causal structures, as many did before and after him.

He had noted in *Genetics and the Origin of Species* the "taxonomic" definition of an "affable taxonomist," C. Tate Regan, that species are what competent systematists consider to be a species, and puts this failure to deliver a universal definition

> ... that would make it possible to decide in any given case whether two given complexes of forms are already separate species or are still only races of a single species[15]

down to the general method of species formation,

> ... through a slow process of accumulation of genetic changes of the type of gene mutations and chromosomal reconstructions. This premise being granted, it follows that instances must be found in nature when two or more races have become so distinct as to approach, but not to attain completely, the species rank. The decision of a systematist in such instances can not but be an arbitrary one.[16]

In short, evolution makes it impossible to determine if species rank has been reached. Nevertheless, the rank itself is real enough: it is the attainment of complete separation. He also notes that

> [w]e find aggregations of numerous more or less clearly distinct biotypes, each of which is constant and reproduces its like if allowed to breed. These constant biotypes

---

[13] Dobzhansky 1935, 353; cf. Dobzhansky 1941, 312.
[14] Dobzhansky 1935, 354.
[15] *Loc. cit.*
[16] Dobzhansky 1941, 310f.

are sometimes called elementary species, but they are not united into integrated groups that are known as species in the cross-fertilizing [*i.e., sexual—JSW*] forms. The term "elementary species" is therefore misleading and should be discarded.[17]

Dobzhansky was influential in this regard, because de Vries' term did largely disappear from the debate. He also notes that biotypes can be cross-specific, and even cross-generic, and that

[w]hich one of these ranks is ascribed to a given cluster is, however, decided by considerations of convenience, and the decision is in this sense purely arbitrary. In other words, the species as a category which is more fixed, and therefore less arbitrary than the rest, is lacking in asexual and obligatorily self-fertilizing organisms. ...

The binomial system of nomenclature, which is applied universally to all living beings, has forced systematists to describe "species" in the sexual as well as in the asexual organisms. Two centuries have rooted this habit so firmly that any thorough reform will meet with determined opposition. Nevertheless, systematists have come to the conclusion that sexual species and "asexual species" must be distinguished. ... In the opinion of the writer, all that is saved by this method is the word "species." A realization of the fundamental difference between the two kinds of "species" can make the species concept methodologically more valuable than it has been.[18]

In the 1951 edition he replaces the final sentence with:

As pointed out by Babcock and Stebbins... , "The species, in the case of a sexual group, is an actuality as well as a human concept; in an agamic complex it ceases to be an actuality."[19]

He also begins to discuss issues of typological thinking, under the influence of Mayr, in the chapter of the "fourth" edition[20] on populations, races, and subspecies, where he notes

[t]he classical race concept [*of human races—JSW*] was typological. ...

Typology is at the bottom of the vulgar notion that any so-called Negro in the United States ... has a basic and unremediable Negroid nature, just as any Jew partakes of some Jewishness, etc. There are no Platonic types of Negroidness or Jewishness or of every race of squirrel or butterflies. Individuals are not mere reflections of their racial types; individual differences are the fundamental biological realities.[21]

There is little hint of discussions of typology or essentialism in the earlier works. Dobzhansky deals extensively with variation within populations; it is the *raison d'être* of the book, and so it might be a case of the wood not needing to be specified when the trees are so well described. By 1970, though, he is well in line with the Mayrian Received View approach.

---

[17] Dobzhansky 1941, 320f.
[18] Dobzhansky 1937, 321.
[19] Dobzhansky 1951, 275.
[20] Dobzhansky 1970.
[21] Dobzhansky 1970, 268f.

AFTER DOBZHANSKY, THE BEGINNINGS OF THE MODERN DEBATE

At the time of the modern synthesis, announced in Julian Huxley's book by that title in 1942,[22] there was little dispute amongst those involved that species were real enough, but there was a wide range of opinion about what that meant. Darlington, for example, explicitly appealed to Ray's dictum (in Latin) that to sort living beings into species we need no more than *"distincta propagatio ex semine,"*[23] but that

> [t]here are many kinds of species and many kinds of discontinuities between species.[24]

He noted that

> We feel we ought to have a 'species concept.' In fact there can be no species concept based on the species of descriptive convenience that will not ensnare its own author so soon as he steps outside the group from which he made the concept. The only valid principles are those that we can derive, not from fixed classes but from changing processes. To do this we must go beyond the species to find out what it is made of. We must proceed (by collaboration) to examine its chromosomal structure and system of reproduction in relation to its range of variation and ecological character. From them we can determine what is the genetic species of Ray, the unit of reproduction, a unit which cannot be used for summary diagnosis, but which can be used for discovering and relating the processes of variation and the principles of evolution.[25]

In contrast, Julian Huxley in his introduction to *The New Systematics* argues that Dobzhansky's 1937 definition "goes far beyond the facts."[26] Huxley notes the constancy of cross-fertilization among plants in particular, and says that therefore "Dobzhansky's definition is untrue, or, if true, taxonomic practice must be so re-cast as to rob the term species of its previous meaning."[27] H. J. Muller's contribution to that volume agrees—there is no fixed rank dividing species from varieties or races, although

> ... divergence goes on very differently, and much more freely, between those which can and do cross, and it is therefore justifiable and useful, even though difficult, to make the species distinction, if it is made in such a way as to correspond so far as possible with this stage of separation. At the same time it must be recognized that the species are in flux, and that an adequate understanding of their relationships can be arrived at only on the basis of an understanding of the relationships between the minor groups and even between the individuals, supplemented by the study of the differences found through observations on the systematics of the larger groups.[28]

---

[22] Huxley 1942.
[23] Darlington 1940, 137.
[24] *Op. cit.*, 158.
[25] *Op. cit.*, 159.
[26] Huxley 1940, 16ff.
[27] Huxley 1940, 17.
[28] Muller 1940, 258.

In the book announcing the synthesis, Huxley later criticized Dobzhansky for underplaying the difficulties that a simple intersterility criterion encountered, particularly in plants.

> The dynamic point of view is an improvement, as is the substitution of incapacity to exchange genes for the narrower criterion of infertility: but even so, this definition cannot hold, for it still employs the lack of interbreeding as its sole criterion. "Interbreeding without appreciable loss of fertility" would apply to the great majority of animals, but not to numerous plants. In plants there are many cases of very distinct forms hybridizing quite competently even in the field. To deny many of these forms specific rank just because they can interbreed is to force nature into a human definition, instead of adjusting your definition to the facts of nature. Such forms are often markedly distinct morphologically and do maintain themselves as discontinuous groups in nature. If they are not to be called species, then species in plants must be deemed to differ from species in animals in every characteristic save sterility...[29]

He gives his own criteria a few pages later, after determining that single-criterion definitions are useless:

> In general, it is becoming clear that we must use a combination of several criteria in defining species. Some of these are of limiting nature. For instance, infertility between groups of obviously distinct mean type is a proof that they are distinct species, although once more the converse is not true.
>
> Thus in most cases a group can be distinguished as a species on the basis of the following points jointly: (i) a geographical area consonant with a single origin; (ii) a certain degree of constant morphological and presumably genetic difference from related groups; (iii) absence of intergradation with related groups. ... Our third criterion above, if translated from the terminology of the museum to that of the field, may thus be formulated as a certain degree of biological isolation from related groups.[30]

After discussing freely hybridizing groups, sympatric ecological forms, plants, polyploidy,[31] and asexuality, Huxley says

> Thus we must not expect too much of the term species. In the first place, we must not expect a hard-and-fast definition, for since most evolution is a gradual process, borderline cases must occur. And in the second place, we must not expect a single or a simple basis for definition, since species arise in many different ways.[32]

---

[29] Huxley 1942, 162f.
[30] *Op. cit.,* 164f.
[31] *Polyploidy* is the state of having three or more complete sets of chromosomes, in contrast to the usual state of diploidy (two sets) in sexual organisms. *Alloploidy* occurs when hybrids are formed through the fertilization of gametes (sex cells) across species. It is usually also polyploidy, in which case it is called allopolyploidy. In plants, fertile individuals often result when chromosomes are duplicated and then separate to form symmetrical diploid chromosome sets. If the hybrid is significantly different from the parental populations, it will not interbreed with them, or will interbreed incompletely, so that eventually the novel karyotype (chromosomal type) will breed true with itself but not with either parental type. Speciation can be achieved in one or only a few generations this way [Grant 1975, 431ff, Ramsey and Schemske 1998].
[32] Huxley 1942, 167.

The new geneticists tended, then, toward an eliminativist view on species, in what they perceived was the tradition of Darwin: sure, the lineages split and this was real, but the rank of the splitting was manifold and had no universally common criteria that could be recognized. There was a division in the way the Synthesis Darwinians approached species, which can be traced back to Darwin's own published ambiguity on the subject. Into this ambiguity of opinion came Ernst Mayr.

ERNST MAYR AND THE BIOSPECIES CONCEPT

Mayr was a German ornithologist who had left Germany well before the war and came to America to the American Museum of Natural History and then to Harvard after spending a number of years in New Guinea and the Solomon Islands studying bird populations and distributions.[33] He was motivated to address the "species problem" because of the publication of another book, opposed to Dobzhansky's approach, by geneticist Richard Goldschmidt, who became Mayr's *bête noire* for many years to come.[34] Goldschmidt proposed that species evolved in a single step, through macromutations involving chromosomal repatterning to form "hopeful monsters"[35] in what is often called "saltation" (i.e., the opposite of *natura non facit saltum* quoted by Darwin). Goldschmidt repeatedly referred to species being separated by "bridgeless gaps,"[36] a phrase he took from Turesson,[37] ignoring the fact that Turesson then went on to give a Darwinian account of species formation. Goldschmidt rejected the Darwinian idea that subspecific races were incipient species entirely, which seems to have motivated Mayr's ire and to have informed his allopatric account of speciation later.

Mayr was invited to give a series of talks on speciation as part of the Jesup Lectures in 1941 at Columbia University's Zoology Department. He was later invited to publish sufficient material to fill an entire volume for Columbia University Press after the other lecturer, Edgar Anderson, fell ill.[38] The result was the single most widely referred to volume of the synthesis.

Basically, this work is a discussion at length of the modes of speciation according to the best knowledge of the day, and much of what Mayr discussed remains valid. In the case of "ring-species" for example, his discussions remain the canonical ones.[39] He also introduced several terms, including *allopatry*[40] and *sibling species*,[41] which

---

[33] Hull 1988, 66f.

[34] Goldschmidt 1940.

[35] *Op. cit.* 390–393; it should be clear now that the term "monster" here refers to a sport or sudden variation from the type.

[36] Cf. *op. cit.,* 143.

[37] Turesson 1927, 100.

[38] Cf. xvii, of the new Introduction to the 1999 reissue of his 1942.

[39] Mayr 1999, 180–185. However, some of the canonical examples are being disputed. Recently, molecular analysis of the *Larus argentatus* (Herring Gull) complex has indicated that they are, in fact, isolated gene pools [Liebers et al. 2004] and the *Parus major* (Great Tit) complex, while it does interbreed to some extent, is a good set of phylogenetic species [Kvist et al. 2003].

[40] Mayr 1999, 149.

[41] *Op. cit.,* 151.

have worn well. But for our purposes, the most important aspect of the book is that here Mayr popularized the definition of species he had already given in 1940:

> A species consists of a group of population which replace each other geographically or ecologically and of which the neighboring ones intergrade or interbreed wherever they are in contact or which are potentially capable of doing so (with one or more of the populations) in those cases where contact is prevented by geographical or ecological barriers.
>
> Or shorter: Species are groups of actually or potentially interbreeding natural populations, which are reproductively isolated from other such groups.[42]

Mayr contrasted this with the "typological" conceptions of taxonomy at the time, especially the "morphological species concept" as Mayr called it (for the first time anyone had, so far as I can tell), as well as the "practical species concept" (a version of Regan's definition), the "genetic species concept" (homozygous populations, which he attributes to Lotsy), and one based on sterility. He called it the "biological species concept"; and so the proliferation of general "species definition" names began. Mayr included Dobzhansky's definition under this rubric, but says of it,

> [t]his is an excellent description of the process of speciation, but not a species definition. A species is not a stage of a process, but the result of a process.[43]

Hence, he proposed the definition above as a practical compromise, allowing the systematist the judgment call, since (as is noted later by critics of the concept) one cannot use reproductive isolation as a test in many cases, not least in allochronic and allopatric populations. Further,

> [t]he application of a biological species definition is possible only in well-studied taxonomic groups, since it is based on a rather exact knowledge of geographical distribution and on the certainty of the absence of interbreeding with other similar species.[44]

Moreover, while it works for "bisexual organisms" (sexual species), it fails, he notes, for "aberrant cases" like protozoans and plants, which are either unisexual (asexual) or freely hybridizing.[45] The remainder of the book discusses speciation processes, and Mayr presents a Wagnerian view that geographical isolation is a precondition for the formation of new species[46] Sympatric species are reproductively isolated absolutely (*"otherwise they would not be good species"*[47]), but the gaps that separate allopatric species are *"often gradual and relative, as they should be, on the basis of the principle of geographic speciation."*[48] He considers sympatric speciation as a possibility, but concludes

---

[42] Mayr 1940, 120.
[43] Mayr 1942, 119.
[44] *Op. cit.*, 121.
[45] *Op. cit.*, 122.
[46] *Op. cit.*, 154–185.
[47] *Op. cit.*, 149, italics original.
[48] *Loc. cit.*

Darwin thought of individuals when he talked of competition, struggle for existence among variants, and survival of the fittest in a particular environment.[49] Such a struggle among individuals leads to a gradual change of populations, but not to the origin of new groups. It is now being realized that species originate in general through the evolution of entire populations. If one believes in speciation through individuals, one is by necessity an adherent of sympatric speciation, the two concepts being very closely connected. However, fewer and fewer situations are interpreted as evidence for sympatric speciation, as it is realized more and more clearly that reproductive isolation is required to make the gap between two incipient species permanent and that such reproductive isolation can develop only under exceptional circumstances between individuals of a single interbreeding population.[50]

This has a whiff of circularity. Mayr defines species as reproductively isolated populations formed in allopatry and absolutely distinct in sympatry. He then claims that sympatric variations cannot form reproductively isolated species because species are formed in allopatry since sympatric "species" have to be absolutely isolated. Of course, there is a lot more to it than this, and Mayr brings in all sorts of impressive empirical evidence, so the charge of plain circularity, once made, must be dismissed. However, this means that the strength of the biological species concept rests *entirely* on the absence of plausible *empirical* reasons to believe that reproductive isolation does not occur in sympatric populations. All it would take to undercut this definition of species, of course, or at any rate Mayr's argument in its favor, is to find an unambiguous case of sympatric speciation. There is a case—cichlid fishes in various lakes in Africa, for instance—where Mayr wonders if these "species flocks" are evidence for "explosive" sympatric speciation.[51] He rejects the idea on the grounds that the closest relatives of each species are not sympatric.[52]

Finally, Mayr considers the factors that cause species, since species are the "effect" of a process. He divides them into internal factors, such as genetic mechanisms, mutation rates, and the like, and external factors. Then he lists a number of subcategories, isolating mechanisms: geographical barriers that restrict random dispersal, ecological barriers, ethological factors, mechanical factors, and "genetic" and physiological factors.[53]

A case for which Mayr does not insist upon allopatry is what he calls "instantaneous sympatric speciation,"[54] such as via polyploidy (the duplication of chromosomes and possible subsequent reduction to a diploid form of differing composition

---

[49] In fact, Darwin occasionally *explicitly* talked of selection and the struggle for existence as occurring between species, as we have seen.

[50] *Op. cit.,* 190.

[51] *Op. cit.,* 215.

[52] Recent work has (unfortunately for Mayr's views) shown otherwise [Mazeroll and Weiss 1995, Albertson et al. 1999, Salzburger et al. 2002, Stauffer et al. 2002, Martin et al. 2015]. Schilthuizen 2000 gives an excellent summary.

[53] Mayr 1942, 237f.

[54] *Op. cit.,* 190ff.

to the parental species), self fertilization, parthenogenesis, and so on, but says this is "apparently rare, even where it is hypothetically possible."[55]

In later publications, Mayr introduced the notion of species being "non-dimensional":

> Noninterbreeding between populations is manifested by a gap. It is this gap between populations that coexist (are sympatric) at a single location at a given time that delimits the species recognized by a naturalist. Whether one studies birds, mammals, butterflies, or snails near one's home town, one finds species clearly delimited and sharply separated from all other species. This demarcation is sometimes referred to as the species delimitation *in a non-dimensional system* (a system without the dimensions of space and time).[56]

He also adds the distinction between the "species category" and the "species taxon"—the former "designates a given rank in a hierarchic classification,"[57] while the taxon is

> ... the concrete object of classification. Any such group of populations is called a *taxon* if it is considered sufficiently distinct to be worthy of being formally assigned to a definite category in the hierarchical classification. *A taxon is a taxonomic group of any rank that is sufficiently distinct to be worthy of being assigned to a definite category.*[58]

By the 1963 version, the biological species as defined by Mayr has become a reproductive community, an ecological unit, and a genetic unit:

> ... species are reproductive communities. The individuals of a species of animals recognize each other as potential mates and seek each other for the purpose of reproduction. A multitude of devices insure intraspecific reproduction in all organisms .... The species is also an ecological unit that, regardless of the individuals composing it, interacts as a unit with other species with which it shares the environment. The species, finally, is a genetic unit consisting of a large, intercommunicating gene pool, whereas the individual is merely a temporary vessel holding a small portion of the contents of the gene pool for a short time.[59]

and by 1970 the definition has changed to read:

---

[55] *Op. cit.*, 192. Again, later work has found sufficient examples, mostly in plants and other gamete broadcasters, to establish this as a real process [Dowling and Secor 1997, Aldasoro et al. 1998, Chepurnov et al. 2002].

[56] Mayr 1970, 14f, italics original. According to Chung 2003, 285, Mayr first began to discuss the dimensionality of species in an address in 1946 [Mayr 1946], where he described the Linnaean conception as having no dimensions. Otherwise his tone is, as Chung remarks, fairly neutral on the difference between the "morphological" species concept and the "biological," "polytypic" species concept at that time. Chung traces Mayr's emerging view of species concepts as differing in their typology and populational nature from 1953 [Mayr et al. 1953] through to 1959 [Mayr 1959] and concludes that he discovered the typological aspect of the prior conceptions at around this time, especially in his 1955.

[57] Mayr 1970, 13.

[58] Mayr 1970, 14, italics original.

[59] Mayr 1963, 21.

*Species are groups of interbreeding natural populations that are reproductively iso-
lated from other such groups.*[60]

Gone is the phrase "actually or potentially" from the 1942 edition; Mayr now
thinks that it is only in sympatry ("with respect to sympatric and synchronous populations") that we can tell for sure that two organisms are distinct species. The definition is "biological," he says,

> ... not because it deals with biological taxa, but because the definition is biological. It
> utilizes criteria that are meaningless as far as the inanimate world is concerned.
>
> When difficulties are encountered, it is important to focus on the basic biological
> meaning of the species: *A species is a protected gene pool. It is a Mendelian popula-
> tion that has its own devices (called isolating mechanisms) to protect it from harm-
> ful gene flow from other gene pools.* Genes of the same gene pool form harmonious
> combinations because they become coadapted by natural selection. Mixing the genes
> of two different species leads to a high frequency of disharmonious gene combinations;
> mechanisms that prevent this are therefore favored by selection.[61]

The text emphasized above is surprising. From being the "effect" of a process in
1942, Mayr now treats species as a *mechanism* of protecting gene pools. Moreover,
selection now plays a role in "protecting" species; previously species were not formed
through selection, and reproductive isolation was a side effect of geographical isolation. Still, isolating mechanisms are still "potentially or actually" active in sympatry
and have to be intrinsic mechanisms of the organisms, and not, for example, "geographic or any other purely extrinsic isolation."[62]

Mayr's view of species seemed not to have changed much since the 1970 volume.[63]
He repeats it in several places.[64] To summarize his final position, let us consider his
second last paper[65] on the species concept, where he claims that species are concrete
describable objects in nature, that they are reproductively isolated even when there is
"leakage of genes," and that the biological species concept (now abbreviated as BSC)
is based on the properties of populations. Although Mayr always stressed the populational nature of species as a result of his insistence on genetic and morphological
polytypy in species, over time there is an increasing emphasis on "populational thinking" in opposition to "essentialism" in his works. For instance, in the introduction to
his history of biology,[66] he discussed this, citing Hull and Ghiselin,[67] dividing western thinking into two phases. The first phase was essentialism deriving from Plato,
and the second, population thinking beginning with Leibniz's theory of monads but
really taking root with the British animal breeders and Darwin and his contemporary

---

[60] Quoted in Mayr 1970, 12.
[61] Mayr 1970, 13, emphasis added.
[62] Mayr 1970, 56.
[63] Apart from flying the "ecological niche" variant mentioned above.
[64] For example, Mayr 1976, 1985, 1988, 1992, 1996.
[65] Mayr 1996. He also provided a chapter in Wheeler and Meier 2000 [Mayr 2000].
[66] Mayr 1982, 45–47.
[67] Ghiselin 1974, Hull 1976.

systematists in Britain. One can take issue with the claim for Leibniz (as a Great Chain thinker, the sort of variation he admitted was scalar rather than distributional), and wonder why de Quetelet has been overlooked as the source of populational thinking and instead been called an essentialist.[68] Most oddly, what is missing here and elsewhere is Karl Popper. Popper had attacked what he called "methodological essentialism" as a malign heritage from Plato,[69] in particular in his *Poverty of Historicism*, §10, where he set up *nominalism*—the doctrine that universal terms are mere labels attached to sets of things—opposed to "realism," or "idealism," which he renames as *essentialism*. Popper ascribes to Aristotle the problem this introduces into science:

> The school of thinkers whom I propose to call *methodological essentialists* was founded by Aristotle, who taught that scientific research must penetrate to the essence of things in order to explain them. Methodological essentialists are inclined to formulate scientific questions in such terms as 'what is matter?' or 'what is force?' or 'what is justice?' and they believe that a penetrating answer to such questions, revealing the real or essential meaning of these terms and thereby the real or true nature of the essences denoted by them, is at least a necessary prerequisite of scientific research, if not its main task. *Methodological nominalists*, as opposed to this, would put their problems in such terms as 'how does this piece of matter behave?' or 'how does it move in the presence of other bodies?' For methodological nominalists hold that the task of science is only to describe how things behave, and suggest that this is to be done by freely introducing new terms wherever necessary, or by re-defining old terms wherever necessary while cheerfully neglecting their original meaning. For they regard *words* merely as *useful instruments of description*.[70]

Popper's sympathies are clearly with the nominalists. Through Hull's seminal essay on essentialism in taxonomy,[71] Popper's distinction came to be widely accepted among taxonomists, and Mayr may be influenced either directly or indirectly by Hull.[72] It is unclear whether he was directly influenced by Popper's *Poverty*, but that work was a *cause célèbre* in its day, and since Mayr and Popper were both leading German-speaking academics in the English-speaking world, it would be surprising if someone as erudite as Mayr had not at least heard of Popper and his ideas.[73] In the 1982 history, Mayr cites Popper only for issues of theory falsification. Why is this?

---

[68] See, for instance, Krüger et al. 1990.

[69] Popper 1957a, 1957b, 1960.

[70] Popper 1960, 28f.

[71] Hull 1965a, 1965b.

[72] Hull (*pers. comm.*) told the story that, as a graduate student, he delivered the talk on which this paper was based in front of Popper, and handed it in at the end of semester. Popper took it upon himself to send the talk to the *British Journal for the Philosophy of Science* without Hull's knowledge, and the author had to ask for it to be returned for revision. Many of his conclusions were not strictly in line with Popper's own ideas, but Popper apparently never read the published work, and so Hull never came under Popper's withering attack himself.

[73] Polly Winsor believes, after discussing the issue with various of Mayr's students and associates, that Mayr did *not* know Popper's work until his attention was drawn to it by Hull's 1965 paper. Mayr does cite Popper in his later work [Mayr 1997, 59f] where he explicitly mentions Popper's attitude to words and essentialism, but prior to that, his ideas on typology seem to be his own, as Winsor calls it, "dragon." According to Winsor, Mayr first used the term "essentialism" as a synonym for "typology" in 1968 [Winsor 2004].

I conjecture that it may be due to the prominence given to Popper by cladists such as Farris, Wiley, Patterson, Platnick, and Nelson as a justification for cladistic senses of naturalness.[74] In any event, this is beyond our scope here, the point being that Mayr is assuming a nominalistic approach to species; the term (as a category) is a useful way of organizing real, concrete objects (the taxa). He particularly criticizes another philosopher, Philip Kitcher,[75] for failing to appreciate the difference between biological populations and classes of non-living (inanimate) objects.[76]

Of particular note is Mayr's version of the history of species concepts. He gives it again in this paper, but he has given various forms of it in his other writings.[77] According to this version,

> [t]he biological species concept developed in the second half of the 19th century. Up to that time, from Plato and Aristotle until Linnaeus and early 19th century authors, one simply recognized "species," eide (Plato), or kinds (Mill). Since neither the taxonomists nor the philosophers made a strict distinction between inanimate things and biological species, the species definitions they gave were rather variable and not very specific. The word 'species' conveyed *the idea of a class of objects, members of which shared certain defining properties*. Its definition distinguished a species from all others. Such a class is constant, it does not change in time, all deviations from the definition of the class are merely "accidents," that is, imperfect manifestations of the essence (eidos). Mill in 1843 introduced the word 'kind' for species (and John Venn introduced 'natural kind' in 1866) and philosophers have since used the term natural kind occasionally for species... [italics original][78]

In many details this account is incorrect, as the preceding chapters show. A distinction *was* made in practice between living and nonliving things by many authors before the nineteenth century, Locke introduced the term "kind" for species, and typological accounts permitted variation in the "essential" characters of type to quite a degree before members of the type became monsters. Mayr seems insistent on finding "forerunners" to his own preferred conception. He then writes of "the morphological, or typological species concept":

> Even though this was virtually the universal concept of species, there were a number of prophetic spirits who, in their writings, foreshadowed a different species concept, later designated [*by Mayr, as it happens—JSW*] as the *biological species concept* (BSC). The first among these was perhaps Buffon (Sloan 1987), but a careful search through the natural history literature would probably yield quite a few similar statements.

This tendency to seek precursors for a favorite personal view is known among historians as the "Whig interpretation of history" or as "presentism," or "creeping

---

[74] Hull 1988, 129, 171, 197, 237–239, 247, 251–253, 268, and references cited therein. Nelson and Platnick's focus on Popper is due to the work of Walter Bock [Bock 1974, Nelson, *pers. comm.*], although Hull [*pers. comm.*] recalled suggesting Popper to one of them. Popper's influence on cladistic taxonomy, including the early cladists, is documented in Rieppel 2003.

[75] Kitcher 1989.

[76] Mayr 1996, 266f.

[77] Mayr 1957, 1970, 1982, 1991.

[78] Mayr 1996, 266f.

precursoritis";[79] it is the importation of modern views into the past. At least since Collingwood, such approaches to history as the progressive leading-up to the modern day or some ideal state have been widely viewed askance by historians:

> Bach was not trying to write like Beethoven and failing; Athens was not a relatively unsuccessful attempt to produce Rome; Plato was himself, not a half-developed Aristotle[80]

and, we might add, the writers on species in the period in question were not precursors to Ernst Mayr.[81] There has *always* been a "biological" (reproductive) component to discussions of species as applied to the living world, which I have called the "generative" notion of species. Moreover, few of the writers adduced by Mayr as forerunners are actually presenting anything much like his view, as most of them include a clear morphological component in their conceptions. Nevertheless, Mayr's claim of E. B. Poulton and K. Jordan as "precursors" would seem to be fair, at least in terms of a similarity of views, and given the number of times he cites them, he may even have them as direct antecedents—that is, they might have directly influenced *him*. Stresemann, as his teacher, clearly did.[82]

In his "point paper" on the biological species concept in the Wheeler and Meier volume, Mayr repeats most of the previous paper. One thing he does add here is that

> [t]he word *interbreeding* indicates a propensity; a spatially or chronologically isolated population, of course, is not interbreeding with other populations but may have the propensity to do so when the extrinsic isolation [is] terminated. [83]

The shift in Mayr's thinking from "actually or potentially interbreeding" in 1942, to "actually or potentially operating isolating mechanisms" in 1963, to "propensity to interbreed" in 2000 is interesting. As a test of species-status, potential anything is obviously useless unless it can be made actual (which is the basis for his insistence that only in sympatry are species fully determinable[84]). But clearly one wants to be able to say that I and the inhabitants of fifteenth-century England are the same species, even if I cannot interbreed with them *actually*, and so Mayr *must* introduce something like a propensity interpretation. However, "propensities to behave" are themselves no easier on the metaphysical eye than potentialities. They remain, in the end, conditional statements: a lump of sugar is soluble if, when immersed in water, it

---

[79] Butterfield 1931; I am indebted to Neil Thomason for the phrase.

[80] Collingwood 1946, 329.

[81] Although this may sound harsh, Mayr *has* referred to his "precursors" as "prophetic spirits" [Mayr 1996, 269], noting "how tantalizingly close to a biological species concept some of the earlier authors had come" [Mayr 1982, 271], and claimed that "Buffon understood the gist of it" and the early Darwin also [Mayr 1997, 130], thus claiming authoritative precursors. Hull 1988, 372–377 and Winsor 2001 discuss the role precursors play in scientific histories. One function for precursors is to give legitimacy to the views of the modern scientist and deflect criticism to dead white males. Similar things happened with Galileo, and also with the "rediscoverers" of Mendel.

[82] Chung 2003, Winsor 2004.

[83] Mayr 2000.

[84] Mayr once told Hull in conversation that not potentially interbreeding was the equivalent of there being isolating mechanisms present (Hull, *pers. comm.*).

dissolves.[85] But if there were no possibility of immersing it in water, would it remain a soluble substance? We want to say so, but if we examine our intuitions, this is because we know of other lumps of sugar that they *have* dissolved, and by analogy with this lump, which has the same composition, and no reason to think otherwise, so it too will dissolve. Propensities may also be interpreted as the average behavior of a reference class in certain conditions,[86] but in this case it makes no sense to talk about the propensity of an individual case (that is, of me and a fifteenth-century woman in England). Either way, Mayr has a problem with potentials or propensities.

Mayr's conception of species is, in the end, one of dispositions to behave in various ways. Henry IV and I are of the same species because we are of the same "substance," and were we in the same population, our genes could freely spread through it. But this is what Mayr wants to object to; this smacks of essentialism, although, as I have argued, it isn't. In the meantime, let us note that Mayr tries to avoid counterfactual claims of *"would* have interbred, if in the same (natural) population" with his insistence on sympatry for full species-hood. It is a problem he makes largely for himself, based on his conflation, I believe, of the epistemic and ontological aspects of being a species. In this, he is not alone.

BIBLIOGRAPHY

Albertson, R. C. et al. 1999. Phylogeny of a rapidly evolving clade: The cichlid fishes of Lake Malawi, East Africa. *Proceedings of the National Academy of Sciences of the United States of America* 96:5107–5110.

Aldasoro, J. J. et al. 1998. The genus *Sorbus* (Maloideae, Rosaceae) in Europe and in North Africa: Morphological analysis and systematics. *Systematic Botany* 23 (2):189–212.

Bock, Walter J. 1974. Philosophical foundations of classical evolutionary classification. *Systematic Zoology* 22:375–392.

Butterfield, Herbert. 1931. *The Whig Interpretation of History*. London: G. Bell.

Chepurnov, Victor A. et al. 2002. Sexual reproduction, mating system, and protoplast dynamics of Seminavis (Bacillariophyceae). *Journal of Phycology* 38 (5):1004–1019.

Chung, Carl. 2003. On the origin of the typological/population distinction in Ernst Mayr's changing views of species, 1942–1959. *Studies in History and Philosophy of Biological and Biomedical Sciences* 34 (2):277–296.

Collingwood, R. G. 1946. *The Idea of History*. 1961 Paperback ed. Oxford: Oxford University Press.

Darlington, Cyril Dean. 1940. Taxonomic species and genetic systems. In *The New Systematics*, edited by Julian Huxley, 137–160. London: Oxford University Press.

Depew, David J., and Bruce H. Weber. 1995. *Darwinism Evolving: Systems Dynamics and the Genealogy of Natural Selection*. Cambridge, MA: MIT Press.

Dobzhansky, Theodosius. 1935. A critique of the species concept in biology. *Philosophy of Science* 2:344–355.

—. 1937. *Genetics and the Origin of Species*. New York: Columbia University Press.

—. 1941. *Genetics and the Origin of Species*. 2nd ed. New York: Columbia University Press.

—. 1951. *Genetics and the Origin of Species*. 3rd rev. ed. New York: Columbia University Press.

—. 1970. *Genetics of the Evolutionary Process*. New York: Columbia University Press.

---

[85] Cf. Sober 1984, 76–78.
[86] Hájek 2003.

Dowling, Thomas E., and Carol L. Secor. 1997. The role of hybridization and introgression in the diversification of animals. *Annual Review of Ecology and Systematics* 28:593–619.

Fisher, Ronald Aylmer. 1930. *The Genetical Theory of Natural Selection*. Oxford, UK: Clarendon Press.

Ghiselin, Michael T. 1974. *The Economy of Nature and the Evolution of Sex*. Berkeley: University of California Press.

Goldschmidt, Richard B. 1940. *The Material Basis of Evolution*. Seattle: University of Washington Press.

Grant, Verne. 1975. *Genetics of Flowering Plants*. New York: Columbia University Press.

Hájek, Alan. 2003. Interpretations of probability. In *Secondary Interpretations of Probability*, ed Edward N. Zalta. Type of Medium. Available at http://plato.stanford.edu/archives /um2003/entries/probability-interpret (accessed 5/30/17).

Hull, David L. 1965a. The effect of essentialism on taxonomy—Two thousand years of stasis. *British Journal for the Philosophy of Science* 15 (60):314–326.

—. 1976. Are species really individuals? *Systematic Zoology* 25 (2):174–191.

—. 1988. *Science as a Process: An Evolutionary Account of the Social and Conceptual Development of Science*. Chicago, IL: University of Chicago Press.

Hull, David Lee. 1965b. The effect of essentialism on taxonomy—Two thousand years of stasis (II). *The British Journal for the Philosophy of Science* XVI (61):1–18.

Huxley, Julian, ed. 1940. *The New Systematics*. London: Oxford University Press.

—. 1942. *Evolution: The Modern Synthesis*. London: Allen and Unwin.

Kitcher, Philip. 1989. Some puzzles about species. In *What the Philosophy of Biology is: Essays Dedicated to David Hull*, edited by M. Ruse, 183–208. Dordrecht: Kluwer.

Krüger, Lorenz et al. 1990. *The Probabilistic Revolution*. 2 vols. Cambridge, MA: MIT Press.

Kvist, L. et al. 2003. Evolution and genetic structure of the great tit (Parus major) complex. *Proceedings of the Royal Society of London—Series B: Biological Sciences* 270 (1523):1447–1454.

Liebers, Dorit et al. 2004. The herring gull complex is not a ring species. *Proceedings of the Royal Society of London—Series B: Biological Sciences* 271 (1542):893–901.

Martin, Cristopher H. et al. 2015. Complex histories of repeated gene flow in Cameroon crater lake cichlids cast doubt on one of the clearest examples of sympatric speciation. *Evolution* 69 (6):1406–1422.

Mayr, Ernst. 1940. Speciation phenomena in birds. *American Naturalist* 74 (752):249–278.

—. 1942. *Systematics and the Origin of Species from the Viewpoint of a Zoologist*. New York: Columbia University Press.

—. 1946. The naturalist in Leidy's time and today. *Proceedings of the Academy of Natural Sciences of Philadelphia* 98:271–276.

—. 1955. Karl Jordan's contribution to current concepts in systematics and evolution. In *Evolution and the Diversity of Life*, edited by Ernst Mayr, 297–306. Cambridge, MA: Harvard University Press.

—. 1957. Species concepts and definitions. In *The Species Problem: A Symposium Presented at the Atlanta Meeting of the American Association for the Advancement of Science, December 28–29, 1955*, Publication No 50, edited by Ernst Mayr, 1–22. Washington, DC: American Association for the Advancement of Science.

—. 1959. Darwin and the evolutionary theory in biology. In *Evolution and Anthropology: A Centennial Appraisal*, edited by B. J. Meggers, 1–10. Washington, DC: Anthropological Society of Washington.

—. 1963. *Animal Species and Evolution*. Cambridge, MA: The Belknap Press of Harvard University Press.

—. 1970. *Populations, Species, and Evolution: An Abridgment of Animal Species and Evolution*. Cambridge, MA: The Belknap Press of Harvard University Press.

—. 1976. Is the species a class or an individual? *Systematic Zoology* 25 (2):192.

—. 1982. *The Growth of Biological Thought: Diversity, Evolution, and Inheritance.* Cambridg, MA: The Belknap Press of Harvard University Press.

—. 1985. The species as category, taxon and population. In *Histoire du Concept D'Espece dans les Sciences de la Vie*, 303–320. Paris: Fondation Singer-Polignac.

—. 1988. The why and how of species. *Biology and Philosophy* 3 (4):431–441.

—. 1991. *One Long Argument: Charles Darwin and the Genesis of Modern Evolutionary Thought.* Cambridge, MA: Harvard University Press.

—. 1992. Species concepts and their application. In *The Units of Evolution: Essays on the Nature of Species*, edited by Marc Ereshevsky, 15–26. Cambridge, MA: MIT Press.

—. 1996. What is a species, and what is not? *Philosophy of Science* 63 (2):262–277.

—. 1997. *This is Biology: The Science of the Living World.* Cambridge, MA: The Belknap Press of Harvard University Press.

—. 1999. *Systematics and the Origin of Species from the Viewpoint of a Zoologist.* New York: Columbia University Press. Original edition, 1942. Reprint, With a new introduction.

—. 2000. The biological species concept. In *Species Concepts and Phylogenetic Theory: A Debate*, edited by Quentin D. Wheeler and Rudolf Meier, 17–29. New York: Columbia University Press.

Mayr, Ernst et al. 1953. *Methods and Principles of systematic zoology.* New York: McGraw-Hill.

Mayr, Ernst, and William B. Provine. 1980. *The Evolutionary Synthesis: Perspectives on the Unification of Biology.* Cambridge, MA: Harvard University Press.

Mazeroll, Anthony I., and Marc Weiss. 1995. The state of confusion in Discus taxonomy. In *Cichlids Yearbook*, 77–83. Cichlid Press.

Muller, Hermann Joseph. 1940. Bearings of the 'Drosophila' work on systematics. In *The New Systematics*, edited by Julian Huxley, 185–268. London: Oxford University Press.

Popper, Karl R. 1957a. *The Open Society and its Enemies.* 3rd ed. London: Routledge and Kegan Paul.

—. 1957b. *The Poverty of Historicism.* London: Routledge and Kegan Paul.

—. 1960. *The Poverty of Historicism.* 2nd ed. London: Routledge and Kegan Paul.

Ramsey, Justin, and Douglas W. Schemske. 1998. Pathways, mechanisms, and rates of polyploid formation in flowering plants. *Annual Review of Ecology and Systematics* 29 (4):467–501.

Rieppel, Olivier. 2003. Semaphoronts, cladograms and the roots of total evidence. *Biological Journal of the Linnean Society* 80 (1):167–186.

Salzburger, W. et al. 2002. Speciation via introgressive hybridization in East African cichlids? *Molecular Ecology* 11 (3):619–625.

Schilthuizen, Menno. 2000. Dualism and conflicts in understanding speciation. *BioEssays* 22 (12):1134–1141.

Sober, Elliott. 1984. *The Nature of Selection: Evolutionary Theory in Philosophical Focus.* Cambridge, MA: MIT Press.

Stauffer, Jay R. et al. 2002. Behaviour: An important diagnostic tool for Lake Malawi cichlids. *Fish and Fisheries* 3 (3):213–224.

Turesson, Göte. 1927. Contributions to the genecology of glacial relics. *Hereditas* 9 (1–3):81–101.

Wheeler, Quentin D., and Rudolf Meier, eds. 2000. *Species Concepts and Phylogenetic Theory: A Debate.* New York: Columbia University Press.

Winsor, Mary Pickard. 2001. Cain on Linnaeus: The scientist-historian as unanalysed entity. *Studies in the History and Philosophy of the Biological and Biomedical Sciences* 32 (2):239–254.

—. 2004. Setting up milestones: Sneath on Adanson and Mayr on Darwin. In *Milestones in Systematics: Essays from a Symposium Held within the 3rd Systematics Association Biennial Meeting, September 2001*, edited by David M. Williams and Peter L. Forey, 1–17. London: Systematics Association.

Section II

Modern Debates

In this section we review the broad species concepts presently in play. These are several classes of fundamental concepts, here divided into *Reproductive Isolation Concepts, Evolutionary Concepts, Phylogenetic Concepts, Ecological Concepts*, and a trashcan categorical of *Other Concepts*. From these a number of subsidiary concepts are composed.

Modern debates follow on from Dobzhansky's posing of the problem and Mayr's advocacy for the Biological Species Concept. I have left out of consideration the social and political aspects of the debate, as it is still too recent, and there is a dearth of considered studies regarding it, leaving Hull's *Science as a Process* to one side, since many of the participants in the "systematics wars" have said to me that they consider it to be one-sided and incorrect in his narrative. I am agnostic about this, as I was neither a participant nor do I have sufficient information to assess the claim.[1]

BIBLIOGRAPHY

Honenberger, Phillip. 2015. Grene and Hull on types and typological thinking in biology. *Studies in History and Philosophy of Science Part C: Studies in History and Philosophy of Biological and Biomedical Sciences* 50:13–25.

---

[1] But see Honenberger 2015 for a consideration of some of the issues.

8 Reproductive Isolation Concepts

They opened Buffon again and went into ecstasies at the peculiar tastes of certain animals.

...

They wanted to try some abnormal mating.

...

They made fresh attempts with hens and a duck, a mastiff and a sow, in the hope that monsters would result, but quite failing to understand anything about the question of species.

This is the word that designates a group of individuals whose descendants reproduce, but animals classified as different species may reproduce, and others, included in the same species, have lost the ability to do so.[1]

The notion that species are kinds of organisms delineated not by the decisions of the classifiers but by the reproductive behaviors and results of the organisms themselves is an old one. As we have seen, it was suggested as part of John Ray's definition in 1688, and also by Buffon in 1748, but it goes back in the form of the generative conception of species to the time of the Greeks. To a greater or lesser degree, it has been a component of nearly all species concepts since Linnaeus (his sexual system implicitly required reproductive isolation), and even now, it is a key component of phylogenetic species concepts and most other operational definitions.[2]

The generative conception of species has been applied to living beings effectively back to Epicurus and the neo-Platonists. That is to say, there has *always* been a requirement not only of constancy of form, but of the reproduction of form, in definitions of living species. This is surprising, since the implication or tacit assumption of many discussions, such as Mayr's[3] or Hull's,[4] has been that species had been for much of the history of the concept arid definitional constructs based on "essences." There can be no doubt that essence *has* played a role in species concepts. However, as we have seen, at least since Locke there has always been a tacit distinction between the *real* essence of a species (that is, what causes a species to *be* a species, its "real

---

[1] Flaubert 1976, 87. Thanks to Neil Thomason for bringing this wonderful passage, first published in 1881, to my attention.

[2] Cracraft 2000.

[3] Mayr, 1982.

[4] Hull 1965a, 1965b, 1988b.

TABLE 8.1

A Classification of Reproductive Isolating Mechanisms (RIMs)[6]

1. Reduction of contact
     (a) temporal
     (b) ecological
2. Reduction of mating frequency
     (c) ethological
     (d) morphological
        PREMATING
        —
        POSTMATING
3. Reduction of zygote formation
     (e) gametic and reproductive tract incompatibility
        PREZYGOTIC
        —
        POSTZYGOTIC
4. Reduction of hybrid survival
     (f) hybrid inviability
5. Reduction of gene flow through hybrids
     (g) hybrid ethological isolation
     (h) hybrid sterility
     (i) hybrid breakdown

Note: That the barriers here are not absolute—the RIMs only produce a reduced *frequency* of successful
            breeding.

constitution") and the *nominal* essence (that is, how we know the species, describe it, define it, and apply a name to it).[5] It is not clear that "essence" in these cases plays a definitional role—the Real Essence is not, by definition, definable.

Modern views of species are widely known and discussed, so we shall be brief in covering them, except for the ways that the most well-known reproductive isolation concept, the class of which we shall call for brevity "isolation concepts" (see Table 8.1), which is of course Ernst Mayr's biological species concept, or *biospecies*, developed. Also, there is considerable confusion and ambiguity in the phylogenetic concepts, so this will need discussion. I have attempted to be comprehensive, though, and provide all the current species concepts on offer since Dobzhansky introduced the issue into the Synthetic debate[7] (see Appendix B).

---

[5] Hull noted (in correspondence) that it is ahistorical and Whiggist in turn for historians such as myself to apply current standards to the Received Historians of the 1960s; he, Mayr, and Cain had what scholarly resources were then available. In large part due to their work, later research has identified the "missing links," and I am certain later work will overturn some of the claims made in this and other modern work too. It should not be thought that I am criticizing him or Mayr, etc., for failing to take into account later scholarship. That *would* be Whiggist.
[6] From Littlejohn 1969, 461.
[7] Building on Mayden 1997.

RECOGNITION CONCEPTS

There have been numerous conceptions of species following Mayr that are fully or partially isolationist. The views of Hugh Paterson in particular[8] have influenced Niles Eldredge and Elisabeth Vrba,[9] among others.[10] Paterson's version of the biospecies concept requires that organisms share a mating system, which he terms the "Specific-Mate Recognition System" (SMRS). He intends this to apply to plants, animals, and other organisms, so the term "recognition" should be taken in the same way the term "selection" is, without voluntaristic or cognitive implications. Paterson defines a species thus:

> We can, therefore, regard a species as that most inclusive population of individual biparental organisms which share a common fertilization system.[11]

He refers to this as the *Recognition Concept*. Mayrian and Dobzhanskyan concepts he calls Isolation Concepts. Since Paterson's version applies, as did previous isolation concepts, only to fully sexual and gendered (anisogamous) organisms, it follows that like them, he does not regard non-sexual organisms as forming species.[12] Paterson's criticisms of the Isolation Concept include its being teleological, since species are seen as "adaptive devices,"[13] but this is hardly fair. Biospecies are not, on Mayr's account, adaptive any more than higher taxa, and on his account, they are formed through isolation and subsequent local adaptation, not *in virtue of* their adaptations. The main difference between the isolation conception of Mayr and that of Paterson is well noted by Scoble, the following contributor to that volume, who notes, "... the BSC [biological species concept] can apply to only *actually*, not potentially, interbreeding groups of organisms. However, if the recognition concept of species ... is accepted ... then we may be able to directly compare at least some of the very characters involved in mate recognition in allopatric and allochronic populations."[14]

However, this does not follow, since we cannot say whether the mate recognition sequences are compatible enough to form viable progeny, either occasionally or repeatedly enough to make them count as the "same" species, until we have actually managed to test it in the lab and the wild. Scoble also notes the problem of uniparental species, and suggests that a homeostatic view might pertain. Since the SMRS is only one kind of homeostatic mechanism, there is no reason to restrict species-hood to sexual organisms only.

GENETIC CONCEPTS

We have several genetic concepts of species, ranging from Dobzhansky's comment that species are "the most inclusive Mendelian population," and Carson's comment

---

[8] Paterson 1985, 1993.
[9] Eldredge 1989, 1993.
[10] Lambert and Spencer 1995.
[11] Paterson 1985, 25, italics original.
[12] *Op. cit.*, 24.
[13] *Op. cit.*, 28.
[14] Scoble 1985, 33.

TABLE 8.2

Classification of Cohesion Mechanisms According to Templeton[17]

I. Genetic exchangeability: the factors that define the limits of spread of new genetic variants through *gene flow*

    A. Mechanisms promoting genetic identity through *gene flow*

        1. Fertilization system: the organisms are capable of exchanging gametes leading to successful fertilization

        2. Developmental system: the products of fertilization are capable of giving rise to viable and fertile adults

    B. Isolating mechanisms: genetic identity is preserved by the lack of *gene flow* with other groups

II. Demographic exchangeability: the factors that define the fundamental niche and the limits of spread of new genetic variants through *genetic drift* and *natural selection*

    A. Replaceability: *genetic drift* (descent from a common ancestor) promotes genetic identity

    B. Displaceability

        1. Selective fixation: *natural selection* promotes genetic identity by favoring the fixation of a genetic variant

        2. Adaptive transitions: *natural selection* favors adaptations that directly alter demographic exchangeability. The transition is constrained by:

            a. Mutational constraints on the origin of heritable phenotypic variation

            b. Constraints on the fate of heritable variation

                i. Ecological constraints

                ii. Developmental constraints

                iii. Historical constraints

                iv. Population genetic constraints

---

that a species is a "field for gene recombination,"[15] through to fully worked-out genetic conceptions like Templeton's and Wu's.

Templeton's view is that a species is

> The most inclusive group of organisms having the potential for genetic and/or demographic exchangeability.[16]

Templeton considers two criteria for being a species—genetic and demographic exchangeability. The first he specifies as "the factors that define the limits of spread of new genetic variants through *gene flow*," and the second as "the factors that define the fundamental niche and the limits of spread of new genetic variants through *genetic drift* and *natural selection*" (Table 8.2). A sexual species requires both criteria. He sees the cohesion concept as sharing a lot with the evolutionary species concept in this respect. These mechanisms generate a cohesive group, and the concept includes asexual taxa (purely in terms of demographic exchangeability) as well as sexual taxa (a mix of both). Nevertheless, Templeton does not indicate by what the *level* of species is indicated. His is more a matter of identifying what mechanisms

---

[15] Carson 1957. This is the phrase that Sewall Wright used, rather than the later "adaptive landscape," in his initial papers [Wright 1931, 1932] that introduced the idea of genetic drift.

[16] Templeton 1989, 25.

[17] From Table 2 in Templeton 1989.

generate species, whatever that level may be. Under both mechanisms, the spread of genetic variants through populations is the *sine qua non* of species rank.

As Ghiselin observes,[18] Templeton's conception resembles Mayr's in that species form as the result of genetic revolutions that constrict exchangeability, but he also favors Paterson's views on mate recognition. "Casting the definition in terms of one gene to be exchanged for another makes it roughly the same as the biological species definition." The requirement for demographic exchangeability means that organisms as units can replace each other in genetic populations, even if they do not do so in actuality.

Mallet has offered up "a species concept for the modern synthesis," the *genotypic cluster species concept* (GCSC) in which species were identified as "identifiable genotypic clusters."[19] Although a genotypic notion, it in many ways resembles the phenetic concept in which instead of morphological variables, the lack of intermediates lies in single genetic loci and ensembles of multiple loci. Like the biospecies concept, the GCSC applies only to populations in sympatry or parapatry, and lacks a nonarbitrary level of similarity or isolation of genetic alleles to specify specieshood.[20] In many ways, it also appears to be a genetic version of the Diagnostic species concept described in the next chapter. It has not found much use outside of this paper, however.

Wu's views are harder to pin down exactly; he may not in fact have a distinct species concept at all.[21] He claims that reproductive isolation (RI) has involved the entire genome under the BSC, but that it instead involves genetic isolation of *adaptive genes and gene complexes*, in particular, what he calls *speciation genes*. Even if there remains gene exchange for the non-adaptive genes, if there is RI for the adaptive genes, good species have evolved. His view relies on speciation itself, the process, resulting in some degree of genome isolation. Species are defined in terms of their genes having reached a particular level of RI, either Stage III—where populations have diverged sufficiently that they will not fuse in sympatry—or Stage IV—where populations are entirely isolated and will not mix at all genetically. Speciation genes have become a topic of interest to researchers with the rise of sympatric speciation models and examples, but Wu's genic conception falls within the traditional genetic and biospecies concepts.[22]

In a major review of research into speciation, Coyne and Orr present a revised version of the BSC in which they make a number of concessions to criticisms and in which they defend against some others.[23] Coyne and Orr's version, which I will here call the *limited BSC*, allows for limited introgression, does not insist upon integration of gene complexes, permits other definitions for asexuals and paraphyly of species, and treats ecological differentiation as necessary to the persistence of species (in sympatry, at

---

[18] Ghiselin 1997, 113.

[19] Mallet 1995, 296; cf. also Mallet 2000, Dres and Mallet 2002, Beltrán et al. 2002, Boenigk et al. 2012.

[20] Brower 2002.

[21] Wu 2001a, 2001b.

[22] Vogel et al. 1996, Bridle and Ritchie 2001, Butlin and Ritchie 2001, Rieseberg and Burke 2001, Van Alphen and Seehausen 2001, Orr and Masly 2004. The mode of speciation as a way to identify the nature of species is the subject of Wilkins 2007.

[23] Coyne and Orr 2004, chapter 1.

any rate) but not to the definition based on reproductive isolation. It allows that non-genetic isolation can form species (for example, intracellular infection by *Wollbachia*), and further that good species might later hybridize to form a new species. They accept Mayr's definition: "species are groups of interbreeding natural populations that are reproductively isolated from other such groups,"[24] although given their qualifying of that concept, it should perhaps read *mostly* interbreeding and isolated.

EVOLUTIONARY SPECIES CONCEPTS

Evolutionary species concepts derive from attempts to deal with the time dimension, something that the biological species concepts of Mayr and Dobzhansky tend to avoid— for them species exist at a given time horizon, and over evolutionary time, of course, they can change and split. Paleontologist George Gaylord Simpson proposed a definition:

> An evolutionary species is a lineage (an ancestral–descendant sequence of populations) evolving separately from others and with its own unitary evolutionary role and tendencies.[25]

Earlier, in a classic paper, he had expressed it slightly differently:

> ... a phyletic lineage (ancestral-descendant sequence of interbreeding populations) evolving independently of others, with its own separate and unitary evolutionary role and tendencies, is a basic unit in evolution.[26]

According to Cain, Simpson characterized the intergrading forms of an evolutionary species as "transients."[27] Species are thus many things: they are *populations* that form *phyletic lineages* through *interbreeding*, and which have *independent evolutionary roles* and *independent evolutionary tendencies*. These properties are reiterated in a later version of the evospecies concept of E. O. Wiley:

> A species is a single lineage of ancestral descendant populations of organisms which maintains its identity from other such lineages and which has its own evolutionary tendencies and historical fate.[28]

By 2000, this has ceased to be a "definition" and is now a "characterization":

> An evolutionary species is an entity composed of organisms that maintains its identity from other such entities through time and over space and that has its own independent evolutionary fate and historical tendencies.[29]

---

[24] Mayr 1995, 5.
[25] Simpson 1961, 153.
[26] Simpson 1951, quoted in Ghiselin 1997, 112f, who suggests the shift from the biospecies emphasis of this definition to the less explicitly interbreeding concept a decade later is due to his falling "increasingly under the spell of the set-theoretical treatment of the Linnaean hierarchy by Gregg ..., who, although mentioning in passing the possibility that species are something else, insisted they are classes." However, a set interpretation does not make it, *ipso facto*, a class interpretation.
[27] Cain 1954, 111.
[28] Wiley 1978, 18.
[29] Wiley and Mayden 2000, 73.

The novel elements here include the "entification"[30] of evospecies in place of the population stipulation, to accommodate asexual organisms,[31] and the inclusion of time and space to indicate the processual nature of the concept. However, one thing that all the evolutionary concepts fail to do adequately is to specify what counts as "independence." If a parasitical species coevolves with its host, which is commonly the case, are they still independent? And of course, there is a metaphorical problem with "fate," given the universal view that evolution is not predetermined, but we can assume that Wiley and his colleagues understand that; this merely specifies that the outcomes of evolution for any one species are unique to that species. In a sense, this is a matter of evolving a unique set of traits.

Evolutionary species concepts have been assumed by some critics to imply a gradual and constant rate of evolution.[32] This need not be the case, however, as Simpson's own classical work on evolutionary rates indicates,[33] but it is also sometimes held that species change over the entire course of their duration. Opposing this, the punctuated equilibrium theorists have argued that species tend to remain stable once evolved.[34] If correct, or rather *when* correct, for it is now accepted to be the case for many species if not all, the "fate" of the species involves the stasis of the unique set of traits once achieved.

Evolutionary species concepts, as with some phylogenetic species concepts, tend to adopt the metaphysical species-as-individuals thesis of Ghiselin and Hull.[35] As Wiley and Mayden note,

Evolutionary species are logical individuals with origins, existence, and ends.[36]

However, if species are correctly thought of as *logical* individuals, they still need not be *historical* individuals. There are three different notions under the term "individual" in the species concept literature, and they ought to be kept distinct:

1. Individuals as *metaphysical particulars* (i.e., not universals or natural kinds)
2. Individuals as *coherent functional objects* like organisms
3. Individuals as *clusterings of properties* in a phenomenally salient manner

I shall expand on this later.

Evospecies, as we might call these entities, are something of a hybrid notion, to my mind. They are phyletic objects, but they are often presented as achieving grades of organization, and often run with evolutionary systematics (itself a hybrid system, conjoining phylogenetic and adaptive conceptions of classification).

---

[30] I owe this useful term to Henry Plotkin (*pers. comm.*).

[31] Wiley and Mayden 2000.

[32] Eldredge and Gould 1972, Ayala 1982, Gould 1982, Ghiselin 1988.

[33] Simpson 1944.

[34] Eldredge and Gould 1972, Gould and Eldredge 1977, Eldredge 1985, Eldredge et al. 1997, Gould 2002.

[35] Ghiselin 1974, Hull 1976, 1978, 1981, Ghiselin 1988, Hull 1992, Ghiselin 1997.

[36] Wiley and Mayden 2000, 74.

LINEAGES

A version of the evospecies is Kevin de Queiroz's *general lineage conception*, sometimes misleadingly referred to as the *Universal Concept*. The key term *lineage* is taken from Hull's metaphysics,[37] based on Simpson's evolutionary species definition,[38] as revised by de Queiroz.[39] It represents the *generative* aspect of species that we saw from early in the history of the concept. It broadly means

> ... a series of entities forming a single line of ancestry and descent. ...
> Lineages in the sense described above are unbranched; that is, they follow a single path or line anytime an entity in the series has more than one descendant. ... Consequently, lineages are not to be confused with clades, clans, and clones—though the terms are often used interchangeably in the literature.[40]
> ... a lineage (an ancestral-descendant sequence of populations) ...[41]

A problem with this is that while every species is a lineage (that is, forms generative sequences), not every lineage forms a species. Essentially a lineage is any part of the evolutionary tree that is unbroken (see Figure 8.1), but we know that species have within them subordinate lineages, of populations, of genes and of kin groups. As we shall see with phylogenetic species concepts in the next chapter, this leads to problems in specifying what level of lineage one wants to call a species.

For example, harbor seals and albatrosses turn out to have extensive within-species variations that form lineages.[42] The decision of the level of diversity of lineages is informed by other considerations, including what the taxonomist considers a "significant" amount of genetic divergence or not. But this means that a lineage, or indeed an evolutionary species, is determined in the end by phenetic criteria: the rank depends on arbitrary (but not necessarily subjective) lines of divergence in the characters or genetic matrix used.

One attempt to deal with this is the Genealogical Concordance species concept of Avise and Ball.[43] De Querioz considers this a phylogenetic conception of species,[44] but in my opinion, it is better thought of as an evolutionary conception. It is basically the view that population subdivisions concordantly identified by multiple independent genetic traits constitute the population units worthy of recognition as phylogenetic taxa. Relying upon genetic coalescence theory, in which genetic lineages are traced back to original populations,[45] a species is considered to exist when it shares a number of genetic coalescents. More recently, multispecies coalescence (MSC) has been used to attempt to delimit species.[46]

---

[37] Hull 1980, 1981, 1984, 1988b, 1988a, 1989, 1992, 1997.
[38] Simpson 1951, 1961.
[39] de Queiroz 1998, 1999.
[40] de Queiroz and Donoghue 1990, 50.
[41] Simpson 1951.
[42] Burg 1999, Burg and Croxall 2004.
[43] Avise and Ball Jr. 1990.
[44] de Queiroz 2007.
[45] Avise and Wollenberg 1997, Wang et al. 1997, Knowles and Carstens 2007.
[46] Carstens and Dewey 2010, Carstens et al. 2013, Yang and Rannala 2014, Lamanna et al. 2016, Leavitt et al. 2016, but see Sukumaran and Knowles 2017 for criticism, arguing that MSC delineates population structure, not species.

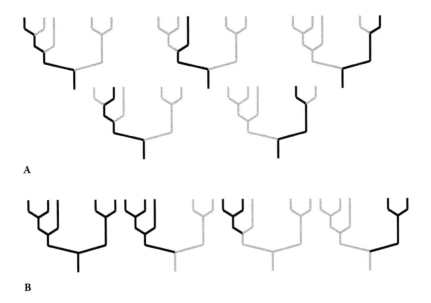

FIGURE 8.1   Lineages, redrawn from de Quierioz. (Redrawn from Figure 3.1 in de Queiroz 1999).

Lineages contrasted with clades, clans, and clones.... . All of the branching diagrams represent the same phylogeny with different lineages highlighted in **A** and different clades, clans, or clones highlighted in **B**. Notice that the lineages are unbranched and partially overlapping, whereas the clades, clans, or clones are branched and either nested or mutually exclusive. Additional (partial) lineages can be recognized for paths beginning at various internal nodes.

BIBLIOGRAPHY

Avise, John C., and R. Martin Ball Jr. 1990. Principles of genealogical concordance in species concepts and biological taxonomy. In *Oxford Surveys in Evolutionary Biology*, edited by Douglas J. Futuyma and Janis Antonovics, 45–67. Oxford: Oxford University Press.

Avise, John C., and Kurt R. Wollenberg. 1997. Phylogenetics and the origin of species. *Proceedings of the National Academy of Sciences of the United States of America* 94 (15):7748–7755.

Ayala, Francisco José. 1982. Gradualism versus punctuationism in speciation: Reproductive isolation, morphology, genetics. In *Mechanisms of Speciation: Proceedings from the International Meeting on Mechanisms of Speciation, sponsored by the Academia Nazionale dei Lincei, May 4–8, 1981, Rome, Italy*, edited by C. Barigozzi. New York: Alan R Liss.

Beltrán, Margarita et al. 2002. Phylogenetic discordance at the species boundary: Comparative gene genealogies among rapidly radiating heliconius butterflies. *Molecular Biology and Evolution* 19 (12):2176–2190.

Boenigk, Jens et al. 2012. Concepts in protistology: Species definitions and boundaries. *European Journal of Protistology* 48 (2):96–102.

Bridle, J. R., and M. G. Ritchie. 2001. Assortative mating and the genic view of speciation. *Journal of Evolutionary Biology* 14 (6):878–879.

Brower, Andrew V. Z. 2002. Cladistics, populations and species in geographical space: The case of *Heliconius* butterflies. In *Molecular Systematics and Evolution: Theory and Practice*, edited by R. DeSalle et al., 5–15. Switzerland: Birkhäuser Verlag.

Burg, T. M. 1999. Isolation and characterization of microsatellites in albatrosses. *Molecular Ecology* 8 (2):338–341.

Burg, T. M., and J. P. Croxall. 2004. Global population structure and taxonomy of the wandering albatross species complex. *Molecular Ecology* 13 (8):2345–2355.

Butlin, R., and M. G. Ritchie. 2001. Evolutionary biology. Searching for speciation genes. *Nature* 412 (6842):31, 33.

Cain, Arthur J. 1954. *Animal Species and Their Evolution*. London: Hutchinson University Library.

Carson, Hampton L. 1957. The species as a field for gene recombination. In *The Species Problem: A Symposium Presented at the Atlanta Meeting of the American Association for the Advancement of Science, December 28–29, 1955*, Publication No 50, edited by Ernst Mayr, 23–38. Washington, DC: American Association for the Advancement of Science.

Carstens, Bryan C., and Tanya A. Dewey. 2010. Species delimitation using a combined coalescent and information-theoretic approach: An example from North American Myotis bats. *Systematic Biology* 9 (4):400–414.

Carstens, Bryan C. et al. 2013. How to fail at species delimitation. *Molecular Ecology* 22 (17):4363–4601.

Coyne, Jerry A., and H. Allen Orr. 2004. *Speciation*. Sunderland, MA: Sinauer Associates.

Cracraft, Joel. 2000. Species concepts in theoretical and applied biology: A systematic debate with consequences. In *Species Concepts and Phylogenetic Theory: A Debate*, edited by Quentin D. Wheeler and Rudolf Meier, 3–14. New York: Columbia University Press.

de Queiroz, Kevin. 1998. The general lineage concept of species, species criteria, and the process of speciation. In *Endless Forms: Species and Speciation*, edited by Daniel J Howard and Stewart H Berlocher, 57–75. New York: Oxford University Press.

—. 1999. The general lineage concept of species and the defining properties of the species category. In *Species, New Interdisciplinary Essays*, edited by R. A. Wilson, 49–88. Cambridge, MA: Bradford/MIT Press.

—. 2007. Species concepts and species delimitation. *Systematic Biology* 56 (6):879–886.

de Queiroz, Kevin, and Michael J. Donoghue. 1990. Phylogenetic systematics and species revisited. *Cladistics* 6:83–90.

Dres, M., and J. Mallet. 2002. Host races in plant-feeding insects and their importance in sympatric speciation. *Philosophical Transactions of the Royal Society B: Biological Sciences* 357 (1420):471–492.

Eldredge, Niles. 1985. *Time Frames: The Evolution of Punctuated Equilibria*. Princeton, NJ: Princeton University Press.

—. 1989. *Macroevolutionary Dynamics: Species, Niches, and Adaptive Peaks*. New York: McGraw-Hill.

—. 1993. What, if anything, is a species? In *Species, Species Concepts, and Primate Evolution*, edited by William H. Kimbel and Lawrence B. Martin, 3–20. New York: Plenum Press.

Eldredge, Niles, and Stephen J. Gould. 1972. Punctuated equilibria: An alternative to phyletic gradualism. In *Models in Paleobiology*, edited by T. J. M. Schopf, 82–115. San Francisco, CA: Freeman Cooper.

Eldredge, Niles et al. 1997. On punctuated equilibria. *Science* 276 (April 18):337c–341c.

Flaubert, Gustave. 1976. *Bouvard and Pécuchet, with the Dictionary of Received Ideas*. Translated by A. J. Krailsheimer. Harmondsworth: Penguin. Original edition, 1881.

Ghiselin, Michael T. 1974. A radical solution to the species problem. *Systematic Zoology* 23 (4):536–544.

—. 1988. Species individuality has no necessary connection with evolutionary gradualism. *Systematic Zoology* 37:66–67.

—. 1997. *Metaphysics and the Origin of Species*. Albany, NY: State University of New York Press.

Gould, Stephen Jay. 1982. Darwinism and the expansion of evolutionary theory. *Science* 216 (4544):380–387.

—. 2002. *The Structure of Evolutionary Theory*. Cambridge, MA: The Belknap Press of Harvard University Press.

Gould, Stephen Jay, and Niles Eldredge. 1977. Punctuated equilibria: The tempo and mode of evolution reconsidered. *Paleobiology* 3 (2):115–151.

Hull, David Lee. 1965a. The effect of essentialism on taxonomy—Two thousand years of stasis. *British Journal for the Philosophy of Science* 15 (60):314–326.

—. 1965b. The effect of essentialism on taxonomy—Two thousand years of stasis (II). *The British Journal for the Philosophy of Science* XVI (61):1–18.

—. 1976. Are species really individuals? *Systematic Zoology* 25 (2):174–191.

—. 1978. A matter of individuality. *Philosophy of Science* 45 (3):335–360.

—. 1980. Individuality and selection. *Annual Review of Ecology and Systematics* 11:311–332.

—. 1981. Units of evolution: A metaphysical essay. In *The Philosophy of Evolution*, edited by Uffe Juul Jensen and Rom Harré, 23–44. Brighton, UK: Harvester Press.

—. 1984. Historical entities and historical narratives. In *Minds, Machines, and Evolution*, edited by Christopher Hookway, 17–42. Cambridge: Cambridge University Press.

—. 1988a. A mechanism and its metaphysics: An evolutionary account of the social and conceptual development of science. *Biology and Philosophy* 3 (2):123–155.

—. 1988b. *Science as a Process: An Evolutionary Account of the Social and Conceptual Development of Science*. Chicago, IL: University of Chicago Press.

—. 1989. *The Metaphysics of Evolution*. Albany, NY: State University of New York Press.

—. 1992. Individual. In *Keywords in Evolutionary Biology*, edited by Evelyn Fox Keller and Elisabeth Anne Lloyd, 180–187. Cambridge, MA: Harvard University Press.

—. 1997. The ideal species concept—And why we can't get it. In *Species: The Units of Diversity*, edited by Michael F. Claridge et al., 357–380. London: Chapman & Hall.

Knowles, L. Lacey, and Bryan C. Carstens. 2007. Delimiting species without monophyletic gene trees. *Systematic Biology* 56 (6):887–895.

Lamanna, F. et al. 2016. Species delimitation and phylogenetic relationships in a genus of African weakly-electric fishes (Osteoglossiformes, Mormyridae, Campylomormyrus). *Molecular Phylogenetics and Evolution* 101:8–18.

Lambert, David M., and Hamish G. Spencer, eds. 1995. *Speciation and the Recognition Concept: Theory and Application*. Baltimore, MD: Johns Hopkins University Press.

Leavitt, Steven D. et al. 2016. Hidden diversity before our eyes: Delimiting and describing cryptic lichen-forming fungal species in camouflage lichens (Parmeliaceae, Ascomycota). *Fungal Biology* 120 (11):1374–1391.

Littlejohn, Murray J. 1969. The systematic significance of isolating mechanisms. In *Reflections on Systematic Biology; Proceedings of an International Conference, University of Michigan*, June 14–16, 1967, 459–482. Washington, DC: National Academy of Sciences.

Mallet, James. 1995. The species definition for the modern synthesis. *Trends in Ecology and Evolution* 10 (7):294–299.

—. 2000. Species and their names. *Trends in Ecology and Evolution* 15 (8):344–345.

Mayden, Richard L. 1997. A hierarchy of species concepts: The denoument in the saga of the species problem. In *Species: The Units of Diversity*, edited by Michael F. Claridge et al., 381–423. London: Chapman & Hall.

Mayr, Ernst. 1982. *The growth of biological thought: Diversity, evolution, and inheritance*. Cambridge, MA: The Belknap Press of Harvard University Press.

Mayr, Ernst. 1995. Species, classification, and evolution. In *Biodiversity and Evolution*, edited by R. Arai et al., 3–12. Tokyo: National Science Museum Foundation.

Orr, H. Allen, and John P. Masly. 2004. Speciation genes. *Current Opinion in Genetics and Development* 14 (6):675–679.

Paterson, Hugh E. H. 1985. The recognition concept of species. In *Species and Speciation*, edited by Elisabeth S. Vrba, 21–29. Pretoria: Transvaal Museum.

—. 1993. *Evolution and the Recognition Concept of Species. Collected Writings*. Baltimore, MA: John Hopkins University Press.

Rieseberg, Loren H., and J. M. Burke. 2001. A genic view of species integration. *Journal of Evolutionary Biology* 14 (6):883–886.

Scoble, M. J. 1985. The species in systematics. In *Species and Speciation*, edited by E. Vrba, 31–34. Pretoria: Transvaal Museum.

Simpson, George Gaylord. 1944. *Tempo and Mode in Evolution*. New York: Columbia University Press.

—. 1951. The species concept. *Evolution* 5 (4):285–298.

—. 1961. *Principles of Animal Taxonomy*. New York: Columbia University Press.

Sukumaran, Jeet, and Lacey L. Knowles. 2017. Multispecies coalescent delimits structure, not species. *Proceedings of the National Academy of Sciences of the United States of America* 114 (7):1607–1612.

Templeton, Alan R. 1989. The meaning of species and speciation: A genetic perspective. In *Speciation and its Consequences*, edited by D. Otte and J. A. Endler, 3–27. Sunderland, MA: Sinauer.

Van Alphen, Jacques J. M., and Ole Seehausen. 2001. Sexual selection, reproductive isolation and the genic view of speciation. *Journal of Evolutionary Biology* 14 (6):874–875.

Vogel, Johannes Christian et al. 1996. A non-coding region of chloroplast DNA as a tool to investigate reticulate evolution in European *Asplenium*. In *Pteridology in Perspective*, edited by J. M. Camus et al., 313–327. Kew: Royal Botanic Gardens.

Wang, Rong Lin et al. 1997. Gene flow and natural selection in the origin of *Drosophila pseudoobscura* and close relatives. *Genetics* 147 (3):1091–1106.

Wiley, E. O. 1978. The evolutionary species concept reconsidered. *Systematic Zoology* 27:17–26.

Wiley, Edward Orlando, and Richard L. Mayden. 2000. The evolutionary species concept. In *Species Concepts and Phylogenetic Theory: A Debate*, edited by Quentin D. Wheeler and Rudolf Meier, 70–89. New York: Columbia University Press.

Wilkins, John S. 2007. The dimensions, modes and definitions of species and speciation. *Biology and Philosophy* 22 (2):247–266.

Wright, Sewall. 1931. Evolution in Mendelian populations. *Genetics* 16 (2):97–159.

—. 1932. The roles of mutation, inbreeding, crossbreeding and selection in evolution. In *Proceedings of the Sixth International Congress of Genetics*, edited by Donald F. Jones, 356–366. Brooklyn, NY: Brooklyn Botanic Garden.

Wu, Chung-I. 2001a. Genes and speciation. *Journal of Evolutionary Biology* 14 (6):889–891.

—. 2001b. The genic view of the process of speciation. *Journal of Evolutionary Biology* 14 (6):851–865.

Yang, Ziheng, and Bruce Rannala. 2014. Unguided species delimitation using DNA sequence data from multiple loci. *Molecular Biology and Evolution* 31 (12):3125–3135.

9 Phylogenetic Species Concepts

Phylogenetic conceptions of species are very similar in some ways to the classifications of the older logic of division discussed in Section I. Hennig's classification is done by division, from synapomorphies to autapomorphies, or in other words, top-down. Hennig distinguished between derived and underived states in taxonomy. *Plesiomorphic* characters are underived in a monophyletic group (a clade) from which the transformations begin, and *apomorphies* are those derived from them in evolution. An apomorphy of a group can be a plesiomorphy of a clade contained within that group. It follows that in Hennig's classificatory scheme that what makes a plesiomorphy and an apomorphy is not absolute rank, but its relative position in a cladogram. In the older terminology, we discussed in Section 1, they are relative to the differentiae, the predicates and the properties those predicates denote which make the differences between taxa. Hennig has no system of absolute taxonomic levels, although he did also attempt to develop such an absolute set of ranks based upon time of evolutionary divergence.[1] There are just taxa, and they are arrayed in a flexible local hierarchy.[2] In the classical logic, the base-level taxonomic rank—the *infimae species*—was a taxic entity that was not itself the genus of any other species and which contained only individuals. Likewise, Hennig has terminal taxa, and these he calls *species*, following Linnaean tradition. Where the *infimae species* and the Hennigian *species* differ from the Linnaean species, however, is that the former are derived from the general group being sequentially divided into subgroups on the basis of characters shared (a single dichotomous key in the early naturalists' practice, on an implied parsimony criterion of many characters for Hennig), while Linnaeus assumed fixed taxon ranks. Linnaean species are an absolute rank, and so also are the higher taxa they comprise.

Species concepts based upon the phylogeny of the groups of organisms are called "phylogenetic," but there are several *phylospecies* concepts, as I will call them, and there are several sub-versions of them in turn. All proponents of phylospecies concepts claim both Darwin and Hennig as their inspirations, but it is arguable how closely each of the modern views relate to those initial expressions of classification of taxa by descent.

In some ways, phylogenetic classification is an outgrowth of the ideas expressed by Haeckel and others during the late nineteenth century. Haeckel coined the term *monophyly*, which Hennig later appropriated and more strictly defined. This is a

---

[1] Hennig 1966, 184–192.
[2] Nelson 1989.

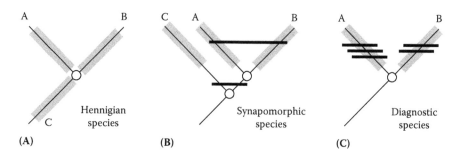

FIGURE 9.1 Phylogenetic species concepts. **A.** Hennigian species are formed at speciation and extinguished at the next speciation event or extinction. The cladogram is a history as well as a classification. **B.** Synapomorphic species are terminal nodes in a cladogram or evolutionary tree specified by synapomorphies. They are recognized by a lack of further discrete patterns of ancestry and descent (i.e., by having no further synapomorphies). **C.** Diagnostic species are the smallest terminal nodes in a cladogram, marked by their autapomorphies (unique sets of characters depending on the authors). All three conceptions may also specify that species are either monophyletic or not.

critical notion in the context of phylogenetic systematics, or, as it is popularly known, cladistics. In Hennig's definition, a monophyletic group is an ancestral species and *all* of its descendent species.[3] Under more formally defined notions of phylogeny, sometimes called pattern cladism, a monophyletic group is a proper subset of some set of taxa, without necessarily implying a particular historical relationship.[4] On this account, a monophyletic group is definable in terms of it having unique characters that are not shared by other taxa. These are called *apomorphies* in Hennig's terminology, or "derived characters." This is a relative term: an apomorphy for one set of taxa is a *plesiomorphy* (an underived, or primitive, character) for a more inclusive set of taxa. If one or more taxa share an apomorphy, it is called a *synapomorphy*, and if only a single taxon carries an apomorphy, it is called an *autapomorphy*.

There are fundamentally three phylospecies concepts (Figure 9.1). The first, defined initially by Hennig, is sometimes called the *Hennigian species concept*.[5] It rests on a conventional decision Hennig made to be consistent with his broader philosophy of classification, and which we shall call the *Hennigian convention*. On this account, if a new species arises by splitting from a parental species, then we shall say that the parental species, no longer being monophyletic (that is, now being paraphyletic), has become extinct, and there are now two novel species.

---

[3] There is a rival conception, *holophyly*, which was coined by Ashlock 1971 (see the response by Nelson 1971) to free up the term *monophyly* to mean a more loose sense of "clade" in which paraphyly could be tolerated; Hennigian monophyly would thus be holophyly on Ashlock's view. In this writer's opinion, this is a concession to the "evolutionary systematics" of Mayr and collaborators, which seeks to group by similarity as well as genealogy [Mayr and Ashlock 1991]. This point was made early on Pratt 1976, 380:

> ...for Mayr the principle of unity in classification is simply evolutionary relationships. He wishes rather to erect groupings on the basis of evolutionary relationships *in so far as these are correlated with resemblances and differences between genotypes.* [Italics original]

[4] Vanderlaan et al. 2013 discuss the development, and many meanings, of the term *monophyly*.
[5] Meier and Willmann 2000.

The second phylospecies concept we might call the *synapomorphic concept.* This further divides into the Monophyletic Concept of Mishler et al. and those who merely apply synapomorphic criteria.[6] A species is, under this account, a taxon in which there is no further division or taxonomic subdivisions—it is the unit of phylogenetic analysis. Mishler's version adds the view that species must be monophyletic.

The third phylospecies concept, which I shall call the *apomorphic species concept,*[7] is derived from the work of Donn Rosen.[8] In various versions, it tends to rely upon the diagnosability of taxa,[9] or, as we may otherwise say, it is a largely epistemological notion of species. All the diagnostic concepts rely upon a species being the "terminal taxon" in a cladogram. Some proponents apply an evolutionary exegesis to this, while others restrict it to a diagnostic relationship, a distinction often referred to as the ontology–epistemology aspects of species.[10] Of course, a phylospecies of one kind can also be a phylospecies of another. The division of the phylospecies into two main kinds, ignoring Hennigian species for the moment, is a reflection of the larger taxonomic debates. These raise the questions (i) should taxonomic classification proceed in terms of descent alone or on the basis of similarity (cladism versus gradism)? and (ii) if classification rests on clades, are homologies (apomorphies) indicators of history, or are they patterns that are evidence in favor of a historical reconstruction but not themselves a model of evolution? Briefly, this is the distinction between ontological and epistemological notions of classification again.

Those who take the epistemological classification position (grouping in terms of synapomorphic relationships only) often fall under the rubric of pattern cladists, although this is not necessarily so. Mishler and Theriot, for example, take synapomorphies as indicating relations at a time, synchronically, to avoid time paradoxes (in which a species becomes paraphyletic after the speciation of daughter species, if it is treated as a sister taxon). Only tokogenetic relations are treated as diachronic relations under this view.[11]

Those who take the diagnosis of monophyly of a group to give a direct hypothesis of evolutionary history are the so-called orthodox, or "traditional" cladists, called "process cladists."[12] Process cladism tends to treat taxa as relationships between organisms, while pattern cladism tends to see taxa as composite entities of which organisms are members. In this regard, the process orthodoxy is more closely allied to some aspects of the evolutionary species concepts of recent times.[13]

---

[6] Mishler and Theriot 2000.

[7] Meier and Willmann 2000, 36–37, call the *Apomorphic Species Concept* the Phylogenetic Species Concept *simpliciter*, and Davis 1997 calls the process or *Synapomorphic Species Concept* the *Autapomorphic Species Concept*, in direct contradiction to my usage in the first edition.

[8] Rosen 1979.

[9] Cronquist 1978.

[10] Wheeler and Platnick 2000. The ontology–epistemology distinction is discussed in Chapter 14.

[11] Mishler and Theriot 2000.

[12] The term "process cladism" was introduced in print in Ereshefsky 2000. It is unclear to me that either pattern cladism or process cladism form monolithic schools, and the differences of opinion on these matters needs to be investigated. Thanks to David Williams for noting this, and catching my transposition of their core ideas.

[13] It is not, in my opinion, true that pattern cladism commits its adherents to an antievolutionary view of taxa. Neither is it true that it is an essentialistic view of taxonomy, as some—notably Mayr—have claimed. It is, however, typological. But then, so is all systematics.

Phylogenetic approaches to species are founded on one of three criteria—synapomorphy, autapomorphy, or prior phenetic identity as operational taxonomic units (OTUs). Roughly, *synapomorphy* acts as the classical logical notion of genera, *autapomorphy* acts as the classical notion of differentia, and in the case of phenetic OTUs, as inputs into a cladistic analysis, while species here are types that are known prior to the formal division. Therefore, it is something of a misconstrual to think that there are only two "phylospecies" concepts besides the Hennigian account.[14]

HENNIGIAN, OR INTERNODAL, SPECIES

In the work that began the cladistic revolution and which remains the source for much cladistic thinking, so clearly and consistently was it expressed, Hennig treated species in an unusual manner.[15] He begins by defining "semaphoronts," or "character bearers" as the elements of systematics. These are effectively stages or moments in the lifecycle of organisms of the taxon. Individuals themselves, or more exactly the ontogenetic lifecycles of individuals, are related through reproductive relationships he called "tokogenetic" relationships. When tokogenetic relationships begin to diverge, they form species, which arise

> ... when gaps develop in the fabric of the tokogenetic relationships. The genetic relationships that interconnect species we call phylogenetic relationships. The structural picture of the phylogenetic relationships differs as much from that of the individual tokogenetic relationships as the latter does from the structural picture of the ontogenetic relationships. In spite of these differences in their structural pictures, the phylogenetic, tokogenetic, and ontogenetic relationships are only portions of a continuous fabric of relationships that interconnect all semaphoronts and groups of semaphoronts. With Zimmermann we will call the totality of these the "hologenetic relationships."[16]

Hennig illustrated this with a now-famous diagram (Figure 9.2).

Hennig is sometimes read as asserting that species cease to exist when they are divided; that is, when the hologenetic relationships cease to be a single set (the species sets are represented in the diagram by the large ellipses). Although Hennig assumes that species are reproductive communities of harmoniously cooperating genes, as Dobzhansky had said, and that species are reproductive groups, as Mayr had said, Hennig nevertheless assumes that species are phylogenetic lineages. In fact, he says that species are defined over spatial dimensions as well, as

> ... a complex of spatially distributed reproductive communities, or if we call this relationship in space "vicariance," as a complex of vicarying communities of reproduction.[17]

---

[14] Brent Mishler pointed out to me that what I had been calling a single class of species concepts, under the term "Monophyletic Species Concepts" (now Phylogenetic Taxon species) was actually pretty diverse, and that some (e.g., Cracraft) did not think species had to be monophyletic. I have attempted to disentangle these views from each other in the text.

[15] Hennig 1966, 28–32.

[16] *Op. cit.*, 30.

[17] *Op. cit.*, 47.

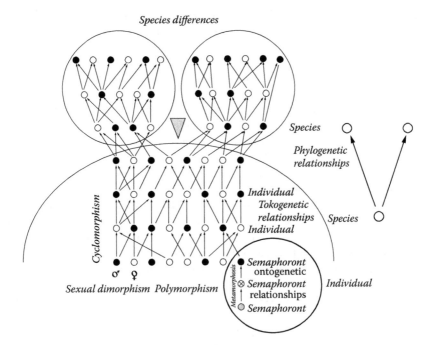

FIGURE 9.2   Hennig's view of systematic relationships. The larger circles are the "hologenetic" relationships that represent species. (Redrawn from Hennig 1966: 31, figure 6.)

The sort of vicariance he has in mind includes trophic replaceability,[18] but also other ecological dimensions, including temperature races.

However, the most significant aspect of Hennig's definition of species lies in the temporal dimension,[19] where he notes that species are to be delimited by events of speciation:

> The limits of the species in the longitudinal section through time would consequently be determined by two processes of speciation: the one through which it arose as an independent reproductive community, and the other through which the descendants of this initial population ceased to exist as a homogenous reproductive community.

Transformations of the species morphology and genetic composition within these two events do not affect the identity of the species, because the species is defined here as a homogenous reproductive community. In effect, Hennig is taking the biospecies isolationist concept to its limits.[20] Species are extinguished at the next speciation event. In his earlier work in German, he was even more concise:

---

[18]   *Op. cit.*, 49–50.
[19]   *Op. cit.*, 58–60.
[20]   Hennig himself said, "the biological species concept, used by me since 1950, does not essentially differ from that of Mayr" [Hennig 1975, 255].

When some of the tokogenetic relationships among the individuals of one species cease to exist, it disintegrates into two species and ceases to exist. It is the common stem species of the two daughter species.[21]

This has become known as the *Hennig Convention*. It has been strongly criticized from all sides, by biospecies proponents, evospecies proponents, and other phyogeneticists,[22] not least because it seems to be an arbitrary way to delimit species taxa. Mayr, for example, takes Hennig to be making a substantive claim on the ways species are formed at speciation, and criticizes it on the basis that the "parent" species can remain unchanged even though it is no longer monophyletic. But it seems to me that the critics have overlooked the most charitable interpretation of the Hennig Convention—it is a convention about *naming* and *denotation*.[23] In short, the *name* of a species is extinguished at speciation. This follows from Hennig's views about the task of systematics. Using (and citing) Woodger and Gregg,[24] and the views of Woodger in particular, about sets in classification,[25] Hennig strives to ensure that there is no ambiguity of reference in the sets named in systematics. Since as soon as a set is divided there is ambiguity about which of the two resultant sets is being referred to by a prior name, Hennig proposes to extinguish the now-ambiguous name and create two new ones. However, he seems to equivocate over whether or not they are new *entities*.

The Hennigian concept of species has been recently expanded and defended by Meier and Willmann,[26] who have proposed a modified Hennigian Species Concept:

> Species are reproductively isolated natural populations or groups of natural populations. They originate via the dissolution of the stem species in a speciation event and cease to exist either through extinction or speciation.[27]

---

[21] Hennig 1950, 102, translated in Meier and Willmann 2000, 30.
[22] See the citations in Meier and Willmann 2000, 31.
[23] Rieppel 2011 argues against my interpretation, and backs it up with a passage from Hennig:

> If one focuses on the genealogical relationships [rather than on morphological or ecological similarity], then one has to accept that a one-to-one relation exists not only between the fossil [i.e., ancestral] species *A* and one of the extant [i.e., descendant] species (*B* or *C*), but also between *A* on the one hand and *B* + *C* on the other. In spite of biological identity between *A* and *B* there exist genealogical relations that also include species *C*, from where it follows that the fossil population (species *A*) cannot be considered identical with only one of the extant populations (with only *B* or *C* respectively) …

Rieppel goes on to say

> What Hennig wanted to say was that the ancestral species (*A*) stands in a one-to-one relation not only with one (*B*) or the other (*C*) of its descendant species but also with the union of *A* and *B*, where $(A \cup B)$ is the monophyletic taxon that emerges from the speciation process.

Rieppel may be correct about Hennig's intentions, but Hennig's phrase "cannot be considered" still sounds to me as if it is about denotation and reference. In any case, the charity offered here is not based on the assumption that Hennig himself was entirely clear on this matter.

[24] See the excellent discussion in Wheeler and Platnick 2000.
[25] Woodger 1937.
[26] Meier and Willmann 2000, 31; cf. Willmann 1985a, 1985b, 1997, 1997.
[27] Quoted from Willmann 1985a, 80, 176.

Like Hennig and most proponents of the biospecies and other isolationist conceptions, they reject the use of a single taxonomic category such as *species* to apply to asexual ("uniparental") organisms. Instead, they call them "agamotaxa."

A set-theoretic model of Hennig's concept has been provided by Kornet (under the title of "internodal species concept") who formalizes the notion of internodality (INT) in a cladogram in terms of the tokogenetic relations being permanently divided, so that any individual in the hologenetic group has a "gross dynastic relationship" (GDYN) with any other individual in that group.[28] Type specimens thus fall out as an appropriate individual to start the GDYN analysis to establish the INT relation. Kornet thus makes the relation of type specimens to the rest of the species scrutable (see below) in set theoretic terms.

The Hennig account is fundamentally a biospecies concept. Hennig himself accepted that species were reproductively isolated, and the criteria used for identifying the relevant phylogenetic edges of the cladogram are simply those of the biospecies. We could therefore say that it is better considered to be considered under that rubric, and that the issue of "extinction" of species at cladogenesis is one of the reference of taxonomic names. In short, the "extinction" is a taxonomic extinction.[29]

PHYLOGENETIC TAXON (SYNAPOMORPHIC) SPECIES

There are many other conceptions that their authors refer to, or which are referred to by others, as "phylogenetic" conceptions of species. It is not clear that these form a natural class of conceptions. I believe there are two basic approaches that derive from the cladistic terminology, but it is not therefore the case that all authors in one or the other classes agree or that cladistic terminology resolves the differences between them.

Synapomorphy-based species are largely due to the advocacy of Brent Mishler, although earlier Joel Cracraft had defined the species taxon as

… the smallest diagnosable cluster of individual organisms within which there is a parental pattern of ancestry and descent.[30]

Mishler and Brandon summarize their Monophyletic version:

A species is the least inclusive taxon recognized in a classification, into which organisms are grouped because of evidence of monophyly (usually, but not restricted to, the presence of synapomorphies), that is ranked as a species because it is the smallest 'important' lineage deemed worthy of formal recognition, where 'important' refers to the action of those processes that are dominant in producing and maintaining lineages in a particular case.[31]

---

[28] Kornet 1993, Kornet and McAllister 1993.
[29] In the context of the debate between Mayr and Hennig over the "correct" way to classify [Mayr 1974, Hennig 1975], Hennig criticizes Mayr for ambiguity in his terms, which lends support to this interpretation. Above all, Hennig wanted clarity.
[30] Cracraft 1983, 170.
[31] Mishler and Brandon 1987, also in Hull and Ruse 1998, 310. See also Mishler and Donoghue 1982.

They redefine *monophyly* in such a way as to be able to include species:

> A monophyletic taxon is a group that contains all and only descendants of a common ancestor, originating in a single event.[32]

De Queiroz and Donoghue, on the other hand, treat species as systems that may not be monophyletic, and indeed may be paraphyletic if a species has split from it, in a parallel with cohesive and functional individuals who lose cells and reproduce.[33] They therefore exclude asexual organisms, and indistinct populations that are not assignable, from being members of species. Mishler and Theriot, extending the monophyletic (i.e., phylogenetic taxon) conception, include asexual organisms largely on the grounds that asexual taxa do not markedly differ in overall phylogenetic nature from sexual taxa—the number of autapomorphies, for example, are similar in both cases.[34]

Synapomorphic species are usually based on the historical, actual, lines of ancestry and descent that are represented in a cladogram. As a phylogenetic taxon, a species is grouped by the synapomorphies shared by organisms that indicates monophyly.[35] It is a phylogenetic or cladistic replacement for the evolutionary species concept of Simpson. Mishler and Theriot give the following broad definition:

> A species is the least inclusive taxon recognized in a formal phylogenetic classification. As with all hierarchical levels of taxa in such a classification, organisms are grouped into species because of evidence of monophyly. Taxa are ranked as species because they are the smallest monophyletic groups deemed worthy of formal recognition, because of the amount of support for their monophyly and/or because of their importance in biological processes operating on the lineage in question.[36]

Here, the monophyletic conception is explicit. The dual nature of the epistemic and the ontological aspects of species are clearly expressed, and the rank of *species* is restricted to "biologically important" lineages. Their version of the synapomorphic concept allows for reticulation as a general problem in classification not merely restricted to species taxa.

De Queiroz and Donoghue, on the other hand, do not think that species have to be monophyletic, because monophyly of populations does not offer a way to specify what the base rank is, and because species evolve from ancestral populations this will leave the species from which the ancestral population derived as paraphyletic.[37] Of course, the monophyly spoken of here is somewhat different from the monophyly of the Mishler et al. version—this one is based on populations as the base entities; the Mishler et al. version is based on phylogenetic lineages. And both conceptions converge on a similar solution—species are regarded as singular phylogenetic *lineages*,

---

[32] *Op. cit.*, 313 in Hull and Ruse 1998.
[33] de Queiroz and Donoghue 1988, 1990.
[34] Mishler and Theriot 2000, 52f. One does wonder, though, if this follows more from the logic of phylogenetic method than anything else.
[35] Mishler and Theriot 2000, 47.
[36] *Op. cit.*, 46–47.
[37] de Queiroz and Donoghue 1988, 1990.

which is de Queiroz's later conception of a cohesive object or group over phylogeny.[38] The actual answer of the earlier paper, however, is that there is no single definition of species that will "answer to the needs of all biologists and will be applicable to all organisms,"[39] although they reject the sort of pluralism I proposed in 2003.[40]

AUTAPOMORPHIC SPECIES

Diagnosis of species has always been involved in the debate, but few if any have until recently suggested that diagnosis is sufficient, apart from (and probably not even there) the so-called Taxonomic Species Concept. Cronquist provided one of the first such conceptions:

> Species are the smallest groups that are consistently and persistently distinct, and distinguishable by ordinary means.[41]

Some accused Cronquist of presenting a "subjective" concept,[42] but it all hinges on what "ordinary means" means. At one time, the use of a microscope was reviled (e.g., by Linnaeus); now assays for specificity ranging from molecular data to morphometric and acoustic traits are considered ordinary practice.

More common phylogenetic diagnostic concepts, though, arise from the recognition that what makes taxa distinct are their apomorphies (or instead, their unique sets of characters; apomorphies are usually single characters, not sets of characters). Species have diagnostic *aut*apomorphies—that is to say, they have unique constellations of characters—while higher taxa (clades) have *syn*apomorphies—shared constellations of characters, which group them together. One instance of this approach is found in the work of Donn Rosen:

> ... a geographically constrained group of individuals with some unique apomorphous characters, is the unit of evolutionary significance.[43]

In his paper on Guatemalan fishes, Rosen defined species in terms of individuals and populations:

> ... a species is merely a population or group of populations defined by one or more apomorphous features; it is also the smallest natural aggregation of individuals with a specifiable genographic integrity that cannot be defined by any set of analytic techniques.[44]

---

[38] de Queiroz 1998, 1999.
[39] Quoting Kitcher 1984, 309.
[40] Wilkins 2003.
[41] Cronquist 1978, 3.
[42] For example, Ghiselin 1997, 106f.
[43] Rosen 1978, 176 quoted in Wheeler and Meier 2000, 55.
[44] Rosen 1979, 277 quoted in Mayr 2000, 99.

He then noted that this means that subspecies are "by definition, unobservable and undefineable," since they have no such apomorphies. Subsequently, Nelson and Platnick proposed in passing that in their book on systematics and biogeography that they would treat species as

> ... simply the smallest detected samples of self-perpetuating organisms that have unique sets of characters.[45]

They, too, noted that this meant that diagnosable "subspecies" (groups identifiable by unique sets of characters) were thus species. This was not intended to be a complete definition but rather a description of their current practice (Nelson, *pers. comm.*). Even so, it was very influential on subsequent discussion. Independently, and almost contemporaneously, Eldredge and Cracraft defined a species as

> ... a diagnosable cluster of individuals within which there is a parental pattern of ancestry and descent, beyond which there is not, and which exhibits a pattern of phylogenetic ancestry and descent among units of like kind.[46]

Cracraft subsequently revised his formulation by removing mention of reproductive cohesion ("like kind") to read

> ... the smallest diagnosable cluster of individuals within which there is a pattern of ancestry and descent.[47]

Quentin (not Ward) Wheeler and Platnick base their definition on this tradition as well.[48] They define species thus:

> We define species as the smallest aggregation of (sexual) populations or (asexual) lineages diagnosable by a unique combination of character traits. This concept represents a unit concept.[49]

This concept is prior to a cladistic analysis,[50] and so unlike Nelson's earlier note that species are taxa like any other level of a phylogenetic tree,[51] Wheeler and Platnick do not worry about apomorphies and homologies when recognizing species, only characters. Species are found wherever characters are fixed and constant across all samples, while traits may be variable.[52] Their willingness to handle and include asexual taxa within their species concept marks them out from most other conceptions, and they bite the bullet on recognizing clones of asexual lineages as species:

---

[45] Nelson and Platnick 1981, 12 quoted in Wheeler and Meier 2000, 56.
[46] Eldredge, 1980, 92 quoted in Wheeler and Meier 2000, 55–56.
[47] Cracraft, 1983, 170 quoted in Wheeler and Meier 2000, 56, and see discussion there.
[48] Wheeler 1999, Wheeler and Platnick 2000.
[49] Wheeler and Platnick 2000, 58.
[50] *Op. cit.*, 59.
[51] Nelson 1989.
[52] See figure 5.1 in Wheeler and Platnick 2000, 58.

If the goal of distinguishing species is thereby to recognize the end-products of evolution, should we seek to suppress naming large numbers of species where large numbers of differentiated end-products exist?[53]

Mishler considers a number of the diagnostic accounts to be phenetically based, including Cracraft's, Platnick's, and Nixon and Wheeler's.[54] Whether this is so (that is, whether they make use of the Cartesian clustering of species in a state space of traits typical of phenetic practice), it is clear that they assume that species are phylogenetically speaking the terminal taxa on a tree. Diagnosis assumes that the traits are specified before the tree is constructed.

WHERE IS THE TAXON LEVEL, OR RANK?

Operationally, phylogeneticists seem to have little operational difficulty in identifying the level of species taxa in their cladograms, but there is a problem that arises if the autapomorphic concept is taken too strictly. Consider cladograms of haplotype groups. The authors have clearly already identified the species, and are considering whether or not the haplotype data taken from, say, mitochondrial DNA support the claim that these populations form subspecies. But on a strictly autapomorphic concept, *each* of the haplotype groups should be considered a separate species (in exactly the way Whately described a logical notion of *species* would do for dog breeds), unless other considerations, such as biogeography and interbreeding, are taken into account, and if they are, then the species concept alone is insufficient to delimit species. And so, it appears that some sort of prior knowledge is required to specify at what level of a cladogram taxa begin and (for example) molecular lineages cease to be subspecific diagnostic criteria.

The claim that species are defined by constant characters, made above by Wheeler and Platnick, is problematic. Either we already know what characters count as species-defining, or we are unable to find a level of species (for there are constant characters for a great many higher-level groups, as well as lower-level groups, than the usual level at which species are identified). Either way, this is not a full concept of what makes a group a species.

BIBLIOGRAPHY

Ashlock, Peter D. 1971. Monophyly and associated terms. *Systematic Zoology* 20 (1):63–69.
Cracraft, Joel. 1983. Species concepts and speciation analysis. In *Current Ornithology*, edited by R. F. Johnston, 159–187. New York: Plenum Press.
Cronquist, A. 1978. Once again, what is a species? In *BioSystematics in Agriculture*, edited by LV Knutson, 3–20. Montclair, NJ: Alleheld Osmun.
Davis, Jerrold I. 1997. Evolution, evidence, and the role of species concepts in phylogenetics. *Systematic Biology* 22 (2):373–403.

---

[53] *Op. cit.*, 59.
[54] Mishler, *pers. comm.*

de Queiroz, Kevin. 1998. The general lineage concept of species, species criteria, and the process of speciation. In *Endless Forms: Species and Speciation*, edited by Daniel J. Howard and Stewart H. Berlocher, 57–75. New York: Oxford University Press.

—. 1999. The general lineage concept of species and the defining properties of the species category. In *Species, New Interdisciplinary Essays*, edited by R. A. Wilson, 49–88. Cambridge, MA: Bradford/MIT Press.

de Queiroz, Kevin, and Michael J. Donoghue. 1988. Phylogenetic systematics and the species problem. *Cladistics* 4:317–338.

—. 1990. Phylogenetic systematics and species revisited. *Cladistics* 6:83–90.

Eldredge, Niles, and Joel Cracraft. 1980. *Phylogenetic patterns and the evolutionary process: Method and theory in comparative biology*. New York: Columbia University Press.

Ereshefsky, Marc. 2000. *The Poverty of Linnaean Hierarchy: A Philosophical Study of Biological Taxonomy*. Cambridge, UK/New York: Cambridge University Press.

Ghiselin, Michael T. 1997. *Metaphysics and the Origin of Species*. Albany, NY: State University of New York Press.

Hennig, W. 1975. "Cladistic analysis or cladistic classification?": A reply to Ernst Mayr. *Systematic Zoology* 24 (2):244–256.

Hennig, Willi. 1950. *Grundzeuge einer Theorie der Phylogenetischen Systematik*. Berlin: Aufbau Verlag.

—. 1966. *Phylogenetic Systematics*. Translated by D. Dwight Davis and Rainer Zangerl. Urbana, IL: University of Illinois Press.

Hull, David L., and Michael Ruse, eds. 1998. *The Philosophy of Biology*. Oxford/New York: Oxford University Press.

Kitcher, Philip. 1984. Species. *Philosophy of Science* 51 (2):308–333.

Kornet, Dina Joanna. 1993. Permanent splits as speciation events: A formal reconstruction of the internodal species concept. *Journal of Theoretical Biology* 164 (4):407–435.

Kornet, Dina Joanna, and James W. McAllister. 1993. The composite species concept. In *Reconstructing Species: Demarcations in Genealogical Networks*, 61–89. Rijksherbarium, Leiden: Unpublished PhD dissertation, Institute for Theoretical Biology.

Mayr, Ernst. 1974. Cladistic analysis or cladistic classification? *Journal of Zoological Systematics and Evolutionary Research* 12 (1):94–128.

—. 2000. A critique from the biological species concept: What is a species, and what is not? In *Species Concepts and Phylogenetic Theory: A Debate*, edited by Quentin D. Wheeler and Rudolf Meier, 93–100. New York: Columbia University Press.

Mayr, Ernst, and Peter D. Ashlock. 1991. *Principles of Systematic Zoology*. 2nd ed. New York: McGraw-Hill.

Meier, Rudolf, and Rainer Willmann. 1997. The Hennigian species concept. In *Species Concepts and Phylogenetic Theory: A Debate*, edited by Quentin D. Wheeler and Rudolf Meier, 30–43. New York: Columbia University Press.

—. 2000. The Hennigian species concept. In *Species Concepts and Phylogenetic Theory: A Debate*, edited by Quentin D. Wheeler and Rudolf Meier. New York: Columbia University Press.

Mishler, Brent D., and Robert N. Brandon. 1987. Individuality, pluralism, and the Phylogenetic Species Concept. *Biology and Philosophy* 2 (4):397–414.

Mishler, Brent D., and Michael J. Donoghue. 1982. Species concepts: A case for pluralism. *Systematic Zoology* 31 (4):491–503.

Mishler, Brent D., and Edward C. Theriot. 2000. The phylogenetic species concept (*sensu* Mishler and Theriot): Monophyly, apomorphy, and phylogenetic species concepts. In *Species Concepts and Phylogenetic Theory: A Debate*, edited by Quentin D. Wheeler and Rudolf Meier, 44–54. New York: Columbia University Press.

Nelson, Gareth J. 1971. Paraphyly and polyphyly: Redefinitions. *Systematic Zoology* 20 (4):471–472.

—. 1989. Species and taxa: Speciation and evolution. In *Speciation and its Consequences*, edited by Daniel Otte and John A. Endler. Sunderland, MA: Sinauer.

Nelson, Gareth J., and Norman I. Platnick. 1981. *Systematics and Biogeography: Cladistics and Vicariance*. New York: Columbia University Press.

Pratt, Vernon. 1976. Biological classification. In *Topics in the Philosophy of Biology*, edited by Marjorie Grene and Everett Mendelsohn, 372–395. Dordrecht: Springer Netherlands.

Rieppel, Olivier. 2011. Against species essentialism. *Metascience* 20 (2):339–341.

Rosen, Donn E. 1978. Vicariant patterns and historical explanation in biogeography. *Systematic Zoology* 27 (2):159–188.

—. 1979. Fishes from the uplands and intermontane basins of Guatemala: Revisionary studies and comparative biogeography. *Bulletin of the American Museum of Natural History* 162 (5):267–376.

Vanderlaan, Tegan A. et al. 2013. Defining and redefining monophyly: Haeckel, Hennig, Ashlock, Nelson and the proliferation of definitions. *Australian Systematic Botany* 26 (5):347–355.

Wheeler, Quentin D. 1999. Why the phylogenetic species concept?—Elementary. *Journal of Nematology* 31 (2):134–141.

Wheeler, Quentin D., and Rudolf Meier, eds. 2000. *Species Concepts and Phylogenetic Theory: A Debate*. New York: Columbia University Press.

Wheeler, Quentin D., and Norman I. Platnick. 2000. The phylogenetic species concept (*sensu* Wheeler and Platnick). In *Species Concepts and Phylogenetic Theory: A Debate*, edited by Quentin D. Wheeler and Rudolf Meier, 55–69. New York: Columbia University Press.

Wilkins, John S. 2003. How to be a chaste species pluralist-realist: The origins of species modes and the synapomorphic species concept. *Biology and Philosophy* 18 (5):621–638.

Willmann, Rainer. 1985a. *Die Art in Raum und Zeit*. Berlin: Paul Parey Verlag.

—. 1985b. Reproductive isolation and the limits of the species in time. *Cladistics* 2 (3):336–338.

—. 1997. Phylogeny and the consequences of molecular systematics. In *Ephemeroptera and Plecoptera: Biology, Ecology, Systematics*, edited by Peter Landolt and Michael Satori. Fribourg: Mauron+Tinguely & Lachat SA.

Woodger, Joseph Henry. 1937. *The Axiomatic Method in Biology*. Cambridge, UK: Cambridge University Press.

10 Other Species Concepts

ECOLOGICAL SPECIES CONCEPTS

Turesson's species concept was not widely adopted, but ecological concepts have been proposed from time to time. Mayr himself ventured one in his 1982 history:

A species is a reproductive community of populations (reproductively isolated from others) that occupies a specific niche in nature.[1]

He did this with neither preamble nor follow-up, and the requirement for the "niche" here seems to have been allowed to quietly drift away, as he does not insist on it elsewhere. A prior instance of another partial ecological concept occurs in Ghiselin, who referred to species as

The most extensive units in the natural economy such that reproductive competition occurs among its parts.[2]

The competition here is for genetic resources, and comes in the context of a strong selectionist account of evolution at various levels. Ghiselin calls it the "hypermodern species concept" and says that species are economic entities analogous to firms.[3]

Van Valen, in a paper that discussed the "odd" reproductive dynamics and evolution of American oaks (gen. *Quercus*), proposed that in these plants at any rate a species was an ecological type. He offered a definition as "a vehicle for conceptual revision, ... not a standing monolith":

A species is a lineage (or a closely related set of lineages) which occupies an adaptive zone minimally different from that of any other lineage in its range and which evolves separately from all lineages outside its range.[4]

This is self-consciously a mixture of the Mayr and Simpson definitions, and Van Valen justifies it in terms of evolution acting on phenotypes, controlled by "ecology and the constraints of individual development." It is therefore a definition founded on a particular view of evolution. He reprises Simpson's notion of a lineage, and defines a population in genetic terms. The novel element here is the idea of the "adaptive zone," which he describes as

---

[1] Mayr 1982, 273, italics original.
[2] Ghiselin 1974, 538.
[3] Ghiselin 1974. See Ghiselin 1997, chapter 9.
[4] Van Valen 1976, 233.

Some part of the resource space together with whatever predation and parasitism occurs on the group considered.[5]

Quercus species are often sympatric, and freely hybridize, and yet they maintain their identity. He refers to these groups as "multispecies," as they exchange genes, but he does not require that they actually form viable hybrids that breed true thereafter. "Multispecies" is defined as a

... set of broadly sympatric species that exchange genes in nature ...[6]

and refers to the syngameon concept of Verne Grant, which, we have seen, is due to Poulton. Multispecies can occur without having component species as such, citing *Rubus*, *Crataegus*, and dandelions. The latter leads to the implication that asexuals are not to be considered species.

Elsewhere Littlejohn has comprehensively reviewed reproductive isolation, and he makes some interesting comparisons between sexual reproduction and asexual isolation.[7] He notes that asexuals (uniparental species; unlike many of his predecessors, Littlejohn does not exclude them from specieshood) must be seen as a "cluster or cloud of individuals representing an adaptive node or adaptive peak," and cites Dobzhansky and G. E. Hutchinson. Hutchinson employs a "taxonomic space" model, and treats species as clusters in that space, formed through adaptation.[8] This is somewhat different to the phenetic concept of an operational taxonomic unit (OTU). For a start, he requires independence of the axes of the space, and that they be adaptive characters. Asexuals, such as bdelloid rotifers, are just as good species as their close sexual relatives, the *Nebalia* class of crustaceans.[9] Similar points about the role of adaptation in delimiting species were previously made by entomologist R. S. Bigelow, who noted that

Reproductive isolation [*in Mayr's 1963 definition—JSW*] should be considered in terms of gene flow, and not in terms of interbreeding, since selection will inhibit gene flow between well-integrated gene pools despite interbreeding.[10]

Recently, the philosopher Kim Sterelny defended an "ecological mosaic" conception of species,[11] based on a reworking of Dobzhansky's metaphor of species as occupying "adaptive peaks" in a Wrightean fitness landscape.[12] Noting that species are typically *ecologically fractured*,[13] Sterelny concludes that most species are geological and ecological mosaics, and are not ecologically cohesive entities. He takes this to explain stasis in the duration of species, because interbreeding between

---

[5] *Op. cit.*, 234.
[6] *Op. cit.*, 235.
[7] Littlejohn 1981.
[8] Hutchinson 1968.
[9] I am indebted to Dr. Littlejohn for providing me with these references.
[10] Bigelow 1965, 458.
[11] Sterelny 1999.
[12] Dobzhansky 1937, 9–10.
[13] Sterelny 1999, 124.

populations within the species' range acts as an inhibitor, which he calls *Mayr's Brake*, on adaptive change over all the entire species. He holds that this reinforces an evolutionary concept of species.

Ecological species concepts are generally partial rather than all-encompassing, and tend to act as adjuncts to more universal conceptions. It is clear that all ecological concepts rely heavily on the pre-eminent role of natural selection to maintain isolation between groups. Selection also plays a critical role in the definition and delimitation of asexual species, to which, among others, we now turn.

"ABERRANT" CONCEPTS

"Aberrant concept" here usually merely signifies that the concept is either not attended to much by zoologists, or that it is not thought by some biologists to be a "true" species concept.

AGAMOSPECIES

Agamospecies are asexual taxa. The term was defined by Cain as the end result of parthenogenesis in animals and apomixis in plants (which is secondary asexuality),[14] but it is now well understood that most unisexual organisms are in fact neither[15]—that is, are primarily asexual, mostly bacteria or algae. Cain defined agamospecies as

> ... those forms to which [the biological species concept] cannot apply because they have no true sexual reproduction ...[16]

This is clearly unsatisfactory. For a start, it is a privative definition—agamospecies are what are *not* something else (in this case taxa comprised of sexual organisms). As we saw also with Fisher, asexuals are considered by Cain to be off the mainstream of evolution, something of a marginal occasional misfiring of the evolutionary process, since sexual recombination permits the more rapid acquisition and spread of favorable mutations, and asexual reproduction is subject to genetic load (the continuation of deleterious mutations). This is no longer the consensus.

So, it has been questioned whether agamospecies are anything more than the morphological species concept applied to asexual organisms.[17] However, there is a notion that applies to asexuals that is, I believe, a more coherent way to understand asexuals, and which avoids privative definition. Originally defined for viruses (which are mostly, but not always,[18] asexual), the concept of the physicist Manfred von Eigen is called *quasispecies* (from the Latin for "*as if, just as, as it were... nearly*"),[19] and

---

[14] Cain 1954, 98–106.

[15] Kondrashov 1994, Schloegel 1999, Taylor et al. 1999.

[16] Cain 1954, 103.

[17] Noted by Dobzhansky 1937, 1941, Sonneborn 1957, 283.

[18] Viruses can crossover genetic material in a superinfected host [Szathmáry 1992, Boerlijst et al. 1996].

[19] Eigen 1993a, 1993b, Stadler and Nuño 1994. See entry "quăsĭ" in Lewis and Short 1879.

it applies equally to uniparental lineages in artificial life simulations as in biology.[20] He notes that

> ...A viral species ... is actually a complex, self-perpetuating population of diverse, related entities that act as a whole[21]

and he defines the quasispecies as a

> ... region in sequence space [which] can be visualized as a cloud with a center of gravity at the sequence from which all mutations arose. It is a self-sustaining population of sequences that reproduce themselves imperfectly but well enough to retain a collective identity over time.[22]

The conception is based on the observation that in a cluster of genotypes of viruses, there will be a mean genotype (the wild type) maintained by selection for optimality for that particular environmental niche (Figure 13.6). Eigen noted that it may eventuate that there is actually no single virus with that "wild type" genotype. Clearly, this makes the agamospecies conception a kind of ecotypical notion, maintained by natural selection; in fact, one might say that it is the purely ecological aspect of the ecospecies conception. Similar observations were made by Hutchinson, but no technical name was applied by him at that time.[23] I therefore propose that "agamospecies" and "quasispecies" be treated as synonyms, and that the inferences made about quasispecies by Eigen be applied to agamospecies hereafter.

MICROBIAL SPECIES

Microbial species are covered by several differing concepts. Bacteriologist Frederick Cohan has proposed a *bacterial species concept*:

> [Bacterial species] are populations of organisms occupying the same ecological niche, whose divergence is purged recurrently by natural selection.[24]

The *polyphasic species concept* has been proposed in which multiple lines of independent evidence are used:

> This polyphasic taxonomy takes into account all available phenotypic and genotypic data and integrates them in a consensus type of classification, framed in a general phylogeny derived from 16S rRNA sequence analysis.[25]

---

[20] Wilke et al. 2001.
[21] Eigen 1993b, 32.
[22] *Op. cit.*, 35.
[23] Hutchinson 1968, 184.
[24] Cohan 2001, 2002, 2006.
[25] Vandamme et al. 1996. See also Varghese et al. 2015.

Another conception, which expands upon Cohan's bacterial species concept, is the *adaptive divergence species concept* of Vos, which employs the McDonald–Kreitman (MK) test:

> Two [microbial] clusters can each be classified as a distinct species when the MK test yields a significant NI value (<1), whereas they can be considered to belong to the same species when the two clusters have not diverged as a result of adaptive evolution and the NI value is nonsignificant.[26]

NI is the Neutrality Index, which measures the selectivity of changes in polymorphisms.

NOTHOSPECIES

The almost exact conceptual opposite of the agamospecies/quasispecies conception is pteridophytologist W. H. Wagner's conception of a *nothospecies*.[27] Effectively, this is a species formed from the hybridization of two sexual species (species formed by the usual method of cladogenesis of sexual species he refers to as *orthospecies*.[28] As Hull notes, "According to recent estimates, 47 percent of angiosperms and 95 percent of pteridophytes (ferns) are allopolyploids"[29] and ferns and their allies continue to cause problems for isolation conceptions of species.[30] The notion is so generally applicable in botany that the concept of nothospecies has been written into the botanical rules for nomenclature.

COMPILOSPECIES

Harlan and de Wet proposed a concept of a species that "plunders" (Latin *compilo*) the genetic resources of another species through introgressive hybridization (where the hybrids preferentially interbreed with only one of the parental species, causing a gene-flow from one to the other, one-way).[31] It is therefore an asymmetric version of the nothospecies concept. It also applies to some plant species.[32]

OTUs AND PHENETICS (PHENOSPECIES)

Phenetics developed from the work of Sokal and Sneath,[33] who trace their attempt to produce a "natural" taxonomy back to Adanson, who used a multivariate classification scheme in *Familles des Plantes*.[34] The phenetic view, also called "numerical

---

[26] Vos 2011, 2.
[27] Wagner 1983.
[28] Wagner, *pers. comm.*
[29] Hull 1988, 103 citing Verne Grant.
[30] Barrington et al. 1989, Paris et al. 1989, Yatskievych and Moran 1989, Haufler 1996, Vogel et al. 1996, Ramsey and Schemske 1998, Wagner and Smith 1999.
[31] Harlan and De Wet 1963.
[32] Aguilar et al. 1999.
[33] Sokal and Sneath 1963, Sneath and Sokal 1973.
[34] Adanson 1763. Winsor 2004 provides a rebuttal for this claim. Initially, Sokal and Sneath referred to their view as "neo-Adansonian."

taxonomy" in virtue of the use of computers for the first time in systematics, basically involves a cluster analysis of continuously varying characters in a Cartesian space of *n*-dimensions (one dimension for each character or principle component) in the belief that species will fall out as clusters that agglomerate in different ways. While Sokal and Sneath generally give priority to the biological species concept, they note

> In the absence of data on breeding and in apomictic groups. ..., the species are based on the phenetic similarity between the individuals and on phenotypic gaps. These are assumed to be good indices of the genetic position, although they need not be. ... In this book it [the term "species"] will be used in the sense of phenetic rank... .[35]

Species are not unique in the phenetic approach; they are operational taxonomic units (OTUs) just as above-species and within-species ranks are.[36] The phenetic view, though, is that species are not monotypic, and Sokal and Sneath define a term, based on Beckner's discussion,[37] *monothetic*, and its antonym *polythetic*,[38] to describe groups based on a single uniform set of characters versus those which vary or have alternative characters; Wittgenstein's family resemblance predicate is explicitly adduced.[39]

SPECIES DENIERS: PURE "NOMINALISM," OR ELIMINATIVISM

It is unfortunate that the alternative to the reality of species has been called "nominalism" by Mayr and others, because nominalism is a philosophical doctrine that asserts that *universals* are not real, and species are not held by many to be universal terms. Strictly speaking, the individuals that "species nominalism" considers real in opposition to species are individual organisms. However, in traditional logic and metaphysics, a universal is not *comprised* of individuals so much as individuals *instantiate* the class.[40] Species *are* comprised of individuals under evolutionary accounts, and so they are real to the extent that their components are (except under ideal morphological views). It would be wrong to call these "nominalistic views" simply because they are founded upon acts of naming. A species nominalism must be directed to the *category* or *rank* of species, and must claim that there is no sense in which that categorical term has any application. I therefore prefer to refer to *species deniers* rather than to species nominalists.

Species deniers include Vrana and (Ward) Wheeler,[41] Pleijel,[42] and Hey.[43] I argued above that Darwin was not a species denier, and that Buffon was only inconsistently one. There are two varieties: *conventionalism*, and *"replacementism,"* the latter

[35] Sokal and Sneath 1963, 30f; but see Sokal and Crovello 1970.
[36] *Op. cit.*, 121f.
[37] Beckner 1959.
[38] *Op. cit.*, 13–15.
[39] *Ibid.*, 14.
[40] See Aaron 1952.
[41] Vrana and Wheeler 1992.
[42] Pleijel 1999, Pleijel and Rouse 2000.
[43] Hey 2001b, 2001a.

involving replacing that term ("species") with another, for example, "deme,"[44] as Pleijel (least inclusive taxonomic units [LITUs]) and Hey (evolutionary groups) do. Given the history of the term, there is no reason to do this except to make it clear that only a particular sort or kind is being referred to, and history shows also that such attempts are always assimilated and subverted. *Species* keeps winning out.

CONVENTIONALISM: THE TAXONOMIC SPECIES CONCEPT

> I object to the term "species concept", which I think is misleading. ... A species in my opinion is a name given to a group of organisms for convenience, and indeed of necessity.[45]

One major stream of thought, particularly among geneticists, with regard to species is what we might call the *conventional* account, although it is sometimes called the "nominalistic" or "cynical" concept.[46] On this widely held account, species are "whatever a competent taxonomist chooses to call a species." The complete quotation from Charles Tate Regan, given by Julian Huxley,[47] is

> A species is a community, or a number of communities, whose distinctive morphological characters are, in the opinion of a competent systematist, sufficiently definite to entitle it, or them, to a specific name.[48]

Huxley goes on to note that the

> difficulty with this definition lies in the term *competent*, which is what we have recently learnt to call the "operative" word. And experience teaches us that even competent systematists do not always agree as to the delimitation of species.

Ghiselin also notes that there are no rules for deciding whether a reproductive community is a species or a subspecies, and that one should wonder whose view to accept when the experts disagree; and such disagreement is common.

In fact, this view is not new, and predates Darwinian evolutionary theory by some time, at least in Britain, for some time. Darwin himself was a member of the drafting panel that proposed just this standard for the new *Rules of Zoological Nomenclature* in 1842—the so-called Strickland Code. In so defining the basic taxon this way, the British Association for the Advancement of Science legislators "consciously and conspicuously distanced themselves from disputes over definitions of species,"[49] and, according to McOuat, established the naming rights of a species to be a delineation of what a competent naturalist was—basically, someone who was accepted by the naturalist community, and primarily, someone with a position at a museum, thus

---

[44] Winsor 2000.

[45] Haldane 1956, 95.

[46] Kitcher 1984.

[47] Huxley 1942, 157. See also Trewavas 1973, Ghiselin 1997, 118.

[48] Regan 1926, 75.

[49] McOuat 2001, 3.

excluding names and species made by bird watchers and gardeners. Darwin himself in several places made this sort of definition, particularly in his *Natural Selection*:

> In the following pages I mean by species, those collections of individuals, which have been so designated by naturalists.[50]

It is interesting to note that Regan refers to "communities." This word typically has been used to apply to ecological communities; that is, ecosystems,[51] but in this case and at this time it is more likely to refer to what we now call a "deme," or a "Mendelian population." If this is so, then Regan is conflating two well-known ideas in our history here: that of a reproductive element and a diagnostic one. In short, Regan might very well have been putting forward an "operative"[52] notion of the generative species conception we have so often encountered. A less biological version of this view, which he calls the "cynical" species concept, is presented by Philip Kitcher:

> Species are those groups of organisms which are recognized as species by competent taxonomists.[53]

Here the operationalist aspect of the concept is primary. Species are made by acts of recognition by experts. Whether or not it includes a strong element of the biological (that is, reproductively isolated) nature of species, conventionalism takes seriously Locke's claim that species are made for communication, and objections to the recent Phylocode proposal (many by strict cladists, no less) are in part founded on the idea that higher taxa, at least, should be convenient, since they are artificial taxa anyway.[54]

REPLACEMENTISM: LITUs (LEAST INCLUSIVE TAXONOMIC UNITS)

Frederick Pleijel, a leading specialist on polychaete worms (bristleworms), has proposed doing away with the notion of species altogether. Instead, he proposes to replace it with the neutral term *least inclusive taxonomic unit*, or LITU (in homage to the OTU of phenetics). Pleijel's and Rouse's "definition" of the LITU runs

---

[50] Darwin and Stauffer 1975, 98, cited by McOuat 2001, 4n10.

[51] Taylor 1992.

[52] The term "operational" is taken from the philosopher of physics (and physicist) Percy Bridgman [1927]. Operationalism, the view named by him, is the view that the only meaning of a scientific concept is the set of measurements and operations that can be done to define it [Chang 2009].

[53] Kitcher 1984, 308.

[54] The Phylocode proposes to replace all Linnaean ranks with rankless strictly monophyletic taxa based on the best cladograms [Cantino and de Queiroz 2010]. It has been supported by eliminativists like Ereshefsky and Pleijel and Rouse [Pleijel and Rouse 2003], but some pattern cladists, such as Norman Platnick and Gareth Nelson (*pers. comm.*) oppose it due to its disruption of scientific communication and meaninglessness, as in their view cladograms are only hypotheses and are subject to revision. Others [Benton 2000, Berry 2002, Bryant and Cantino 2002, Forey 2002, Carpenter 2003, Gao and Sun 2003, Nixon et al. 2003, Kojima 2003, Keller et al. 2003] attack the proposal for a range of reasons, ranging from a personal distaste to a rejection of cladism. Most think that named monophyletic higher taxa are not going to be stable as new results come in.

... named monophyletic groups which are identified by unique shared similarities (apomorphies) ... which are at present not further subdivided. ... Identification of taxa as LITUs are statements about the current state of knowledge (or lack thereof) without implying that they have no internal nested structure; we simply do not know if a given LITU consists of several monophyletic groups or not.[55]

However, as the historical reviews above make abundantly clear, the notion of a least inclusive taxonomic unit historically *is* a species in both logic and biology. What Pleijel has done here is rediscover the past way of looking at things. However, LITUs are ontological units, and not merely epistemic ones.

SPECIES CONCEPTS IN PALEONTOLOGY (PALEOSPECIES)

The problem of applying any concept of species in paleontology has long been understood, and has spurred many discussions.[56] The difficulty lies in the way the data are presented. In neontology (the study of living organisms), the behaviors of organisms, both sexual and ecological, can be observed in some detail and repeated if the initial observations are inadequate. Under some circumstances, the organisms can be experimentally mated. Molecular evidence, especially that of DNA, can be harvested and assayed, meaning that where a traditional taxonomist might have used at most around 40 characters, the molecular systematist has more like 40,000.[57] Also, polytypy can be investigated in extant species, enabling the investigator to delineate the populations and subspecific variants and to tell whether, on the concept used, these still are included in the specific group.

Not so the paleontologist. The information in that case is usually restricted to several individual specimens (except when the bulk of the population leaves fossils, as in the case of silaceous forams, whose shells are fossilized in sediments on the seafloor, and which when they are found provide information about distribution, variety in populations, and changes over time). The famous *Tyrannosaurus rex*, for example, is known from around 50 specimens, none complete. Many hominid species are known from a single individual. In cases where the ancestors of a lineage lived in conditions unconducive to fossilization—the study of which is known as taphonomy—will have entire series of species unknown to science. For example, few ape fossils have been found for those species that lived in forest and jungle environments, where decomposition in the acidic soils, scavengers, and plants will dispose of the carcass relatively quickly.

So, paleontologists rely almost exclusively on morphological data. This means that there is pressure on them to lump stratigraphic specimens together (since fossil taxa are often used as stratigraphic markers by geologists) if there is some subjectively

---

[55] Pleijel and Rouse 2000, 629. See also Pleijel 1999, 2003.

[56] Simpson 1943, Sylvester-Bradley 1956, Schopf 1972, Smith 1994.

[57] A point made at the Melbourne Systematics Forum during 2002. I did not catch the name of the person making the point, but it is important. In the end, unless we know the ways in which DNA is expressed developmentally in all the species being analyzed, DNA is just a richer source of "morphological" data. However, few of these nucleotides are likely, in practice, to be informative (D. Williams, *pers. comm.*).

acceptable similarity between specimens.[58] In the rare cases when many specimens are found, problems due to variation can cause taxonomic complications. If many locales are involved, that is if the specimens are allopatric, it is unclear whether they form a single species or allopatrically isolated but closely related sister species. The problem is critical in the case of *Homo erectus*, which has a range from southern Africa to east and southeastern Asia. If all that remains is skeletal information, can we really be sure if these are the same biospecies, for example, or even whether *H. erectus* is one morph of a broader biospecies, possibly including *H. sapiens*? The now-amalgamated species *Canis lupus* includes morphs like the timber wolf, the Pekinese pug, and the Great Dane, all interfertile or fertile along a series of intermediate forms. Any paleontologist would split them into distinct taxa on morphological grounds alone. In the case of many birds, on the other hand, skeletal changes between good species are minimal, and only behavioral and ecological differences will tell them apart.

CHRONOSPECIES (SUCCESSIONAL SPECIES)

A problem that used to be common in discussions of paleontology and species concepts was that of chronospecies, where speciation occurs over time such that at the starting and end points of a time series, the morphs are different species. This was discussed by Simpson, for example, in his book on evolutionary tempo and mode.[59] However, the notion of chronospeciation is no longer thought by many to be an operational one—for a start, many specialists think that for animal species, such changes do not occur without lineage differentiation—in short, anagenesis is accompanied by cladogenesis—and that in most cases species remain largely unchanged after their initial period of adaptation until they go extinct.[60]

Chronospecies are formed, according to the author, T. N. George, who proposed the notion,[61] when a lineage changes sufficiently to be given a new name. In this respect, it is a temporalized version of the taxonomic species concept, or even an interpretation of the paleospecies concept.[62] A chronospecies is effectively an arbitrary division of a gradually evolving lineage,[63] and seems to have been come to prominence as the "species concept" of "phyletic gradualism," the target of criticism by advocates of the theory of punctuated equilibrium, although it is not often used in the interim, so far as I can tell.[64]

---

[58] An excellent discussion and a proposed resolution to this problem is Polly 1997. See also Simpson 1943, George 1956.

[59] Simpson 1943.

[60] Eldredge and Gould 1972, Gould and Eldredge 1977, Eldredge 1985, Gould 2002.

[61] George 1956, 129.

[62] Cain 1954, 106f, Simpson 1961, 166.

[63] Eldredge 1989, 98.

[64] Texts such as Mayr's and Simpson's [Simpson 1961, Mayr 1963] refer instead to "chronoclines" as the directional change of characters in the paleontological record, akin to geographical clines.

BIBLIOGRAPHY

Aaron, Richard Ithamar. 1952. *The Theory of Universals.* Cambridge, UK: Cambridge University Press.

Adanson, Michel. 1763. *Familles des plantes: I. Partie. Contenant une Préface Istorike sur l'état ancien & actuel de la Botanike, & une Téorie de cete Science.* Paris: Vincent.

Aguilar, Javier Fuertes et al. 1999. Molecular evidence for the compilospecies model of reticulate evolution in *Armeria* (Plumbaginaceae). *Systematic Biology* 48 (4):735–754.

Barrington, David S. et al. 1989. Hybridization, reticulation, and species concepts in the ferns. *American Fern Journal* 79 (2):55–64.

Beckner, Morton. 1959. *The Biological Way of Thought.* New York: Columbia University Press.

Benton, Michael J. 2000. Stems, nodes, crown clades, and rank-free lists: Is Linnaeus dead? *Biological Reviews* 75 (4):633–648.

Berry, Paul E. 2002. Biological inventories and the PhyloCode. *Taxon* 51 (1):27–29.

Bigelow, Robert S. 1965. Hybrid zones and reproductive isolation. *Evolution* 19 (4):449–458.

Boerlijst, M. C. et al. 1996. Viral quasispecies and recombination. *Proceedings of the Royal Society of London, Series B* 263:1577–1584.

Bridgman, Percy W. 1927. *The Logic of Modern Physics.* New York: Macmillan.

Bryant, H. N., and P. D. Cantino. 2002. A review of criticisms of phylogenetic nomenclature: Is taxonomic freedom the fundamental issue? *Biological Reviews* 77 (1):39–55.

Cain, Arthur J. 1954. *Animal Species and Their Evolution.* London: Hutchinson University Library.

Cantino, Philip D., and Kevin de Queiroz. 2010. *Phylocode: A Phylogenetic Code of Biological Nomenclature.* Available at http://www.ohio.edu/phylocode/ (accessed 7/6/17).

Carpenter, J. M. 2003. Critique of pure folly. *Botanical Review* 69 (1):79–92.

Chang, Hasok. 2009. "Operationalism." In *Secondary Operationalism*, edited by Edward N. Zalta. Stanford, CA: Metaphysics Research Lab, Stanford University. Available at https://plato.stanford.edu/archives/fall2009/entries/operationalism (accessed 9/16/2017).

Cohan, Frederick M. 2001. Bacterial species and speciation. *Systematic Biology* 50 (4):513–524.

—. 2002. What are bacterial species? *Annual Review of Microbiology* 56:457–487.

—. 2006. Towards a conceptual and operational union of bacterial systematics, ecology, and evolution. *Philosophical Transactions of the Royal Society B: Biological Sciences* 361 (1475):1985–1996.

Darwin, Charles, and Robert Clinton Stauffer. 1975. *Charles Darwin's Natural Selection: Being the Second Part of His Big Species Book Written from 1856 to 1858*, edited from manuscript by R. C. Stauffer. London: Cambridge University Press.

Dobzhansky, Theodosius. 1937. *Genetics and the Origin of Species.* New York: Columbia University Press.

—. 1941. *Genetics and the Origin of Species.* 2nd ed. New York: Columbia University Press.

Eigen, Manfred. 1993a. The origin of genetic information: Viruses as models. *Gene* 135 (1–2):37–47.

—. 1993b. Viral quasispecies. *Scientific American* July 1993 (32–39).

Eldredge, Niles. 1985. *Time Frames: The Evolution of Punctuated Equilibria.* Princeton, NJ: Princeton University Press.

—. 1989. *Macroevolutionary Dynamics: Species, Niches, and Adaptive Peaks.* New York: McGraw-Hill.

Eldredge, Niles, and Stephen J. Gould. 1972. Punctuated equilibria: An alternative to phyletic gradualism. In *Models in Paleobiology*, edited by T. J. M. Schopf, 82–115. San Francisco, CA: Freeman Cooper.

Forey, Peter L. 2002. PhyloCode—Pain, no gain. *Taxon* 51 (1):43–54.

Gao, Keqin, and Yuanlin Sun. 2003. Is the PhyloCode better than Linnaean system? New development and debate on biological nomenclatural issues. *Chinese Science Bulletin* 48 (3):308–312.

George, T. Neville. 1956. Biospecies, chronospecies and morphospecies. In *The Species Concept in Paleontology*, edited by Peter C. Sylvester-Bradley, 123–137. London: Systematics Association.

Ghiselin, Michael T. 1974. A radical solution to the species problem. *Systematic Zoology* 23 (4):536–544.

—. 1997. *Metaphysics and the Origin of Species*. Albany: State University of New York Press.

Gould, Stephen Jay. 2002. *The Structure of Evolutionary Theory*. Cambridge, MA: The Belknap Press of Harvard University Press.

Gould, Stephen Jay, and Niles Eldredge. 1977. Punctuated equilibria: The tempo and mode of evolution reconsidered. *Paleobiology* 3 (2):115–151.

Haldane, John Burton Sanders. 1956. Can a species concept be justified? In *The Species Concept in Palaeontology: A Symposium*, edited by Peter C. Sylvester-Bradley, 95–96. London: The Systematics Association.

Harlan, Jack R., and J. M. J. De Wet. 1963. The compilospecies concept. *Evolution* 17 (4):497–501.

Haufler, Christopher. 1996. Species concepts and speciation in pteridophytes. In *Pteridology in Perspective*, edited by Josephine Camus et al. Kew: Royal Botanic Gardens.

Hey, Jody. 2001a. *Genes, Concepts and Species: The Evolutionary and Cognitive Causes of the Species Problem*. New York: Oxford University Press.

—. 2001b. The mind of the species problem. *Trends in Ecology and Evolution* 16 (7):326–329.

Hull, David Lee. 1988. *Science as a Process: An Evolutionary Account of the Social and Conceptual Development of Science*. Chicago, IL: University of Chicago Press.

Hutchinson, G. E. 1968. When are species necessary? In *Population Biology and Evolution*, edited by Richard C. Lewontin, 177–186. Syracuse, NY: Syracuse University Press.

Huxley, Julian. 1942. *Evolution: The Modern Synthesis*. London: Allen and Unwin.

Keller, Roberto A. et al. 2003. The illogical basis of phylogenetic nomenclature. *The Botanical Review* 69 (1):93–110.

Kitcher, Philip. 1984. Species. *Philosophy of Science* 51 (2):308–333.

Kojima, Jun-Ichi. 2003. Apomorphy-based definition also pinpoints a node, and PhyloCode names prevent effective communication. *Botanical Review* 69 (1):44–58.

Kondrashov, Alexey S. 1994. The asexual ploidy cycle and the origin of sex. *Nature* 370 (6486):213–216.

Lewis, Charlton T., and Charles Short. 1879. *A Latin Dictionary Founded on Andrews' Edition of Freund's Latin Dictionary*. Revised, enlarged and in great part rewritten by C. T. Lewis and Charles Short. Oxford: Clarendon Press.

Littlejohn, Murray J. 1981. Reproductive isolation: A critical review. In *Evolution and Speciation: Essays in Honor of M. J. D. White*, edited by William R. Atchley and David S. Woodruff, 298–334. Cambridge, UK: Cambridge University Press.

Mayr, Ernst. 1963. *Animal Species and Evolution*. Cambridge, MA: The Belknap Press of Harvard University Press.

—. 1982. *The Growth of Biological Thought: Diversity, Evolution, and Inheritance*. Cambridge, MA: The Belknap Press of Harvard University Press.

McOuat, Gordon R. 2001. Cataloguing power: Delineating 'competent naturalists' and the meaning of species in the British Museum. *The British Journal for the History of Science* 34 (1):1–28.

Nixon, Kevin C. et al. 2003. The PhyloCode is fatally flawed, and the "Linnaean" system can easily be fixed. *Botanical Review* 69 (1):111–120.

Paris, Cathy A. et al. 1989. Cryptic species, species delimitation, and taxonomic practice in the homosporous ferns. *American Fern Journal* 79 (2):46–54.

Pleijel, Frederik. 1999. Phylogenetic taxonomy, a farewell to species, and a revision of *Heteropodarke (Hesionidae, Polychaeta, Annelida)*. *Systematic Biology* 48 (4):755–789.

Pleijel, Frederik, and Greg W. Rouse. 2000. Least-inclusive taxonomic unit: A new taxonomic concept for biology. *Proceedings of the Royal Society of London—Series B: Biological Sciences* 267 (1443):627–630.

—. 2003. Ceci n'est pas une pipe: Names, clades and phylogenetic nomenclature. *Journal of Zoological Systematics and Evolutionary Research* 41 (3):162–174.

Polly, Paul David. 1997. Ancestry and species definition in paleontology: A stratocladistic analysis of Paleocene-Eocene Viverravidae (Mammalia, Carnivora) from Wyoming. *Contributions from the Museum of Paleontology, The University of Michigan* 30 (1):1–53.

Ramsey, Justin, and Douglas W. Schemske. 1998. Pathways, mechanisms, and rates of polyploid formation in flowering plants. *Annual Review of Ecology and Systematics* 29 (4):467–501.

Regan, C. Tate. 1926. Organic evolution. *Report of the British Association for the Advancement of Science*, 1925:75–86.

Schloegel, Judy Johns. 1999. From anomaly to unification: Tracy Sonneborn and the species problem in protozoa, 1954–1957. *Journal of the History of Biology* 32 (1):93–132.

Schopf, Thomas J. M. 1972. *Models in Paleobiology*. San Francisco, CA: Freeman Cooper.

Simpson, George Gaylord. 1943. Criteria for genera, species and subspecies in zoology and paleontology. *Annals New York Academy of Science* 44:145–178.

—. 1961. *Principles of Animal Taxonomy*. New York: Columbia University Press.

Smith, Andrew B. 1994. *Systematics and the Fossil Record: Documenting Evolutionary Patterns*. Oxford/Cambridge, MA: Blackwell Science.

Sneath, Peter Henry Andrews, and Robert R. Sokal. 1973. *Numerical Taxonomy: The Principles and Practice of Numerical Classification, A Series of Books in Biology*. San Francisco, CA: W. H. Freeman.

Sokal, Robert R., and T. Crovello. 1970. The biological species concept: A critical evaluation. *American Naturalist* 104 (936):127–153.

Sokal, Robert R., and Peter Henry Andrews Sneath. 1963. *Principles of Numerical Taxonomy, A Series of Books in Biology*. San Francisco, CA: W. H. Freeman.

Sonneborn, Tracy M. 1957. Breeding systems, reproductive methods and species problems in Protozoa. In *The Species Problem: A Symposium Presented at the Atlanta Meeting of the American Association for the Advancement of Science, December 28–29, 1955*, Publication No 50, edited by Ernst Mayr, 155–324. Washington, DC: American Association for the Advancement of Science.

Stadler, Peter F., and Juan Carlos Nuño. 1994. The influence of mutation on autocatalytic reaction networks. *Mathematical Biosciences* 122 (2):127–160.

Sterelny, Kim. 1999. Species as evolutionary mosaics. In *Species, New Interdisciplinary Essays*, edited by Robert A. Wilson, 119–138. Cambridge, MA: Bradford/MIT Press.

Sylvester-Bradley, Peter Colley, ed. 1956. *The Species Concept in Palaeontology: A Symposium*, Publication of the Systematics Association. London: Systematics Association.

Szathmáry, Eörs. 1992. Viral sex, levels of selection, and the origin of life. *Journal of Theoretical Biology* 159 (1):99–109.

Taylor, J. W. et al. 1999. The evolution of asexual fungi: Reproduction, speciation and classification. *Annual Review of Phytopathology* 37:197–246.

Taylor, Peter D. 1992. Community. In *Keywords in Evolutionary Biology*, edited by Evelyn Fox Keller and Elisabeth A. Lloyd, 52–60. Cambridge, MA: Harvard University Press.

Trewavas, Ethelwynn. 1973. What Tate Regan said in 1925. *Systematic Zoology* 22 (1):92–93.

Van Valen, L. 1976. Ecological species, multispecies, and oaks. *Taxon* 25 (2/3):233–239.

Vandamme, Peter et al. 1996. Polyphasic taxonomy, a consensus approach to bacterial systematics. *Microbiology Review* 60 (2):407–438.

Varghese, Neha J. et al. 2015. Microbial species delineation using whole genome sequences. *Nucleic Acids Research* 43 (14):6761–6771.

Vogel, Johannes Christian et al. 1996. A non-coding region of chloroplast DNA as a tool to investigate reticulate evolution in European *Asplenium*. In *Pteridology in Perspective*, edited by J. M. Camus et al., 313–327. Kew: Royal Botanic Gardens.

Vos, Michiel. 2011. A species concept for bacteria based on adaptive divergence. *Trends in Microbiology* 19 (1):1–7.

Vrana, Paul, and Ward Wheeler. 1992. Individual organisms as terminal entities: Laying the species problem to rest. *Cladistics* 8 (1):67–72.

Wagner, Warren H. 1983. Reticulistics: The recognition of hybrids and their role in cladistics and classification. In *Advances in Cladistics*, edited by Norman I. Platnick and Vicki A. Funk, 63–79. New York: Columbia University Press.

Wagner, Warren H., and Alan R. Smith. 1999. Pteridophytes of North America. In *Flora of North America*. Oxford/New York: Flora of North America Association.

Wilke, Claus O. et al. 2001. Evolution of digital organisms at high mutation rates leads to survival of the flattest. *Nature* 412 (6844):331–333.

Winsor, Mary Pickard. 2000. Species, demes, and the Omega Taxonomy: Gilmour and The New Systematics. *Biology and Philosophy* 15 (3):349–388.

—. 2004. Setting up milestones: Sneath on Adanson and Mayr on Darwin. In *Milestones in Systematics: Essays from a Symposium Held within the 3rd Systematics Association Biennial Meeting, September 2001*, edited by David M. Williams and Peter L. Forey, 1–17. London: Systematics Association.

Yatskievych, George, and Robbin C. Moran. 1989. Primary divergence and species concepts in ferns. *American Fern Journal* 79 (2):36–45.

11 Historical Summary and Conclusions

It is time to sum up the major historical claims of this book. We have considered several historical and several philosophical claims, each in the light of one another. The biological conclusions are not mine to draw, but I can opine, and do in Section III.

From Aristotle through to the end of the Middle Ages and the Renaissance, the notion of species has not remained static; Aristotle's conventions and notions have been modified. Mostly, they were modified by the neo-Platonists and especially by Porphyry, who made Aristotle's top-down classification scheme dichotomous after the manner of Plato. Aristotle had opposed "privative" classification, classifying in terms of what things are *not*, although the grander schemes of later writers up to Lamarck had little problem with this, and happily classified groups such as Invertebrata.

In the medieval scheme, the notion of genus and species did not involve fixed ranks; a species might in turn be a genus on its own. The only "absolute" ranks were the *summum genera*, which represented in the Aristotelian tradition the universal categories (topics, or *topoi*) from which all things were to be divided, and individuals. The nominalist issue whether these general terms were merely aspects of mental categories or were real was alive and active well into the scientific period.

We find in the Epicureans a *generative conception of species* as early as the fourth century BCE, and this recurs throughout the remaining discussions until the biological tradition begins in the seventeenth century, and beyond. Essences, on the whole, were not themselves thought of as necessary and sufficient criteria for membership in species, and almost all writers admitted that there were deviations from the type. The classical scheme was, however, almost always based on a top-down classification with a large admission of apriorism, until the collapse of the Universal Language Project, and the rise of corpuscular philosophy, which rendered species secondary qualities, or unknowable.

The Great Chain of Being meant that, depending on what emphasis was given to the principle of plentitude and the principle of continuity, species were arbitrary divisions in a plenum—sometimes logical, sometimes substantial and actual. The specific nature of a member of a species was thought by Cusa to be a contraction of the essence in that individual, but the ultimate reality, according to Cusa, is the individual. A continuing battle between nominalists and realists (idealists in modern terms) meant that there was a field of alternate opinions. Variation is recognized to be a fact within taxa from the fifteenth century onward. Some, such as Ficino, held that there was a species that was most representative of a genus, since other species within the genus could play on variations of the generic theme, as it were.

The role of the Noachic story of the Ark effectively set up the problem of a lowest kind in zoology, and as a result the term *species* (or *genus*) became anchored as a basic term for such kinds. However, it is worth noting that in order to accommo-date the rising number of discovered species during the age of exploration (or rather the age of colonial expansion), they permitted Aristotle's notion of new species by hybridization to have greater play, along with spontaneous generation for species without lungs such as insects and worms.

With Bacon, we move from an immediate generalization of the universals from observed instances, and a subsequent top-down division of things, to the inductive construction of increasingly broader generalizations. He too allows for deviations and variations in species and other taxa, and Locke proposes that not only is there biological variation, but that species (sorts) themselves are conventional names we use to communicate easily. Nevertheless, he did allow for a real essence, but it is one that cannot be defined or even discovered. Leibniz was more optimistic; Kant even more pessimistic.

Kant rejects Leibniz's view of the law of completion (principle of plenitude) and argues that nature is discretely divided, and he too uses a generative conception of species. Although the traditional logic survives until the institution of set theory in the nineteenth century, even Mill is able to twist it to serve biological realities. Both the logical tradition before Darwin and after, in his own country, allow for a differ-ence between essentialist logical species and typological biological ones.

There is everywhere a remarkable lack of the sort of essentialism that Mayr and others believe permeated this period and its philosophy. While we see typology, when it comes to dealing with biological organisms, most of the time there is no insistence upon essences, and sometimes there is an explicit exemption for biological species of any knowable essence.

The major break with the classical notions of species from the tradition of "universal taxonomy," as I have called it, came with the Baconian insistence that instead of beginning classification with the universals, species of living things are found by a process of ascending abstraction and generalization. Species have been understood to be propagative forms, which I have here termed the *generative concep-tion*, and in various degrees and emphasis this remained the basic conception until the modern era. For this reason, the focus has been on seed, the fructative apparatus, and the reproductive behaviors, according to then-current views on generation.[1] The reason for a focus on species as units of biology derived from the medieval tradition and the neo-Platonic revival of the seventeenth century, but this was richer than the Received View supposes.

Many conceptions of species depended on, or were confounded by, the Great Chain. In one respect, a species was whatever was lowest in generality, but in another, species were conventions, or artificial divisions, since not all variant forms are to be found in a locality. Plenitudinous views expected that intermediate forms would be found, and, right into the nineteenth century with Macleay and Swainson, some held that while species might be discrete, there would be a continuity of form itself. There was often a species regarded as most like the vanilla generic characters, the *type*

---

[1] Hull 1967, 312 did quote Aristotle (*De Anima* 415a26) on the generation of like forms. He notes that Aristotle and Theophrastus did deal with cases where breeding true did not result in like forms, though. This was not emphasized in Hull's later work.

species, which relied upon the Great Chain conception. There was disagreement about the level of organization, too. Buffon thought that local forms were variants of the true species, the *premiere souche*, while Linnaeans tended to name any persistent variant form as a species. Bonnet and the early Buffon thought that specific forms were mere abstractions, a nominalist view that gradation imposed also on Lamarck. Even so, all these writers made use of a generative conception. Species may not exist, but if they did they were the result of heredity on the reproduction of forms.

Species realists in the later biological tradition tended to be fixists, such as Cuvier and Agassiz, but of course such a contrast was not possible until the possibility of transmutation over time was mooted. In the period before fixism was introduced, typology was widespread, but species were at best only mutable, not transmutable, excepting the case of spontaneous generation, in which species could sequentially change in outward form, rather like the metamorphosis of insects. The process by which species came to be in the first instance, however, was irrelevant to how they were maintained, and a generative notion was used by realists as well, especially those who stressed form as an identifying or diagnostic factor. Most of them differed from the transmutationists as to whether the formal diagnosis was a mark of the real essence; fixists thought that the generative process was the reason for the form, transmutationists that the generative process (Lamarck's *feu éthéré*, for example) was the reason only for the form in a given time or place or conditions. In short, for fixists, the form gave the essence, while for transmutationists, the "essence" meant that the form would not remain constant. Some like Buffon and A.-P. de Candolle bridged the two camps, and declared that the real essence of a species was modified to a certain degree by local conditions and hybridization.

In the period before the *Origin*, in Britain and presumably elsewhere, species were declared to be the "property," as it were, of competent and recognized taxonomists, particularly in museums, as McOuat has discussed. A number of experts doubted the permanence of species so that by the mid-century, several writers had made the suggestion in reputable forums that species did change.

Darwin's own views changed—at first, he was quite comfortable with the standard generative notion, and made notes on the idea that it was either the physical impossibility of interbreeding or the "repugnance" of species to interbreed in the wild. Diagnostic issues followed from the facts of the matter (the nominal essence did not give the real essence, in other words). But early on he started to toy with the notion that species were meaningless and conventional, and at the time of the *Origin* and shortly after, he seems to have taken this as his position—species are the outcomes of the evolutionary process acting on varieties, and although real entities individually, the rank of species is not real. It doesn't seem to have made much impact on his own systematic practice, as it was not to have made much impact on anyone else either, for some time to come. The *rank* of species is arbitrary in the *Origin*. However, species could "do" things—they could even compete with each other.

Darwin first thought that geographical isolation formed species most of the time but by the *Origin* had shifted to the view that selection against intermediates and hybrids was the major force. The geographic view of speciation, though, was to become the majority position in opposition to Darwin except for a few of his most devoted followers. However, while he may have been a conventionalist in his view of identification, he was not a nominalist, unlike several of his followers. He was a pluralist as to the degree of difference between, and causes of, species. It is significant that he thought that sexual constitutions were what caused the isolation of species, not form or adaptive traits. Darwin's evolving views are significant, as his is perhaps the first and the most complete attempt to deal with the implications of the transmutation of species.

The adaptationists such as Wallace, Romanes, and several others, however, were monists. Species were formed solely by selection. Weismann is an interesting exception here—he appears to have allowed for stochastic causes as well as deterministic selectionist causes of species. Poulton's essay, discussing these turn-of-the-20th-century arguments, effectively set the stage for the modern debates by introducing the notion of a genetic population as the boundary of species. Several contemporary writers such as Lotsy, Karl Jordan, and Turesson elaborated on that theme under the recent introduction of Mendelian genetics, with Turesson reintroducing the older notion of ecological habitat affecting form (albeit in a much more limited way). Several authors treated asexuals as a different kind of entity than species; Fisher even thinks asexuality is not a real phenomenon. This will become significant later in the century.

One thing that is obvious, even at this early stage in the debates over the "species problem," is that there is a distinction to be made between "universalist" species concepts, which are intended to apply to all organisms, especially in regard to speciation mechanisms after Darwin, and those proposing more limited notions, which apply only to some sorts of organisms. In the twentieth century, most conceptions of species are universalist conceptions, and the properties that mark them out are those derived from the proponent's preferred mechanisms of speciation. Biospecies are formed through the acquisition of reproductive isolating mechanisms in allopatry or sympatry according to the preference of the author. Saltationists claim that all species are formed through macromutations. Punctuated equilibrium theory later treats speciation as a rapid process followed by a period of stasis, and so species are universally delimited by this "sudden" event, thus finding the individualism thesis and the evolutionary conception of species congenial, and so on. Only a few people proposed limited conceptions, such as ecological conceptions of agamospecies, which had no general application.

So let us list the historical conclusions of this book in summary form.

Genus and *species* and their cognates in Greek and other languages are vernacular terms, logical terms, and biological terms. It is important to read each text in the right context and understand that when a classical author, including Aristotle, uses these terms (or their Greek equivalents), they may be using them in a vernacular or technical logical sense. The biological notion of *species* did not develop until the end of the sixteenth and through the seventeenth centuries CE. Consequently, it is anachronistic to read the classical writers as making claims about biological species.

Throughout the history of biological thought, *species* has always been thought to mean the generation of similar form. That is, a living kind or sort is that which has a generative power to make more instances of itself. The generative conception of species was the common view from the Greeks to the beginnings of Mendelian genetics around 1900. Prior to this, there had been a *"species question"*—that is, the question of the origination of new species. After this, there was a *"species problem,"* in which various attempts were made to identify the genetic substructure of species. I date the changeover to Poulton's essay in 1903.

There never was a morphological species tradition as such, apart from the use of morphology to *identify* species, and this continues today. Moreover, Idealism and Morphology are distinct programs—not all morphologists are idealists (and hence not all who rely on morphology are Platonists).

Species have always been understood to involve deviation from the type, from Aristotle through to the modern era.

Type and *essence* are distinct ideas in biological history. Types can have variation, while essences cannot. Systematics has always used the type concept and a "method of exemplars," but rarely has it used essences as anything but a useful set of diagnostic keys or as an aid to identification. *Essence* can be understood as a *formal* notion—one used in description, definition, or identification—or it can be understood to be a *material* notion—in which the essence *causes* the thing to be what it is. Many biologists followed Locke in supposing that the Nominal (formal) Essence was not the Real (material) Essence. There is no substantial material essentialism in natural history or biology until *after* Darwin, if then. What there is, is probably a reaction to Haeckel and the "evolutionists" who followed Schopenhauer and Nietzsche, and largely relies upon the revival of Thomism after Vatican I. There is a minor tradition, scientifically speaking, of Neo-Thomist-inspired scientific material essentialism from around 1870 to the end of the 1960s.

The Synthetic Darwinians generated and promoted a history of *species* before Darwin as essences based upon a misreading of the (mostly logical and metaphysical) sources and applying them incorrectly to biological cases. Scientists often use history as part of a program to promote their current scientific views, either by demonizing their opponents in proxy or by demonstrating that they are the culmination of a progressive historical process of discovery. This is not restricted to either side of any debate. Whiggism is rife in textbook histories.

The overall problem of species derives from its neo-Platonic history as a top-down category of the logic of classification.[2] Modern taxonomy works in the opposing direction, beginning with the organisms, the individuals in the medieval system, and thence to lineages, populations, and then species. Species in biology are the result of inductively generalizing from individuals, rather than dividing general conceptions into subaltern genera to reach the infimae species. We still desire to treat *species* as a natural kind term, and hence to find essential features that define all and only those taxa. Between-species synapomorphies are not like this; all they have in common is that they keep lineages distinct (either causally or cognitively, the ontological and the epistemic sides of this issue). They necessitate a bottom-up classification logic, or perhaps better, an *in media res* logic.[3] The reason I have made such play with cladistic conceptions of classification in this work is that cladistic classification is, depending how you interpret the matter, either a prolegomenon to induction, or an act of induction in itself. Bottom-up classification involves projectable inductive

² Boodin 1943.
³ Ghiselin 1997, 182ff.

inferences and predicates, and "being a species" is one of these predicates. We predicate of some group that it is a species, and mean by that, that it is held distinct from other groups; this is all that the biological taxon concept has in common with the medieval conception of classificatory categories. We need to resist the tendency to fall back into the older way of thinking about classification. There is no universal grammar or language of nature. John Ray was right, and John Wilkins (the *other* John Wilkins) was wrong.

One thing that ought to be clear from this book so far is that the standard stories and assumptions from the architects of the Modern Synthesis are often simply incorrect. What implications might the loss of the essentialist myth have for scientific research? We might no longer see it needful to deny there is a human nature, for example, while remaining true to the understanding that there is no human *essence*. This could affect our approach to such topics as evolutionary psychology. We might begin to see that formal considerations and biological considerations do not immediately inter-translate, and so defuse a good many arguments about classification. We might see that species can be real phenomena in, say, a local ecosystem, without requiring of them that they play the same explanatory role in every ecotype. We might stop trying to overgeneralize species concept[ion]s or speciation mechanisms to all species. This would reduce the heat in a number of biological forums. And we might just value conceptual clarity and stop trying to employ the dead in support of modern views, while not overvaluing modern views at the expense of a strawman of the past. Perhaps this will help biologists appreciate older work without the polemics and caricatures currently in use. And Suidae may evolve feathered forelimbs for locomotion.

BIBLIOGRAPHY

Boodin, John Elof. 1943. The discovery of form. *Journal of the History of Ideas* 4 (2):177–192.
Ghiselin, Michael T. 1997. *Metaphysics and the Origin of Species.* Albany: State University of New York Press.
Hull, David Lee. 1967. The metaphysics of evolution. *British Journal for the History of Science* 3 (12):309–337.

Section III

Philosophical Discussions of the Species Concept

12 Philosophy and Species
Introduction

Having covered as best we can the *history* of the notion of species, one may wonder what this has to do either with the *philosophy* or the *science* of *species*. My justification for such a long prelude is that the essentialism story has impeded both philosophical reflection on the matter, through a lack of correctly understood examples, and scientific consideration of pre-Darwinian ideas and solutions. Moreover, philosophy of science must be based on actual histories, rather than any kind of armchair intuition or scientific mythology. In this Section I will do what Marjorie Grene and David Depew did for the philosophy of biology in general, and use the historical context to arrange and discuss the philosophical ideas that are posed and debated by philosophers and biologists alike.[1]

There are many ways that a philosophical topic can be divided, but the three basic questions of philosophy are—What exists? (*Ontology*), What can be known? (*Epistemology*), and What is it worth? (*Axiology*). When considering the problem of species in biology, all three have been approached in detail. The ontology of species has been the subject of metaphysical questions such as those we have seen developed both within and without natural science—are they classes, sets, individuals? The epistemology of species has covered such matters as how they are individuated, explained, or what role they play in theory. The axiology of species is central to issues of environmental ethics and policy, and so on.

We will not cover the axiological issues here, as the ethical and metaethical issues are too vast, especially in ethics and conservation biology.[2] This leaves us with the ontology and epistemology of species. The distinction has been made many times, especially by Cracraft,[3] but as we have seen, similar issues have appeared in the pre-Darwinian literature and often since. Hence I give here a short historical (of course) introduction to the *philosophy* of species, followed by my interpretations of the issues, and then my own dissolution to it.

LITERATURE ON THE PHILOSOPHY OF SPECIES

This is not intended to be a full discussion of the philosophical issues. For that, there are many books in recent years that have introductions to the problem of species. The most significant are Richard Richard's[4] and David Stamos',[5] and several collections on

---

[1] Grene and Depew 2004; however, I do not thereby announce my unqualified agreement with their conclusions or arguments. Hull 1989 also argued in favor of the use of actual evidence in philosophy of science.
[2] See, for example, Maclaurin and Sterelny 2008.
[3] Cracraft 1987, 1989.
[4] Richards 2010.
[5] Stamos 2003.

the subject have also been published.[6] However, the best overview of which I know of the entire philosophical issues, as well as the scientific ones, is Frank Zachos' book, which is written for non-philosophers. I strongly recommend it.[7] For those who are new to the philosophical issues, Peter Godfrey-Smith's introduction has a worthwhile chapter, and Kim Sterelny's and Paul Griffiths' now dated but still worthwhile book offers a more extensive discussion.[8] Matt Slater has an extensive and clear argument for a view he calls Populationism, and discusses the philosophical aspects of essentialism.[9]

THE THREE SPECIES PROBLEMS

The species problem is actually a number of problems that biologists have dealt with since the term was first applied to biological organisms by Aristotle. I call the three main problems the *grouping problem*, the *ranking problem*, and the *commensurability problem*. It will benefit us to clearly distinguish these at the beginning of our discussion and to bear them in mind as we consider the philosophy of species.

THE GROUPING PROBLEM

Our first problem is, How are members of species to be grouped together and excluded from other groups? No matter whether species are constructs and arbitrary, or natural and objective, the issue remains how we can include some and exclude other organisms from a species. During the Middle Ages, the notions of *differentiae* and *relata* were brought to bear. A species was defined by characters, and those that differentiated organisms in some relevant way enabled classifiers to exclude some character bearers from a species. Likewise, characters that related organisms were thought to be sufficient to include organisms within species despite their individual differences. In short, the grouping problem is one primarily of *operationality*. But not entirely—organisms can be seen as *self-classifiers*, especially in the reproductive isolation concepts of species. So long as organisms are able, it is said, to differentiate between themselves and find related enough organisms to mate with, then they form species even if we humans cannot identify what it is about them that the organisms themselves find salient in the process. Maynard Smith, for example,[10] notes that one reason we have so little trouble identifying some kinds of species (e.g., birds) is that they are of the same general size as we are, and use much the same criteria in self-differentiating: sound, appearance, and behavior. It follows, some think, that other groups equally salient to us must also have some causally significant differentiae and relata. Likewise, it follows that organisms that are (reproductively) distinct but are otherwise indistinguishable by us must have some differentiae and relata that we cannot recognize (e.g., biochemical differences) without specialist equipment. So our grouping problem reduces to the issue: what is appropriately seen (by us or by organisms themselves) as grounds for distinguishing some and grouping others into species?

---

[6] Most recently, Pavlinov 2013, an online open source book.
[7] Zachos 2016. Previous works include Claridge et al. 1997, Hey 2001.
[8] Sterelny and Griffiths 1999, Godfrey-Smith 2014.
[9] Slater 2016.
[10] Maynard Smith 1975.

THE RANKING PROBLEM

Is species a fixed—that is to say, absolute—rank in taxonomy (in contrast to the so-called "artificial" higher taxa)? Arguments about whether some or all taxonomic ranks are natural or not go back a very long way. Linnaeus clearly thought that species and genera were natural ranks, while Buffon felt that they were not (at one time), and even today some present alternative views that species themselves are not natural, but only individuals, or presently terminal taxa, are. Much of the species debate has been over what it is that species are that makes them a natural (that is, real) rank. Discussions of reproductive isolating mechanisms, for example, or ecological niches, geographic replaceability, and so forth take it for granted that there does exist a real rank of taxa, that it is appropriate to call that rank the species-level, and that what we are all disagreeing about is what to use to define that level, or rather what it is that makes some taxon that level. Other problematic cases, such as ring species, well-defined geographic races, facultatively interfertile species, superspecies, sympatric species swarms, and of course asexuals, both quasispecies and secondarily asexual taxa, lead some to think that the notion of species is a homonym for many distinct concepts. Moreover, species seem to be constituted in different ways in different groups of organisms. Many plant species are formed, for example, by hybridization, a process that is often ignored or glossed over when discussing animal species. Some fungi have multiple sexual morphs instead of the regulation two, and so reproductive isolation becomes a much harder concept to qualify in their case than in animals.

Consequently, if we are trying to compose a general concept of species that applies to all living things, or to compare and contrast the concepts that are relevant only in particular cases, two questions arise: one is whether there is warrant for thinking that there *is* indeed a rank of species in all taxonomic hierarchies or whether it is relative to the discriminating criteria used in a given group of taxa; the other is whether a species taxonomic level is required at all.

THE COMMENSURABILITY PROBLEM

The most common and popular species concept is, of course, the biological species concept championed by Mayr since 1940. It is typically the definition taught to undergraduate zoology students and often the one taught to botany students. It is not, however, taught to bacteriology students, nor is it the preferred definition of lichenologists, as the taxa that are formed by most bacteria do not rely on constant sexual reproduction if any, and lichens are obligate symbionts formed by a mutualistic association between blue-green algae and fungi.[11] Immediately the commensurability problem rears its head. Species among animals are not commensurate with species among many single-celled organisms or species among symbionts. Some, such as Dobzhansky,[12] simply denied, as we saw, that asexual organisms *can* form species (since by definition reproductive isolation is what makes species, and every asexual individual is reproductively isolated). This is surely putting the definitional cart before the evolutionary horse. If species are outcomes

---

[11] Purvis 1997, Leavitt et al. 2016. A recent paper suggests that lichens are the result of *three* symbionts, with an ascomycete yeast completing the triad [Spribille et al. 2016].

[12] Dobzhansky 1937, 320ff.

of evolution, and asexual taxa—forms, morphs, types, niche occupiers, whatever—are the outcome of evolution, then we need to be able to justify the special status of taxonomic units like species over other taxonomic units. Simple familiarity through tradition and acquaintance in one domain of biology is insufficient—zoological hegemony has been a tendency in evolutionary and taxonomic theory for most of the twentieth century. In part, the species problem arises because categorical concepts that applied well in zoology failed to generalize outside it, or even those that applied well to some particular group such as birds, mammals, or insects failed to generalize even to other animals, let alone plants, fungi, lichens, algae, and so on. Perhaps, some commentators—both biologists and philosophers—suggest, the term "species" is a trashcan categorical and should be replaced altogether. At least one major group of polychaete worms has been described recently without mention of species, except to explain why species are not mentioned. Perhaps "taxon" is sufficient. This will be discussed in Chapter 15.

MONISM VERSUS PLURALISM

Philosophers of biology, along with biologists themselves, have divided into two broad camps: *monists* about species concepts, and *pluralists*. Monists hold that there is one conception of species that is correct; pluralists that there are many. Philosophers may be monists or pluralists about a metaphysical view of species as well as about which particular conception of biological species it is right to adopt and champion. David Hull set up the philosophical species concept debate:

> One reason why philosophers find the monism-pluralism debate so interesting is its apparent connection to the dispute over realism and antirealism. Of the four possible combinations of these philosophical positions, two seem quite natural: monism combined with realism, and pluralism combined with antirealism. ... The other two combinations ... are somewhat strained. It would seem a bit strange to argue that one and only one way exists to divide up the world, but that groups of natural phenomena produced on this conceptualisation are not "real." They are as real as anything can get! ... A combination of pluralism and realism seems equally peculiar. ... The world can be divided up into kinds in numerous different ways, and the results are all equally real![13]

So Hull identified four philosophical options (Table 12.1) and argued that pluralist realism is untenable. A great many philosophers followed him in this.

Given that few now think that there is one universal species conception, many have argued for pluralist antirealism. Antirealism comes in two kinds: either there simply are no such things as species (*ontic* antirealism) or there is no such category of "species" that will suffice to capture the complexity of the living world (*epistemic* antirealism). The latter view ties in with the commensurability issue mentioned above.

However, neither monist realism nor pluralist realism are incoherent. If there is one kind of *species*, anything not covered by this category is simply not a species, no matter what else it may be (a population, a type, a subspecific clade, or an agamospecies). This would be the view of Mayr and Simpson, as discussed above. Moreover, pluralist realism is not unknown either. Kitcher stated:

---

[13] Hull 1999, 25.

TABLE 12.1
Monism, Pluralism, Realism, and Antirealism

Species...	Monist	Pluralist
Realist	Monist realist	Pluralist realist
Antirealist	Monist antirealist	Pluralist antirealist

The species category is heterogeneous because there are two main approaches to the demarcation of species taxa and within each of these approaches there are several legitimate variations. One approach is to group organisms by structural similarities. The taxa thus generated are useful in certain kinds of biological investigations and explanations. However, there are different levels at which structural similarities can be sought. The other approach is to group organisms by their phylogenetic relationships. Taxa resulting from this approach are appropriately used in answering different kinds of biological questions. But there are alternative ways to divide phylogeny into evolutionary units. A pluralistic view of species taxa can be defended because the structural relations among organisms and the phylogenetic relations among organisms provide common ground on which the advocates of different taxonomic units can meet.[14]

I have defended a similar view myself:

In one class of species concepts—the reproductive isolation concepts of species—being a species depends upon a natural kind: sex. We tend to generalise from zoological, and particularly mammalian, sexual modalities, but the biological reality is that sexuality is pluriform. Nothing that we can set up as necessary and sufficient criteria captures all and only sexual reproduction. On the cladistic account, the only natural group is a monophyletic group, known as a clade (a single taxon plus all and only its descendants). Sex is not a trait of only a single monophyletic group, and hence is polyphyletic (and indeed paraphyletic). ... However, the many different modalities of sexual reproduction are due to the many monophyletic origins of these modalities. It follows that biospecies are evolved modes, and that there are therefore many different modes of biospecies. Since biospecies have independent origins, on the cladistic convention different modes of sexual species must be seen as distinct groups within the tree of life, and the more general kind of species is the more inclusive class of asexual species, or agamospecies. Hence, there is no natural group of biospecies. But the actual processes that isolate the many kinds of sexual species are empirically determinable, and so just in terms of this one concept we are limited pluralists, constrained by the evidence. By induction, other concepts that aim to capture biological realities, will be likewise constrained.[15]

In short, ways of being species evolve, just like ways of being vertebrates, or mammals, or angiosperms.[16] This of course raises the issue of what it is that *makes* some group a species, to which I shall return (see Chapter 14, Species Realism).

---

[14] Kitcher 1984, 309.
[15] Wilkins 2003, 625.
[16] See also Ereshefsky 1998.

BIBLIOGRAPHY

Claridge, Michael F. et al. 1997. *Species: The Units of Biodiversity*. London/New York: Chapman & Hall.

Cracraft, Joel. 1987. Species concepts and the ontology of evolution. *Biology and Philosophy* 2 (3):329–346.

—. 1989. Speciation and its ontology: The empirical consequences of altering species concepts for understanding patterns and processes of differentiation. In *Speciation and its Consequences*, edited by D. Otte and J. A. Endler, 28–59. Sunderland, MA: Sinauer.

Dobzhansky, Theodosius. 1937. *Genetics and the origin of species*. New York: Columbia University Press.

Ereshefsky, Marc. 1998. Species pluralism and anti-realism. *Philosophy of Science* 65 (1):103–120.

Godfrey-Smith, Peter. 2014. *Philosophy of Biology, Princeton Foundations of Contemporary Philosophy*. Princeton: Princeton University Press.

Grene, Marjorie Glicksman, and David J. Depew. 2004. *The Philosophy of Biology: An Episodic History, The Evolution of Modern Philosophy*. Cambridge: Cambridge University Press.

Hey, Jody. 2001. *Genes, Concepts and Species: The Evolutionary and Cognitive Causes of the Species Problem*. New York: Oxford University Press.

Hull, David Lee. 1989. A function for actual examples in philosophy of science. In *What the Philosophy of Biology is: Essays Dedicated to David Hull*, edited by Michael Ruse, 309–321. Dordrecht: Kluwer.

—. 1999. On the plurality of species: Questioning the party line. In *Species, New Interdisciplinary Essays*, edited by Richard A. Wilson, 23–48. Cambridge, MA: Bradford/MIT Press.

Kitcher, Philip. 1984. Species. *Philosophy of Science* 51 (2):308–333.

Leavitt, Steven D. et al. 2016. Hidden diversity before our eyes: Delimiting and describing cryptic lichen-forming fungal species in camouflage lichens (Parmeliaceae, Ascomycota). *Fungal Biology* 120 (11):1374–1391.

Maclaurin, James, and Kim Sterelny. 2008. *What Is Biodiversity?* Chicago, IL: University of Chicago Press.

Maynard Smith, John. 1975. *The Theory of Evolution*. 3rd ed. Harmondsworth/Baltimore: Penguin.

Pavlinov, Igor Ya, ed. 2013. *The Species Problem—Ongoing Issues*. Rijeka, Croatia: InTech.

Purvis, Ole William. 1997. The species concept in lichens. In *Species: The Units Of Biodiversity*, edited by Michael F. Claridge et al., 109–134. London: Chapman & Hall.

Richards, Richard A. 2010. *The Species Problem: A Philosophical Analysis, Cambridge Studies in Philosophy and Biology*. Cambridge: Cambridge University Press.

Slater, Matthew H. 2016. *Are Species Real? An Essay on the Metaphysics of Species*. Basingstoke, New York: Palgrave Macmillan.

Spribille, Toby et al. 2016. Basidiomycete yeasts in the cortex of ascomycete macrolichens. *Science* 353 (6298):488–492.

Stamos, David N. 2003. *The Species Problem: Biological Species, Ontology, and the Metaphysics of Biology*. Lanham, MD: Lexington Books.

Sterelny, Kim, and Paul E. Griffiths. 1999. *Sex and Death: An Introduction to Philosophy of Biology*. Chicago/London: University of Chicago Press.

Wilkins, John S. 2003. How to be a chaste species pluralist-realist: The origins of species modes and the synapomorphic species concept. *Biology and Philosophy* 18 (5):621–638.

Zachos, Frank E. 2016. *Species Concepts in Biology: Historical Development, Theoretical Foundations and Practical Relevance*. Switzerland: Springer.

13 The Development of the Philosophy of Species

There are two major questions about the analysis of the species category: First, to what general ontological category do species taxa belong? Second, what distinguishes species taxa from other members of this category?

Philip Kitcher[1]

All philosophical arguments have antecedents, and all special topics (such as the species problem) are broadly influenced by philosophical ideas and arguments from outside the specialty in which they occur. This is equally true for the philosophy of biology, and multiple resources are employed from the three broad areas of philosophical enquiry.

Scientists, especially of recent decades, however, are often not so comfortable with philosophical treatments of their specialty. For instance, Van Valen argued that the resources of philosophy (at the time he wrote) are not really fit for the problem of species, and lists the heterogeneous criteria that working taxonomists use to identify a species:

First, though, it may be useful to list some properties which, I trust, most biologists agree apply to most species:

(1) A single origin and final extinction, with reproductive (informational) continuity between these events.

(2) Origin taking many, rather than several or one, generations.

(3) Limited but real extension in both time and space.

(4) Origin by transformation of a population of individuals, not from one or two parents.

(5) Capacity to evolve.

(6) Capacity to act as a unit in evolution.

(7) Occurrence in spatially disjunct populations.

(8) Potential reproductive continuity among all included populations; compatibility for development and fitness of offspring as well as for mating and fertilization.

(9) A mechanism for recognizing other individuals or gametes of the same species.

(10) Reproductive isolation from other species.

(11) Being composed of individuals.

(12) Capacity to speciate, either with or without phyletic branching.

(13) Capacity to remain the same species while, and after, part becomes another species after phyletic branching.

(14) Possession of phenotypic, genic, and genotypic characters jointly distinct from those of any other species.

---

[1] Kitcher 1987, 185.

(15) Occupancy of a perhaps broad and flexible niche (in the sense of a perhaps arbitrarily bounded part, or even disjoint parts, of the [biotic and abiotic] environmental hyperspace) different from that of any sympatric species.

(16) Ultimate regulation of population or metapopulation density being causally different from that of any sympatric species.

All of these properties are of species taxa, not of the species category, which is merely the set of all species …. The list is very heterogeneous, as it should be, but all the attributes are ontic rather than epistemic.[2]

In recent years, microbiology and virology have added various issues to the *explananda* of the species problem, such as lateral genetic transfer, endosymbiosis, community formation, and so forth. These, then, are the key issues that the problem of species must address. As science has observational data that it seeks to explain, philosophy has criteria that it needs to explain under unified concepts, or to dissolve the concepts as being non-unifiable.

Van Valen goes on to argue that Beckner's notion of a *polytypic* set[3] provides the best resources for addressing the problem, but this is getting ahead of ourselves. For now, let us note that polytypy, recast by Sneath and Sokal as a *polythetic* set, is a philosophical notion. Philosophy is unavoidable even by scientists.

In fact, Beckner's book began what is now understood as the modern discipline of the philosophy of biology. Hull was influenced by him, as was Ruse. Nevertheless, philosophical treatments of biology are of a long lineage,[4] including Dobzhansky's 1935 paper "A critique of the species concept in biology," published in *Philosophy of Science*.[5] That noted, though, the modern discipline begins in the 1960s, and the keystone paper that began it is Hull's 1965 paper, written a mere five years after Beckner's book was published, "The effect of essentialism on taxonomy—two thousand years of stasis," and the key issue was species.

As we have seen, the essentialism story was instigated by Hull taking Popper's view of Aristotle, together with Cain's 1958 paper, as being history. This found its way into orthodoxy for the next fifty years through the advocacy of, among others, Ernst Mayr. And once introduced in that fashion, philosophy became crucial to the species problem, as essentialism was a long-standing topic in metaphysics and epistemology.

It is now consensus among philosophers of biology, and many biologists, that essentialism is contrary to biological reality. In large part, this is because of Mayr's authority and his notion of *population thinking*.[6] Divorced from the essentialism *story*, however, the arguments in favor of populations as the "units" of Darwinian

---

[2] Van Valen 1988, 51.

[3] Beckner 1959, 23–25. Beckner was directly influenced by Wittgenstein's "family resemblance" notion in the *Philosophical Investigations* [Wittgenstein 1968, §67], but also the notion of a polytypic species in biology [Van Valen 1988, 53].

[4] For example, Smellie 1790, Whewell 1840, Oken 1847, Goodsir 1868, Papillon 1875, Spencer 1884, Bailey 1896, Johnstone 1914.

[5] Dobzhansky 1935.

[6] For an excellent treatment of the notion of populations in a Darwinian context, see Godfrey-Smith 2009. See also Queller 2011.

evolution and against the essentialist approach to species whether or not that was ever offered in biology, remain. However, in order to explain why species exist, and what they and other taxonomic kinds are, both unitary causal accounts and epistemic criteria for judging that the kinds share that unitary causal account are required. For instance, not all populations form a species. And arguably, not all species are metapopulations.

THE PHILOSOPHICAL BACKGROUND

The species problem is often discussed in the context of developments in philosophy generally. For instance, Mill and Venn introduced the notion of *natural kinds* into the philosophy of science, and since then philosophers have returned to this as the basis for species in natural history/biology.[7] With the invention of "essentialism" in the 1950s in philosophy, natural kinds were identified with essential classes, sets, or sums.[8]

When Hull began his work on essentialism in taxonomy in 1965, he was greatly influenced by the work of Michael Scriven on explanations and predictions.[9] Under the dominant theory at the time of explanation, the *nomological-deductive* (ND) model, in order for any kind of entity to play a role in an explanation, that kind had to be subject to laws of regularity, from which, with the appropriate boundary and initial conditions, one could deduce the state to be explained. Since species were not subject to laws,[10] species did not play an explanatory role in biology and hence were not something that fulfilled the appropriate role for a natural kind, in Hull's view:

> On the traditional view, the species category is a class of classes defined in terms of the properties which particular species possess ... and particular organisms are individuals ... The relation between organisms, species and the species category is membership. An organism is a member of its species and each species is a member of the species category. On the view being urged [by Ghiselin and Hull], both particular species and the species category must be moved down one category level. Organisms remain individuals, but they are no longer members of their species. Instead an organism is part of a more inclusive individual, its species, and the names of both particular organisms (like Gargantua) and particular species (like *Gorilla gorilla*) become proper names. The species concept is no longer a class of classes but merely a class.[11]

Considerable ink has been employed to argue whether species are classes or individuals. The debate continues even today.[12] Ghiselin and Hull argued that species cannot be sets, because sets cannot be historical; that is, they cannot change over

---

[7] I have discussed this history in Wilkins and Ebach 2013.
[8] Khalidi 2013 offers a nice discussion of these issues.
[9] Scriven 1959.
[10] Under the view that selection occurred at the population level, natural selection was not a law of *species* evolution.
[11] Hull 1976, 174f.
[12] Stamos 1998.

time, and species do.[13] Kitcher, on the other hand, argued that sets can, and do, evolve.[14]

When people say, as do Kitcher and Ruse,[15] that species are classes, they typically mean that species are natural kinds in a Kripkean sense of having some intensional definition or set of properties.[16] On this account, species are things that share some set of criteria or microstructures that function as the essence of the species. It might be a genome, a wild-type, or just some simple morphological criterion, but all members of the species must have that set of properties. Hence, says Ruse, we can regenerate *Tyrannosaurus rex* because if we have the genome and a suitable egg, we have the essence of that terrible lizard. According to Hull, though, if we have recreated a dodo by breeding it from its ancestral pigeon stock, it is not the same species as *Dodo ineptus*, destroyed by sailors on the Indian Ocean islands where it once lived. Species, like people, can only be born once—everything else is a knockoff, like a Tiffany-style lamp is a knockoff of an actual Tiffany lamp.[17]

A problem for Hull is that species *have* been observed, under certain conditions, to evolve more than once—in plants which can repeatedly duplicate their chromosomes to "instantly" form new species, or in fish which independently evolve deepwater morphs in different lakes, and which will breed preferentially with similar morphs in other lakes.[18] Let us call this the *Respeciation Problem*, after Turner's title. Moreover, the new individuals formed this way are interfertile with their predecessors, and fold their genetic complement into that of the established lineages. How can Hull deal with this?

One way might be to take the Hennigian approach and say that the two new species are extinguished when they merge and form a new one. This is a purely nomenclatural solution to an ontological problem. Yes, the new species has a somewhat different character (unless the modifications are infinitesimally distinct from each other), but if, in order to save the individuality thesis one has to say that every time a change occurs you have a new individual, the thesis is bordering on banality. Anyway, it is not Hull's solution; he wants to be able to restrict species in time, but not to reject their mutability. In a discussion using the Hobbesian Ship of Theseus example, which being rebuilt ends up entirely replaced and then duplicated from the older parts, Ghiselin, the originator of the species as individuals thesis, proposed that there is no fact of the matter which is the "same" ship, but that this does not mean Theseus' ship is a universal.[19]

So perhaps here, Hull might wish to treat species as having vague boundaries in time—there is a point x at which there is no species A, and a point y where there is, and in between we have a vague border where it is ambiguous whether or not A

---

[13] Hull 1980, Ghiselin 1997.

[14] Kitcher 1984. Kitcher's argument was based on the extensionality of sets—if the composition of the set changed over time, so too did the set change (to a new species).

[15] Ruse 1987. See also LaPorte 2004.

[16] Mellor 1977 discusses this sense of natural kind, but see Hacking 2007b, 2007a for discussion of the modern essentialist tradition of natural kinds, and Slater 2016 for further elaboration.

[17] With a difference in the price tag to match, Hull 1988, 78.

[18] Turner 2002, Martin et al. 2015.

[19] Ghiselin 1997, 52. See also Hull's discussion of theoretical traditions [Hull 2002].

exists, or whether the two still-distinct lineages *M* and *N* are *A* or not. After all, this is the case when we discuss nascent species anyway. But this seems counterintuitive as well—if ontology is what we are discussing, then each lineage, say *M*, is identical physically to *A*, and all that is missing is a date indexical. This seems odd.

For this reason, Ruse wants to treat *A* as a class which can be instantiated in actual historical entities or processes as many times as need be. It avoids the respeciation problem. The disagreement lies in different answers to the question, "what do we need to capture in our metaphysics for things that can arise more than once?" Consider a different problem—that of the identity of Individual[20] persons. We typically say of an Individual that he or she has some terminus a quo at a particular point—it might be conception, or nativity, or some point of development. In taxonomic terms, we would say that this Individual is monophyletic, being the ancestral cell population and all descendent cell populations. But it is possible, in biology, for an Individual fetus to be a mixture of two genetically distinct fertilized cells or zygotes; these are called chimera after the mythical monster. Are *these* Individuals, in the ordinary sense? And to make matters worse, what if one zygote had been commenced in vitro some years before the second and frozen in nitrogen before it was fused with the other? What *is* the *terminus a quo*?

If we want to capture this case in our definitions of what it is to be an Individual, then we have the same sorts of problems for identity of organisms that we have for *species*, and how we deal with them will depend largely on how we think meaning inheres in such terms. We might say that to be an Individual, say Socrates, to use the classic example, is to instantiate Socratic properties—that is, to treat Socrates as a class that can be instantiated in many different individuals courtesy of some science fiction replicating machine. That would be what Ruse would say about species. Or, we could say that there is only one Socrates, who was born, perhaps chimerically, at a relatively exact period, only in this case the chimeric Socrates was "born" or "begun" in an extended process unlike the monophyletic Individuals we usually see. Even if the unfused zygotes were viable and developing organisms, we might say that they ceased to be upon being fused; and so on for all the exotic cases we might imagine.

Chimeric species are not a counterexample to our intuitions of what species are, metaphysically speaking; they may extend our conception, or we might (arbitrarily) exclude them, but I see no reason to reject the individuality thesis because of these cases and be forced to instead adopt a "natural kind" class conception of the species categorical term. What we choose as the boundary for the term may be arbitrary or it may be forced by the biological realities, but species are still, as Caplan phrased it,[21] déclassé, I think.

---

[20] I am capitalizing the word "Individual" to avoid confusion with the metaphysical notion being discussed here.

[21] Caplan 1980.

INDIVIDUAL, COHESIVE, OR CONCRETE

There is a major equivocation on the term "individual" that causes some confusion in the literature. There are three kinds of individual that one finds in the discussions, and they are not always disambiguated by the authors.

Metaphysical or logical individuals ("Particulars"): These are entities that are not classes; that is, which are particulars. A class is defined (in my preferred metaphysics[22]) as the denotation of a non-restricted predicate. The metaphysical aspect of "individual" *sensu* Ghiselin/Hull is that a taxon (in particular a species) is a *concrete* object, not an abstract one. Concrete objects are spatiotemporally restricted, or, to put it another way, have an index space–time coordinate or range of coordinates. Abstract objects do not have such an index.

Functional or cohesive individuals ("Systems"): These are objects which happen to have some common systematic set of interactions (either temporal or formal). This is the sense of "individual" that applies to a Mayrian biospecies—his notion of gene flow as a homeostatic mechanism suggests that he thought of species as functional units.

Phenomenal individuals ("Appearances"): These are objects which can be observed in their entirety as entities. This is very often scale-dependent (a colony can look unified at one scale and appear as a multiplicity of objects at a finer scale).

Now, this gives us a field of eight options: a putative taxonomic entity can be one of those seen in Table 13.1. Shading in the table indicates the possible kinds of classes (top) and individuals (bottom), with names for each option in italic. The question is then what sort of entity/individual a species is (this can be different for different species). Leaving out the metaphysically abstract species (A), which is what Ghiselin and Hull do (but Ruse does not), working from Strawson's *Individuals*[23] for obvious reasons (as no individual organism would ever be a member of its species, only a "realization" of it), and the appearance-based kinds (B–D) and we are left with four options: a species is a historical entity without causal cohesion or phenomenality (E); a species is a historical entity with causal cohesion but no phenomenality (F); a species is a historical individual with no causal cohesion but phenomenality (G), and a species is all three (H). Option E fits any spatiotemporally delimited group of organisms (including any lineage or clade); Option F and G fit Mayrian species (the second being cryptic species), and Option E represents any "taxonomic" species that lacks a mechanism as such for keeping it distinct.

Consequently, I can't agree with the premise that to be a metaphysical individual *requires* causal cohesion or functional integration, as is sometimes expressed.[24] Paradigmatic individuals (i.e., organisms) do have this cohesiveness and are usually phenomenally distinct, but this is not an argument that all *metaphysical* individuals must be. A hive is functionally cohesive but sometimes not phenomenally distinct, for example (of course, as phenomenal distinctiveness is scale-relative, a hive might

---

[22] Zalta 1988.

[23] Strawson 1964, Chapter 8.

[24] For instance, by Ghiselin 1997. Gould 2002, 602–603, for example, offers four conditions for selectable individuals: *change, discreteness and cohesion, continuity,* and *functionality or organization.* He needs these for species to be subjected to species selection.

TABLE 13.1
Types of Individuals

Taxon	Metaphysical	Causal	Phenomenal
A. Pure abstraction [*Idea*]	✓	✓	✓
B. Phenomenal individual [*Group*]	✓	✓	✗
C. Integrative individual [*Effect*]	✓	✗	✓
D. Phenomenal, Integrative individual [*Apparent Effect*]	✓	✗	✗
E. Historical individual [*Particular*]	✗	✓	✓
F. Historical, Phenomenal individual [*Apparent Particular*]	✗	✓	✗
G. Historical, Integrative individual [*System*]	✗	✗	✓
H. Full individual	✗	✗	✗

behave very distinctly in, say, time-lapse photography, as also would other vague individuals like fungal mats, etc.).

Moreover, the distinction between species as a time-slice (*synchronic*) entity and lineages as a time-extended (*diachronic*) entity also seems to me to fail as a sharp distinction. All species are lineages, as de Querioz has it, but not all lineages are species. A synchronic species is just a very foreshortened lineage in a slight time horizon. It is never entirely "non-dimensional," and so I do not see the need for a complex and in my opinion unsupportable metaphysics of "potentials" or "propensities" such as Mayr offered. Calling on the potential to interact in a given manner is begging the question; this is only known *ex post facto*. In the end, a species must be seen logically as an individual lineage, extended over time and space.

CLOUDS, CLADES, AND GRADES: NATURAL KINDS OR NATURAL GROUPS?

Throughout the natural kinds debate, philosophers have assumed that kinds of organisms—in particular animals like tigers, zebras, and domestic animals—are exemplars of natural kinds, or presented arguments why they are not, without stating exactly what it is they *are*. Although species have been given a metaphysical treatment as individuals, or particulars, in contradistinction to the "default" or Received View that species are classes or universals, the nature of those individuals and of all higher taxa they comprise has not been fully explored.

LaPorte has revived the claim, made also by others in recent years, that organisms do form natural kinds.[25] In this, he focuses on the *kinds*—it is taken for granted they are *natural*. LaPorte describes the sort of kinds organisms form as having a "historical essence," one that is not based upon the microstructure of the organisms in each kind being shared, but rather on a shared historical origin. Griffiths has also offered a similar account.[26]

---

[25] LaPorte 1997, 2000, 2004, 2018.
[26] Griffiths 1999.

I do not consider that these are either essences or that they form kinds, despite the extensive defense of that view LaPorte presents in terms of naming groups and what their necessary referents are thereafter. In contrast, I want to make the following claims and defend them:

1. Species and other evolutionary taxa do not form natural kinds; they form natural groups.
2. Natural kinds in biology are model-dependent and are neither inductively projectable nor informative about anything not contained in the model[s] in which they occur.
3. Natural groups are both inductively projectable and informative.
4. While natural kinds are not historical or concrete objects, but are timeless abstractions, natural groups are concrete and historical.
5. Natural kinds are not natural in the sense that they have objective existence independently of cognition; natural groups are natural in that way.

These are broad and contentious claims, so I shall attempt to make them plausible and provide support for them. To achieve this, I shall rely on historical practice of biologists, and in particular of recent taxonomy.

TAXA AND KIND TERMS

A *taxon* is any class of objects in a classification in modern biology, and species are taxa. The term has been shrouded in ambiguity since it was coined, due exactly to the problem we confront now. It was unclear whether a taxon was a kind that was defined or whether it was a concrete group. In biological systematics, there have been three main ways of conceiving of classes since evolutionary theory became dominant in biology, and in particular since the end of the World War II. These three are, in historical order: classes are *grades*, classes are *clades*, and classes are *clouds*. We use the word "class" in this case simply because we use the term "classification"—it is not meant to imply any conclusions about the nature of classifications. Sometimes in logic a class is understood to be an *intensionally defined* grouping. This is not something we must accept in biology from the outset. In fact, a "universal" term, which is how classes are often described when discussing kinds in biology, is under the Aristotelian system merely *any* term for which there are two or more particulars under its rubric.

Terms in the biology debate, and to a lesser extent in the broader context of philosophy of language and essentialism in science, tend to be part of the problem. We are so used to seeing kind terms as intensional classes that it has become part of the Received View that prior to some threshold point, usually the publication of the *Origin of Species* in 1859, we all thought that taxa in biology were definitional and essential kinds. This is historically untrue, as we have seen, but in any event, we must take care not to import these assumptions into the topic with our terms.

A *grade* can be defined as a defined kind. It is some state that may be acquired or achieved or entered in virtue of acquiring, achieving, or arriving at some set of jointly sufficient and severally necessary properties (JSSNPs), which form the *essence* of the class. Such a conception has several implications I shall attempt to bring out.

A *clade* is, as the Greek root word *klados*, meaning "branch," suggests, a historical class.[27] This notion involves a causal history, which may or may not result in a set of JSSNPs, and in the leading school of biological classification named after it (*cladism*) a clade is understood to be uniquely isolated from the rest of the tree by a single stem or "cut."[28] Such clades have the property of being *monophyletic*, which is usually defined as the stem (taxon, or species) of the branch and all of its dependent branches. Clades nest within more inclusive clades. The diagram of relationships between taxa forms a *cladogram*; there is a dispute over whether or not a cladogram is a direct or indirect representation of the evolutionary history of these groups. Those who think it is a direct tree usually refer to the internal nodes of the cladogram as earlier taxa; those who do not refer to internal nodes of the tree as formally equivalent to inclusive sets; that is, each branch represents a set of smaller sets, or of individual organisms.

A *cloud* is just a cluster in some abstract morphometric or phase space of characters of organisms. On the well-founded assumption that organisms vary over distribution curves for any measurable character, numerical taxonomy, known later as *phenetics* (in contrast to cladistics), attempted to find "natural kinds" of organisms— that is, of taxa—in a purely empirical and atheoretical way. Unfortunately, organisms vary in ways that are often uncorrelated with other traits they bear (due to mosaic evolution), and so it transpired that what appeared as a phenomenal cluster on one set of principal components was divisible in different ways on another set. Many of the analytic techniques of phenetic analysis have been assimilated by both the cladistic and the gradistic schools of classification, but the underlying philosophy of pure operationalism is pretty well a dead issue in biological systematics.

The difference between these schools was not over the task of classification, for each thought that it was to find the "joints between natural things." The dispute largely turned on what was acceptable in the way of theory-dependence in classification. Pheneticists were extreme empiricists; a classification should reflect no prior theoretical commitments. Gradists are comfortable with the notion that we can use theoretical kinds to distinguish objects, in ways I shall explore below. Cladists assumed that we are recovering some signal of the history of the organisms through the employment of a formal and logically coherent methodology, but that some general knowledge of the organisms and the taxa in which they exist is required to separate the informative from the uninformative characters.

---

[27] The term was coined by Julian Huxley, and defined as "delimitable monophyletic groups" [Huxley 1957b] and "*groups* of common ancestry" [Huxley 1957a, 90]. He also used the term *grade* to mean an advance "sometimes independently achieved, sometimes in common." I am deliberately not adopting the rather insulting terms employed by Mayr and Bock [2002, 180] to distinguish between "classification," which they define as, or rather assert, is based on similarities, and "cladification," or genealogical "ordering." It appears that Mayr and Bock wish to restrict the nature of classification to grouping by similarities (in conjunction with phylogenies). This is contrary to historical antecedent, and restricts, deliberately, the role of systematics or taxonomy to the collation of grades as defined here. Alternatively, should one wish to assert that cladification is one kind of classification, which I do, the other kind might be better termed "gradification," or the making of grades. I first used this term in Wilkins 2002. See also Hennig 1975.

[28] Sober 2000, 164.

Each of these has a philosophical analogue or equivalent. Gradism is essentialism, phenetics is (quite overtly) based on Wittgensteinian family resemblance, and cladism is either an Aristotelianism or a nominalism. It is odd that while essentialism is considered fallacious in biology and especially in taxonomy, largely due to the advocacy of Ernst Mayr, it is the received view in philosophy of language. Recent work, neatly summarized by Susan Gelman,[29] suggests that we are all born essentialists, which therefore raises the question: must we remain essentialists? If biology is not just the essences of kinds, then of course once we commence our investigations into biology we need not remain essentialists; but oddly Mayr, who stridently opposes the idea that species and populations are essentialistically definable, nonetheless adopts a gradist perspective when classifying taxa in addition to genealogical "ordering." Let us consider why his "evolutionary systematics" adopts this chimeric approach.

According to Mayr and Simpson, the genealogical aspect of evolution, which Darwin called "common descent," is paramount, but nevertheless, we can overlay on that evolutionary history further divisions based on the evolutionary "niche" or "grade" that some branches achieve. For example, there is some similarity of lifestyle and body plan that groups crocodiliforms and reptilians together, even though in cladistic—that is, evolutionary genealogical—terms reptiles and birds are more closely related, which is to say they share a more recent common ancestor. The "avian grade" is sufficiently distinct from the "reptile-like" body plans that we can ignore "propinquity of descent" as a grouping criterion in this case. A similar case is put for humans: we are sufficiently different from other apes and primates in general that we must form our own named group. In fact, Julian Huxley once suggested, half-seriously, that we ought to form a new *kingdom*, *Psychozoa*, of cultural animals with mind and language.[30] This kind of grade/clade hybrid is a paraphyletic group, if more than one species attains it.

In large part, such claims depend critically on the importance of the factors used in defining the grade. Perhaps there is a grade of parrots that can act cute and mimic language—if so, the only thing saving science from making this a grade, and from a host of similarly arbitrary grades, is the fact that scientists do not think "acting cute and mimicking language" forms a natural kind, while "feathered flying animals," and "mentally endowed language users," do form two kinds. If they do, though, it is clear in an evolutionary sense that such grades can be achieved more than once by different genealogical routes. In other words, feathered birds and talking brainy apes might evolve more than once. Cuteness is a subjective grade, which depends on the criteria of the observer. We shall not consider subjective grades as worthwhile kinds here, as to do so amounts to adopting conventionalism.

The cladistic account is more particularistic in the ontological sense—each branch of the tree is a singular historical object. What the scientists think is important is not of any real relevance to the "naturalness" of the groups. What happened to them in their history is. It may be that one can identify "kinds" such as "warm-blooded"—once

[29] Gelman 2004.
[30] Huxley 1957a, 91.

a candidate taxon, *Homothermia*[31]—which clades can attain through evolution, but that is a purely theoretical name. To define it you require some reason to think it is an important kind and not simply a subjectively interesting one. You must have some prior commitment to the significance of warm-bloodedness (and of course it was important in the minds of those who "discovered" it because *they* were warm-blooded). In short, once has to have some sort of model from which to draw significant and salient criteria to define the grade out of an infinite range of *possible* grades.

Natural Boundaries

In the cladistic approach to classification, the classes are causal, not descriptive. One often cannot define a clade, only identify (or diagnose) it and point to it. Theory, and theoretical terms such as "warm-blooded," "intelligent," "predator," "parasite," can only be applied or not to a clade according to the standards of a theoretical structure and interpretation, and they are no more informative than the model on which the grades are based. If you only know that something is a predator, you only know whatever "predator" implies (that it is a heterotroph, that it is ambulatory in most cases, that its population size will co-vary with its prey's in Lotka-Volterra cycles under some *ceteris paribus* conditions[32]); you know little else about it. On the other hand, if you know that something is a member of a particular clade, you know a *lot* about it—you know most of: its dental structures, skeletal structures, cell types, developmental sequences, genetic features and functions, overall lifecycle, and so on. On the cladistic account, these things are affinities, as they might have said in the nineteenth century, while ecological and other universalizing theories only provide information about analogies (in the biological sense of the similarity relation that is not due to common descent, opposed to homology).

Natural kinds have always been an attempt to draw natural boundaries among universals. The thing about universals, though, is that they are timeless and abstract. A universal is something that any historical thing may instantiate, but which continues to exist if nothing instantiates it. A historical entity, though, is a particular, an individual. When nothing of that particular exists any more, the particular is gone, extinct. At best, particulars instantiate, or subsist as examples of, universals. So we must ask ourselves a number of questions—some historical and some formal. One question is, how is classification in biology to proceed, via universals or via particulars, on the basis of laws or actual processes? Another is, why do we insist that classification must be based on universals? There are others, which we shall encounter as we proceed, but these two offer a starting point.

Because Aristotle's logic, in the *Categories* and elsewhere, begins with predicates, it is perforce an essentialistic (definitional) logic, although pretty well everybody from him through to Locke understood that the definitional account of essence did not apply in a simple manner to living things. Moreover, it was understood after the

---

[31] Huxley, *loc cit.*

[32] Mark Colyvan noted in commenting on an earlier draft that the Lotka-Volterra equations also apply to the relationship between grazers and the plants they graze, so perhaps it turns out that the naïve category of "predator" in fact is artificial, or an arbitrary partition of a natural kind (see Colyvan and Ginzburg 2003).

introduction of a biology-alone species concept that the *biological* species was not identical to the Aristotelian species, as I have shown above. Although the received view is historically false—biologists before Darwin and logicians knew that living species could not be defined and were decomposable, unlike logical species which were both definable and indivisible—there is sufficient truth in it to set out a philosophical problem addressing the nature of kind terms like "species" in biology and their relation to natural kinds overall. I shall argue that biology does not have natural kinds (NK) of the sort that physics and the other ahistorical sciences do, and that the appropriate sortal in biology is a natural *group* (NG).

CLASSES IN BIOLOGY

A natural kind is defined by Wilkerson[33] as a real essence; an intrinsic property or set of properties that make a thing the kind of thing it is, irrespective of any system of classification. It is not important here whether this commits a Natural Kindist to scientific essentialism, modal realism, or the logical or causal necessity of laws of nature. What matters is that NKs imply that class terms involve a *de re* commitment to a fixed set of properties intrinsic to any particular that is a member of, or falls under, the natural kind.[34] I use the term "class" in this section for any sortal concept covering more than one particular, without prejudice with respect to essentialist definitions or not, as Aristotle did with universals. Here, clouds, clades, and grades are all class concepts, and data organized using them are classifications. Effectively, a classification is the organization of any set of data or objects under general terms.

It is clearly the case that if biological taxa are NKs, they cannot evolve (despite LaPorte's contrary assertion). What has to occur in evolution, if taxa are NKs, is that the actual ancestor-descendent, or genealogical, lineages move *out* of one species/ NK and into another. Species would become, on the NK account, timeless abstract classes, leaving the dynamic and historical entities of evolution to be the lineages themselves.[35] Moreover, it would then also be the case that "being a species instance" would be sharply defined. If there were a microstructural essence, variation elsewhere in the genome or phenotype other than in what is essential would not affect the essence (would be *accidental* in the old terminology), but as soon as *the essence* varied beyond its limits the lineage would no longer *be* of that species; allowing, with LaPorte and Aristotle, that the essence is within a range, so that a member of a species can vary right up to the critical threshold of being a species member, and that any further divergence from the typical mode will push it over into a novel species

---

[33] Wilkerson 1995.

[34] LaPorte's claim that his kindism, or essentialism, is historical seems to me to be another example of the metaphysical chimerism that is found in Mayr.

[35] LaPorte discusses this objection [2004, 9–10] but fudges this point. He shifts from discussing change of species to change of *members* of species. If species are NKs, they do not change, which he admits, but if they are, then their members are actually only *instances* of the species. There then needs to be a connection or commensuration between instances of the first and second species in an evolutionary sequence, and hence the need for lineages. That said, Ghiselin's objections to which he is responding that species have *no* NK essences must be contingently false. If a species consists entirely of clones, or the species is reduced to a single closely inbred population, it can easily have properties that are unique to all and only its members. Most species will vary (be *polytypic*) but not all need to.

class. It would therefore necessarily follow that a member of one species would literally give birth to a member of another just at the point at which the intrinsic property set is modified or lost, and another gained. Any organism that retained the essential characters of the ancestral species and gained the essential characters of a new one would simultaneously be a member of both species. Add to this the fact that most species are polytypic for the majority of their genes and morphological characters, and it becomes likely there is no intrinsic property set for all and only members of all species. Some views of species, such as some phylogenetic conceptions ("Phylogenetic Taxon" conceptions) have indeed defined species as differentiable or diagnosable groups of organisms which do *not* vary (thus increasing the number of species beyond reproductively isolated groups), but typically, species are thought to be delineated *de re* rather than *de dicto*, and to vary in any candidate physical "essence" or JNSSPs. (What is true of species is true, *mutatis mutandis*, of higher and lower taxa.) As NKs are tied to universals, and in particular laws of nature, the question must arise, and has repeatedly done, whether there are laws in biology, and natural kinds of living things. It may pay therefore to compare the attributes of the object classes of physical NKs with the attributes of candidate object classes of biology, such as organisms, taxa, or ecotypes. How do they critically differ, if they do?

INDISCERNIBLES

Universal objects in physics are typically indiscernibles. One electron differs only from another in exogenous properties such as location and velocity. Its endogenous properties, such as charge and mass, are identical to every other electron (this argument is put clearly by Hull and Ghiselin). The same is true, *pace* isotopes, of atoms of gold, *pace* isomers, of molecules of amino acids and electromagnetic radiations of identical wavelengths. With biological classes, though, it is different. In physics, laws apply universally because the same entities are covered by that law in each application of it. But in biology, each organism, each molecular gene, each cell is distinct, even when these entities fall under "the same" class, taxonomically speaking. Like snowflakes,[36] each biological object is intrinsically an individual. Unlike snowflakes, the similarity relations between organic entities are statistical rather than the result of deterministic causal processes under boundary conditions. It is arguable whether this is true even for snowflakes. Physical objects might never, in fact, be rigidly determined by causal laws. But whether there is a qualitative difference or not, biological organisms, at least, do fall into distributions over a sample set or population. Organic systems are also subject to a complex range of boundary conditions, as well as having unique beginner states, but if this is the only difference, the indiscernibles involved are still only those of physics (and chemistry), and as such biological organisms remain subject to physicochemical laws such as the laws of thermodynamics. There appear to be no universal laws of biology that are not laws of physics over

---

[36] So far as we know...

indiscernible entities of physical theory.[37] So far as we can tell, *biological* objects are nearly always discernible.

Without laws, biological kinds are in sharp distinction from other natural kinds, so we shall consider shortly if there *are* laws in biology. Another difference lies in the historical aspects of biological entities—an electron at t behaves as an electron at $t + n$ does. A radioactive atom has the same probability of decaying at one moment as another, although in large samples of such atoms a fixed proportion will have decayed after a set period has elapsed. However, biological kinds are highly sensitive to prior states. Genetically individual organisms develop differently based on differences in their point of departure, such as parental antibody inheritance in placental mammals, or inheritance of methylation patterns. Likewise, gene trajectories in a population depend on the initial structure of the gene pool, and the evolutionary history of a new species is tightly constrained by the states of the ancestral species, its gene frequencies and developmental processes. Horse descendants will never[38] evolve wings, for instance. No vertebrate will evolve into a centaur. The tetrapod body plan is too restricted by the developmental sequence. And a trait once lost will rarely be regained in an evolutionary sequence, and almost certainly not in the same manner, since, as the number of steps of mutation and recombination of genes increase, the likelihood of complete reversion drops astronomically (Cope's Rule).[39]

Must we therefore deny that biology has NKs of its own, and that it has no laws, however many law-like generalizations it may develop?[40] Must we say that theoretical biology is not actually scientific?[41] Or are all biological NKs actually the NKs of physics or chemistry? To answer this (the last query in the affirmative) we need to understand the role of models in biology, and their relation to contingent historical individuals.

Species and other phylogenetic—that is to say, historical—taxa do not form NKs just in virtue of their phylogeny, but they do form NGs. More exactly: taxa *can* form NKs, but in a sense not usually intended by NK advocates: there *are* laws in biology, and they do involve NKs, but as NKs are not historical entities, they are timeless entities, *sub specie aeternitatis*. Grades are states that may be attained in many different ways and at many different times, and which need not share either the same physical substrates or historical causal processes. The universals of grades are supervenient universals, which, while they are *de re* kinds, get picked out of the world as kinds in virtue of the models we apply to a more or less restricted domain. If nothing

---

[37] A common response is to claim ecological laws as laws of biology that are not reducible to physics, but it is my opinion that ecological laws are merely biological and complex cases of the least action principle. As such, they are physical laws. Further, by "law" here, I simply mean a generalization or model that covers the entire explanatory domain.

[38] Never? Hardly ever. The probability is so low as to be unlikely to be realized within the duration of the universe. But Hull's Law states that there is nothing so strange that there is not at least one example of it in biology.

[39] One caveat: organisms may be (genetically, behaviorally, or ecologically) "identical" in the sense that substituting one for another makes no major difference to the outcomes in a given model. This is not unlike Templeton's "demographic or genetic indiscernibility."

[40] Nagel 1961, Armstrong 1978, Dupré 1993, Rosenberg 1994, Woodward 2001, Ao 2005.

[41] Murray 2001.

else, these laws are known only to apply to terrestrial organisms. In short, biological NKs are universals in a very small and domain-specific universe.[42]

Biological generalizations, usually ecological ones such as the Lotka-Volterra predator–prey relationship, define the universals that are active under that regime. "Predator" is defined as a grade of organism that consumes other organisms, and there is an implied relationship between population size and structure of the one and the other. Predators must be fewer in number than their prey, simply due to the energy-relationships implied by organisms subsisting on other organisms. When they over-predate, there is a population crash of prey followed by a population crash of predators shortly after. What, from this model, do we know about the instances of predators and instances of prey? We know they will evolve in an arms race if they are tightly linked trophically. We know that there is a general ratio of energy budgets where the energy budget of the predator is greater than the energy budget of the prey, and so on. All these things are implied by the model. But should we discover that predator A has some feature—say a claw structure—can we therefore conclude that another predator, B, will? Without further information, no. One predator might be an eagle, while another might be an ant-lion. In fact, without phylogenetic information, nothing else apart from the definition of the grade "predator" can be inferred. Model-based grades are not inductively projectable, phylogenies are.[43] Ecological laws apply in any case where there are "transactions" of energetic capture, limited space, and resources. They are laws of economics irrespective of whether the objects over which they range are biological or not. The model (that is, in this case, the Lotka-Volterra equation plus some rules for interpretation, and a possibly tacit delimitation of the explanatory domain) offers only a description of the relations between two populations. It might equally describe an inorganic system; say, an economic market, a computer model, or an engineering process.

However, for all that, a clade, or a monophyletic group, *may* instantiate a grade so long as a number of conditions are met. It must of course be monophyletic. Any taxonomic group that is only partial, or which is formed by the independent attainment of the grade, is considered non-natural in cladistic classification. The so-called "evolutionary systematists," such as Mayr and Simpson, believed that this was perfectly acceptable so long as the grade was salient, as illustrated above. However, cladists rejected this on the grounds that it returned classification to an arbitrary state. So long as the individual systematist considered the grade significant, *any* grade-based classification can be proposed and defended equally as well. But in a cladistic system, the grade, which is an intensionally defined class, and the clade, which is an extensionally defined class, must coincide for the grade to be (contingently or historically) natural. As soon as any other lineage attains the grade, or a lineage of the clade leaves it, the grade is no longer representative of a natural group. In short, NKs may be NGs and vice versa, but they *need* not. A grade may be a polyphyletic

---

[42] Richard Dawkins has claimed [1989, 191f] that natural selection must apply to any living thing irrespective of it having evolved from terrestrial antecedents. But I am unsure that natural selection is a law as such. It is a formal model, but it need not apply in every case or even most cases of living organisms even on earth, as stochastic processes like genetic drift indicate.

[43] Nelson 1989, Griffiths 1994. I discuss this further in Wilkins and Ebach 2013.

(independently or convergently evolved) or a paraphyletic (privatively defined as a clade minus some branches) group.

So what are the differences between NKs and NGs? I claim that these are the following.

1. NKs are timeless abstractions that are only instantiated or realized, while NGs are concrete objects that change when their constituents do.
2. NKs are realized by indiscernibles, while NGs are collections of discernible particulars.
3. NGs are historical objects, with a commencement and cessation point and a location.
4. NGs comprise objects that form causal chains, or lineages in biology, over time.

These are the features of biological classes. How do they apply outside biology? What distinguishes *all* NKs from *all* NGs? It is my opinion that the difference lies in one being defined by a model, and the other comprising causal relations between objects. And at the foundation, the root, of an "ultimate physics," they may very well coincide.

NGs are formed in biology by the transmission of genes, epigenetic systems, environments, learning, and behaviors. These things cause members of a species to resemble each other and to have much the same physical properties, or phenotypes. Out of a universe of physical properties they *might have* had, only those that are passed on to them and which they are competent to achieve directly as a result of what is passed onto them are those they do have. There is no definite microstructure for an NG, although genotypes are often mostly the same for most members of a species.

NKs, on the other hand, are always exemplified in the same manner by anything that is an exemplar of that NK. It is definitionally required, for example, that all predators are heterotrophs. There are many physical ways by which this can happen, but all NK-objects have only one way to be that NK (by consuming other organisms, for example). In short, NKs are fully populated possibility spaces, while NGs are only sparsely populated. Every possible state in the possibility space allowed for by the intension of the NK is a full and total member of the NK, an indiscernible. But of an NG, only those that actually *do* result from the shared causal processes are members of the class. Hence, members of an NK must have a shared microstructure. This is not a *causal* requirement, but one of definition—it is to be (*esse*) that kind to have that microstructure (*essence*).

When Locke made the point about the difference between relation of the linguistic and familial kinds, he was in effect noting that causal or generative relations between entities—the *natural relations*, as he calls them—form groups that are subsequently named.[44] The implication is that, as Locke notes elsewhere, these names need not

---

[44] This appears to contradict Locke's claim elsewhere in the *Essay* (II.xx.4, II.xxv.8) that relations exist only in the mind. His usual exactness of expression slips here. It is possible that he did think the relations between dam and foal were merely conceptual, but genealogies are formed on observation, not comparison of ideas naming the *relata*, as he claims relations are (II.xxv.2). See Stewart 1980 for a discussion (reprinted in Thiel 2002).

have a Nominal Essence—that is, they need not be definable—but they have, in Locke's metaphysics, a Real Essence, in the sense that they have something in common that *causes* them to be a group. The essences are, of course, the pedigrees of these things so related. A pedigree is a real causal relationship. People, and horses, have "a community of blood," or as we would say today, a genetic or genealogical relationship.[45] Darwin himself later wrote that all true classification was genealogical, in the *Origin*. This was taken up, but seemingly not further developed, by Kant. His "scholastic," or "academic" classification, based on similarity, is a class-based or gradistic classification. The "natural system" is based on the actual causal relations of the organisms, as with Locke.

Such groups are causally generated. And they are as natural as anything is capable of being. But they are not NKs, as they have no definitions, unless it happens that accidentally with respect to the propinquity of descent (Darwin's term) that they all share some character or trait, in which case it is better thought of as a diagnostic trait. If all members of the *A* family have red hair, they are not a family by virtue of their auburnity but by virtue of being related by (real) parent–child lineages that are more closely connected in a network graph representation than they are with the outgroups used to diagnose it.

Edward Zalta, in his interpretation of metaphysics,[46] introduces a useful distinction between *abstract objects*, which are spatiotemporally unrestricted, and *concrete objects*, which are indexed as to time and space, which is to say they are bounded in time and space (though they need not be contiguous or even connected). A species exists at a time and in a location, even if it happens to vary over ranges (and even if it re-evolves). A natural kind does not. Once an NK-object, always an NK-object. If a species is an NK, then it can come and go into and out of concrete existence many times. It can be caused by different physical processes. For example, an NK *T. rex* can evolve by ordinary evolution once, and once extinct, might be reformed by Jurassic Park scientists. But an NG *T. rex* can evolve only once. The latter "*T. rex*" is only a copy and deserves its own name, and it will have quite distinct causal relations to the earlier "example."[47] In the case of a respeciation event such as described for some species of fish, there is a less clear-cut "birth" in evolutionary terms for the species, but the "speciation event" is still singular, and the species is still a metaphysical particular, and thus a concrete object. Vagueness is expected in evolution, and presents no direct challenge to the reality of the outcomes of its processes.

How does this generalize beyond biology? Are there NGs in other disciplines and domains? In physics, a particle has the same properties no matter where it appears. An electron always has the same charge, components, and propensities as

---

45 Interestingly, Stamos 2003, 40–47 reads Locke's views on classification quite differently, seeing him as a gradist pure and simple. I think, however, that this is due to taking Locke's views on classification purely from his discussion in the *Essay* on logical species. It is true, though, that Locke does not treat causal relations as the basis for [logical] species. Stamos' own relational account of [biological] species could be traced to this passage in Locke. That Locke was a gradist is also argued, not in those terms, by Ayers 1981.

46 Zalta 1988.

47 Gradists such as Ruse and possibly even Darwin of course, challenge this, although I believe Darwin was merely unclear about the sole necessity for genealogical classification, and not confused, as Padian claims [1999]. However, see the discussion in Stamos 2003.

any other electron. Physics space is, so to speak, fully populated (or the states are highly restricted in a physical NK to only a few possible occupiable coordinates). Chemistry, being fully determined by physics, likewise has all kinds fully occupied, as the periodic table indicates (although we may not yet have found some of the higher atomic weight elements, and they may not even exist in the universe—but if they do or will, they will occupy the location in the periodic table fully). But there are several sciences that do not have indiscernible units, and biology is one of them, and these sciences also do not have fully occupied state spaces. Historical processes will not typically explore all possible states allowed by the properties of the substrates; hence it is feasible that a planet might never have metamorphic rocks, or a star might never go nova, although these states are permitted by the underlying physics, and are required if the initial and boundary conditions are satisfied. Geology, astronomy, and biology, universalized to cover all such instances in the knowable universe, are the historical sciences. In these cases, NGs are to be expected. In some cases such as stellar evolution or ecology, there may be a generalized classification based on NK physics, but to the extent that the objects are the special objects of these disciplines—organisms or astronomical bodies—they form NGs.

In such disciplines classes must still be differentiated, and it is a mistake to think that these classes *must* be differentiable in terms of definitions or essences. But the classes *will* turn out to be those formed by common causal chains, and they will be historical individuals as well. So NGs are not restricted to biology. Historical sciences study and classify differentiable particulars.

I propose that we should conceive of biological groups as NGs rather than as NKs unless they happen to coincide with each other. NKs are model-dependent, and they are timeless and abstract classes, while real biological groups are time-indexed and spatially located. This applies equally to other disciplines. It follows that while there may *be* NKs in a historical discipline, and hence we may say they have laws, these laws occur as generalizations within restricted or occasional models. The physical systems covered by these models or theoretical structures will not always exemplify them completely, and rarely in the same way as another instantiation. Thus we may find that the metamorphic rocks of one region contain more or less of a trace element than those of another, or that selection acts on the gross morphology of one species, but on the metabolic processes of another, though the theoretical structure or explanation is identical—the same selection "pressures," the same "fitnesses," and so forth (Sober's point about fitness being supervenient[48]).

NGs can be generalized beyond biology, to cover any collective particular which is caused by a unique set of causal processes at a time and place, and which has similar, but not necessarily identical in any respect, members. For example, any aggregate, such as a sand dune, in which the properties depend on the nature of the members (the shape and size of the grains, for example) is a natural group. Waves, hills, continents, and clouds are all NGs, although not every cloudlike formal entity is—phenetic groups are only NGs if they are formed by unique causal processes.

It is no accident that Wittgenstein's family resemblance predicates, so widely discussed, describe natural groups to the extent that the things predicated have the same

[48] Sober 1984.

general causal underpinnings. A family resembles each other from the *causes* of heredity acting according to Mendelian segregation and assortment through actual causal relations, which point, ironically, Wittgenstein failed to appreciate (see below, "Family Resemblance"). But just because we see a pattern does not make it a *natural* class, either of a kind or a group. Hence the phenetic program failed, as it did not restrict resemblance to causally linked objects (and thus it fell into the error of confusing the recognition of patterns with the recognition of actual objects).

THE "NEW" ESSENTIALISMS

A "revival" of essentialism has occurred since the 1990s, associated with *process structuralism*, which is basically a neo-Thomistic account of structure determining function,[49] *causal essentialism*, and a couple of philosophical attempts to revise essentialism as *developmental resources* shared by members of taxonomic groups.

Origin Essentialism

The first new essentialism is Paul Griffiths' "historical essentialism."[50] Griffiths accepts that variation within species is not a matter of what Sober[51] referred to as deviation from a natural state, where the natural state, or "intrinsic nature," plays the role of essence. Instead, he sees the developmental system inherited by members of a taxon as an *extrinsic*, or relational, essence: nothing that is not descended from a common ancestor is a member of that taxon-kind. Griffiths accepts the cladistic notion of monophyly as the only good license for projectability in induction, and assigns that license to the shared developmental resources—the developmental system and environmental and social resources of each organism—which are passed on historically in a relatively, but not rigidly, invariant manner. This is not a "natural state" essence so much as it is a phylogenetically entrenched essence. Some developmental states are more effective or deeply entrenched in the lineage than others, and act as a brake on variation, licensing inferences that have a law-like nature, albeit with exceptions. Historical, or *origin* (as the approach has come to be called) essentialisms have also been proposed by Okasha[52] and Rieppel[53] and criticized by Pedroso and Ereshefsky.[54]

---

[49] Goodwin et al. 1983 and Goodwin 1994 offer examples of the process structuralist approach. See the discussion in Griffiths 1999.
[50] Griffiths 1999.
[51] Sober 1980.
[52] Okasha 2002.
[53] Rieppel 2010, see also Rieppel 2008.
[54] Pedroso 2013, Ereshefsky 2014.

INTRINSIC BIOLOGICAL ESSENTIALISM

The second new essentialism is Michael Devitt's,[55] which he called *Intrinsic Biological Essentialism* (IBE[56]). Devitt defines essential properties thusly:

> A property *P* is an *essential property* of being an *F iff* anything is an *F* partly in virtue of having *P*. A property *P* is the essence of being an *F iff* anything is an *F* in virtue of having *P*. The essence of being *F* is the sum of its essential properties.[57]

This is one of a number of definitions of what makes an essential set of properties given in the literature. More broadly, Brian Ellis defines it as

> The real essence of any natural kind is a set of properties or structures in virtue of which a thing is a thing of this kind, and displays the manifest properties it does.[58]

The difference between intrinsic and origin essentialism is that the properties that define IBE are, as the name indicates, intrinsic to the species. However, they need not be intrinsic to each *member* of the species; that is, the essential properties need not be microstructural essences. The essentialist argument is nicely summarized by Wilson and co-authors:

> ... a kind's essence is *universally instantiated* by members of the kind, is *causally responsible* for that kind's typical traits, and is *explanatorily salient* in accounts of those traits. If biological kinds are intrinsically heterogeneous in the sense described above, *then there are no such universally instantiated traits*, and so the causal and explanatory roles played by putatively corresponding essences do not exist.[59]

Devitt offers the argument that "being a member of a certain taxon is more than informative, it is *explanatory*," and "when biologists group organisms together under some name on the basis of observed similarities, they do so partly *on the assumption that those similarities are to be explained by some intrinsic underlying nature of the group*."[60] In other words, biological practice is essentialist both methodologically and epistemologically.[61] Devitt distinguishes between the question of what it is for *F* to be a species, and what it is for an organism to be a member of *F*. He rejects the relational (origin) account of species that Griffiths presented, as it makes all members of several species with a shared common ancestor members of a single species.

Devitt's essentialism permits both gradual change and arbitrariness of delimitation, so long as there are cases where species are clearly distinct in their underlying

---

[55] Devitt 2008, 2011.

[56] Not to be confused with Inference to the Best Explanation, Chapter 14, this book.

[57] Devitt 2008, 345. See Dumsday 2012 for a discussion of these definitions.

[58] Ellis 2001, 54. Ellis' essentialism applies far more broadly than just biology; he intends it to apply to physical, dynamic, kinds in general.

[59] Wilson et al. 2007, 193.

[60] Devitt 2008, 352f.

[61] This is similar to the views of Fitzhugh 2009.

properties. However, he holds that Linnaean taxa are defined by their intrinsic properties:

> Intrinsic Biological Essentialism: Linnaean taxa have essences that are, at least partly, underlying, probably largely genetic, intrinsic properties.[62]

The partial nature of his definition is because the nature of the species as taxa is due to relational, as well as intrinsic, properties, while IBE is a claim about the explanations of the properties of *organisms* and their membership in those taxa, not the category of species. The "largely genetic" component includes developmental programs.[63]

Devitt and others have appealed to a causal essence. This is to say, the essential traits (which is the biological instance of "properties") must be causally efficacious in maintaining the species. The causal account that he and many others indicate underpins a realistic biological essentialism is our third option: Richard Boyd's *Homeostatic Property Cluster* account, to which we now turn.

HOMEOSTATIC PROPERTY CLUSTER KINDS

One of the early (1960s) arguments against scientific essentialism (outside of physics) was that natural kinds do not have uniquely shared properties among all their members. In short, organisms vary. As we have seen, variation within species arose as a concern in the early nineteenth century, and by the mid-twentieth century, species are known widely to be polytypic, with variants and multimodal distributions of traits common. One essentialism proposal by Richard Boyd took this variation to be real and normal, and sought to account for kinds in the face of their statistical nature.

Boyd argued that natural kinds do not, generally, in science require JNSSPs.[64] His "accommodation thesis" is that natural kinds had better account for scientific practice, and that the archetypical natural kinds—species—did not require the "traditional" essentialism, but were cluster concepts. However, Boyd said, the clusters were not merely social constructions, but *causal* clusters that kept the kinds in homeostasis with which the social constructive practices of science engage. This is called the *Homeostatic Property Cluster* (HPC) concept of natural kinds. Like Devitt's essentialism, who in part follows him here, this is a *causal essentialism*.[65] Boyd's exemplar is species, and this has been taken up by numerous others.[66] He has extended the HPC account to genealogically based higher taxa (i.e., monophyletic groups) in the process redefining monophyly as a conserved developmental modularity.

---

[62] Devitt 2008, 378.

[63] Devitt does not argue in favor of a genetic definition here, but merely appeals to such things as DNA barcoding and genome comparisons between species. As I have argued above, barcoding does not define or even with certainty identify species. See also DeSalle et al. 2005, Wheeler 2005, Will et al. 2005 for critiques. As to cross-genome comparisons, Devitt is conflating identification of a "typical" organism within a species with what makes the species bear typical traits, in this case.

[64] Boyd 1991. See also Boyd 1999b, Boyd 1999a.

[65] Although Boyd's HPC kinds view has gained wide currency among philosophers and biologists, it is not without its critics. Ereshefsky and Reydon 2015 argue that HPCs include some non-scientific kinds, and excludes some scientific kinds, for example.

[66] Brigandt 2009, Rieppel 2009, Barker and Wilson 2010, Martínez 2015, Neto 2016.

Boyd defines the three core ideas of HPC in this way:

> Three ideas characterize the early literature on HPC kinds. First, there are important natural kinds such that membership in them is a matter of participating in (types or tokens of) natural kinds of processes. Second, in many important cases the relevant processes will involve imperfect homeostatic clustering, so the kinds defined in terms of those processes will have indeterminate boundaries.
>
> Finally, because the HPC conception was developed so as to apply to species definitions, there was an emphasis on the dynamics of HP cluster(ing) (so that the relevant homeostatic clustering mechanisms were to be counted as elements in the HP cluster) and on *variability*: on the possibility that the components of a single HP cluster would vary over space and time. Such HP cluster(ing)s were seen as historically individuated in ways that reflected the contributions that reference to them makes to the reliability of inductive and explanatory practices (i.e., in order to make the corresponding kinds causally grounded...[67]

This sense of "essence" is, as I have argued before, relatively benign. Assuming that HP clusters can change over time by substitution (i.e., that the homeostasis is not rigid all the time), HPC kinds function both as ampliative, or projectable, foundations for inference and explanation, and overcome some of the issues that modern evolutionists have about essentialism. This is not the place to do an in-depth analysis,[68] but one thing we might say is that, like the monophyly-based phylospecies definitions, HPC kinds do not specify how much clustering or what kind of properties must cluster in order to identify a species (or for that matter, higher taxa) as that category of kind. In short, one already needs to know what is a species, in order to causally account for the homeostasis that makes it one. How that is done I shall argue below.

Another observation: If the developmentally entrenched properties, HPC or not, that causally generate species are the basis for being a species, and I have no reason to doubt that this is *one* of the property clusters that does so, then the developmental system itself is an autapomorphy of that species (or, if species hybridize in that group, a synapomorphy of the clade). In short, the HPC kind account doesn't solve the phylospecies problem.

However, as a metaphysics of biological kinds, the HPC account has much to recommend it. For a start, the account of kinds is *drawn* from biology (and social kinds) rather than from logic and language. It deals with, as it was intended to deal with, polytypy. If one can discover an HPC kind, then one can make projectable inferences about unobserved members of that kind. Indeed, that is one of the virtues of phylogenetic classification—because traits are phylogenetically conserved on the whole, monophyletic groups share properties of all sorts. Hence inferences from one species to another, or one species member to another, are warrantable. But if species themselves are not monophyletic, and organisms do vary, that license is reduced accordingly. This is why monophyly or coalescence are often proposed as a *definiendum* of species.

[67] Boyd 2010, 690.
[68] For which, see the treatments by Richards 2016, Slater 2016.

PHILOSOPHICALLY SPEAKING, HOW MANY SPECIES CONCEPTS ARE THERE?

Before I argue my positive case for species being phenomena rather than theoretical ranks or groups, we need to clarify how many conceptions of species there are, and what they mean.

It is my view that there are six "basic" species conceptions: *biospecies* (reproductively isolated sexual species), *ecospecies* (ecological niche occupiers), *evolutionary species* (evolving lineages), *genetic species* (common gene pool), *morphospecies* (species defined by their form, or phenotypes), and *taxonomic species* (whatever a taxonomist calls a species) (see Figure 13.1). There are also those conceptions defined privatively, by the absence of some feature or property, such as agamospecies (lacking recombination) or nothospecies (lacking reproductive isolation).

Notice that some of these basic concepts are not concepts of *what* species *are*; that is, what makes them species (the *ontology* of species), but instead are concepts based on how we *identify* species: by morphology, or the practices of taxonomists. Others are roughly equivalent (the *epistemology* of species). A gene pool is defined as a population of genomes that can be exchanged, and so a genetic species is basically a reproductive species. Evolutionary species are not what species *are* so much as what happens when some processes (such as ecological adaptation or reproductive isolation) *make* them species that persist over a long time. One common "concept" of species, the so-called *phylogenetic species concept*, is likewise a mix either of morphospecies, biospecies, or evospecies, or all of them. The "polyphasic" concept is also based upon a method for identifying species through many kinds of evidence. Agamospecies are species that *lack* some property: sex. An agamospecies is a not-biospecies species (although some, like Simpson and Mayr, simply denied they were species, which is a problem given that sex is a relatively rare property in the universal tree of life; it means most biological taxa do not come in species). It is a combination of taxonomic (or diagnostic) and genetic elements.

So, what *makes* an agamospecies a species? It can't be reproductive isolation, for obvious reasons, so it must be the only thing that we have left on the list: ecological niche adaptation. It could be chance, but if grouping happens by chance it is unlikely to be maintained by chance. In the absence of sex, therefore, we need ecological niche adaptation to keep the cluster from just randomly evaporating. Of course, few if any microbial species are purely asexual in the sense that they don't *ever* exchange genes; microbes have several mechanisms to do this even if they lack genders or mating types and fail to reproduce by any other means than division. Some genetic material can be exchanged through viral insertion, through DNA reuptake in the medium after a cell has lysed, or by deliberate insertion of small rings of DNA, plasmids, through pili. "Horizontal" or "lateral" genetic transfer is probably as old as life itself. While this might introduce some genetic variation into a population, it is selection for a local fitness peak that makes the asexual genome not stray too far from the "wild type."

As sex becomes more frequent, rising from near zero recombination per generation up to the maximum of 50% exchanged for obligatorily sexual organisms, another factor comes into play. Increasingly, the compatibility of genomes, reproductive

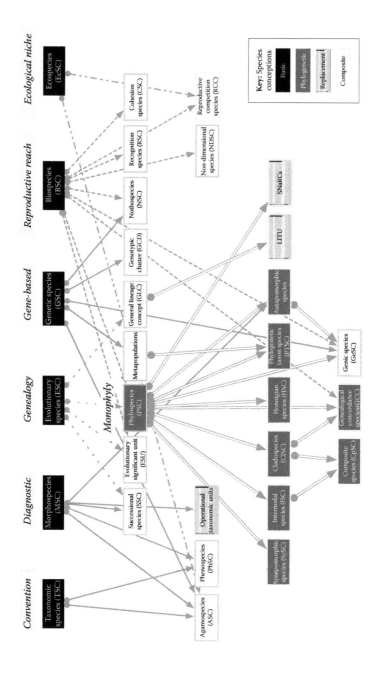

FIGURE 13.1 Basic and derived conceptions of species. LITU = Least Inclusive Taxonomic Unit. SNaRC = Smallest Names and Registered Clade. See Appendix B for details.

processes at the cellular, organ, and physiological level become important. In organisms with behavioral signaling (that is, with nervous systems and sensory organs), reproductive behaviors like calls and movements become important. Sex acts to ensure that the organisms that can interbreed tend to be those whose genome and anatomy are consistent enough. I call this *reproductive reach*: the more closely two organisms are related, the more likely they are within each other's reach as potential mates, and so the species is maintained by reproductive compatibility, and of course *some* ecological adaptation. This is very similar to a definition of the geneticist Alan Templeton, who said that species were "the most inclusive population of individuals having the potential for phenotypic cohesion through intrinsic cohesion mechanisms," "that defines a species as *the most inclusive group of organisms having the potential for genetic and/or demographic exchangeability*."[69] "Genetic" exchangeability here means the ability to act in the same manner in reproduction—any two members of the species are (more or less) interchangeable. "Demographic" exchangeability means that any two members of the species behave the same, ecologically, behaviorally, and so forth, and are interchangeable (more or less). With these two *causes* of being a species, we can now narrow down the number of concepts to two: ecospecies or biospecies.

There's a philosophical matter to clear up. These causal explanations are just that: *explanations*. They are not the *concept* of species. There was a concept of species before we had any clear idea of what they might be. We identified species in the fifteenth century that are still regarded as species, and there wasn't the slightest whiff of a theoretical biological explanation in the air at the time.[70] There is also a difference between the use of species definitions as aids to *discovery*, and *explanation*. A number of species conceptions have been formulated as operational aids to identifying a new species. Some of these are genetic or molecular[71] and some are morphological,[72] but many conceptions have attached to them species delimitation techniques or protocols.[73] Mostly these are protocols of the sub-discipline or field rather than natural classifications. Many of them are phenetic or phylogenetic conceptions, or both.[74] Theory-based concepts presume the universal applicability of that theory outside the groups on which it was formulated.[75] Discovery techniques that are based upon explanatory concepts are hostage to empirical fortune.

And it is an old concept, too, although the first simply *biological* definition of "species" waited until 1686 when John Ray defined it. Ray's definition was based on a simple observation: progeny resemble their parents. Species are those groups of organisms that resemble their parents. Versions of it go back to the Greeks. As I have argued, this presumes there is some power, a *generative capacity*, to make progeny resemble parents, and it seems to rely upon seeds. I call this the *generative conception of species*, and it was not only the default view before Darwin, but Darwin

---

[69] Templeton 1989.
[70] Aristotelian accounts of generation notwithstanding.
[71] Sites and Marshall 2003, Caron et al. 2004, Hanage et al. 2005, Staley 2006, Birky et al. 2010.
[72] For example, John and Maggs 1997, Purvis 1997.
[73] DeSalle, Egan and Siddall 2005, Knowles and Carstens 2007, Raxworthy et al. 2007, de Queiroz 2007.
[74] Templeton 1989, Mallet 1995, 2000, Beltrán et al. 2002.
[75] For example, Wu 2001a, 2001b.

himself held it, as do nearly all modern biologists. It is what it is that the explanations explain. So technically there is only *one* species "concept," of which all the others, the 2 or 6 or 27+, are "concep*tions*." The idea that there is one generic category into which there are put many "concepts" is a mistake made by Ernst Mayr.[76] In ordinary philosophical usage, it is the *concept* that is the category, and the definitions define, in various ways, that concept. Another mistake often made by biologists is to think that if there is a concept/category, there has to be a specified *rank* or "level" at which all species arise.[77] This seems to rely on the idea that because Linnaeus took Ray's concept of species and made it the lowest rank in his classification scheme, there must be something that all and only species have as properties, and this assumption has caused no end of confusion. That species exist does not imply that all species share some essential property (any more than because we can usually identify what an organism is implies there is something that all and only organisms share). This philosophical error is essentialism itself about categories, and it is a supreme irony that Mayr, the opponent of essentialism about individual species, was held in thrall to essentialism about taxonomic concepts.

Some people think that there *are* no species. Moreover, they wrongly think this view is a consequence of evolution and that Darwin himself denied there were any.[78] Now what Darwin thought 150 years ago is of no real consequence to modern biology, but he didn't think species were unreal constructs; he thought there was no single set of properties species had to have. He was not a taxonomic essentialist. But neither is it the case that species are unreal because they shade into each other. In modern philosophy there is an ongoing debate over whether one can have vague and fuzzy sets or kinds,[79] but for science we need only a little logic and metaphysics: if we can identify mountains, rivers, and organisms, we can identify species, and they will tend to have a "family resemblance." What is a species among primates will tend to be like species in all other close relatives. What is a species among lizards will (usually) be like what a species is in close relatives (for instance, some lizards are parthenogens, and have no males, where their nearest relatives are sexual, but in that case, they are like their sexual cousins ecologically and morphologically). But some think that species do not exist except in the minds of biologists and their public.[80] So for them, zero. Our final score is: 22–28, 6, 2, 1 or 0.

My solution is this: There is *one* species concept. There are *two* explanations of why real species are species: *ecological adaptation* and *reproductive reach*. There are *seven* distinct definitions of "species," and *numerous* current variations and mixtures (listed in Appendix B, A Summary List of Species Definitions). And there are $n + 1$ definitions of "species" in a room of n biologists.

---

[76] Mayr 1963.
[77] Baum 2009.
[78] Mallet 2010.
[79] Graff and Williamson 2002.
[80] As we have seen, this is not new: both Buffon and Lamarck also thought species had no independent reality from taxonomists' conceptions.

PHILOSOPHICAL TERMINOLOGY

Epistemology: The study of the conditions under which knowledge is gained and justified. *Adj.* epistemic.

Ontology: The study of the metaphysical kinds of objects that exist, such as number, sets, classes, or relations. *Adj.* ontic.

Metaphysics: The study of the fundamental nature of reality and being. This is typically focused upon issues that cannot be resolved through science, and includes concepts like universals, individuals, substance, space, time, mind, reality, and so on. ONTOLOGY is a part of metaphysics.

Essence: Either what makes something what it is (the constitutive version), or something that is necessary for the thing to be what it is (the modal version). A fixed nature. One or all of the necessary properties of a class of things. In older (scholastic) writings, the essence is the form of objects, independently of their substance.

Essentialism: The view that some objects have essential properties. Usually, the view that classes of things are defined by their essential properties, or essences. Sometimes expressed modally as the [jointly] *necessary and* [severally] *sufficient* properties (JNSSPs) of a class; which is to say these properties must be properties of all members of the class (necessity) and all of them jointly are enough to specify the class.

Class: A set of things that fall under the defining concept of the class. The extension of a set's INTENSION. *Alternative definition*: that which is classified, a taxon.

Set: A collection of things or ideas, however formed. Sets with definitions are *intensional sets*.

Intension/Extension: The *intension* (not to be confused with *intention*) is the meaning or connotation of an expression. The *extension* is the state of affairs, or objects, that fall under the intension of an expression, or denotation. Roughly, intension marks out the defining properties, while extension marks out the things that are defined.

Monothetic/Polythetic: Monothetic groups are classified by sharing a single set of properties, the absence of any one of which excludes an individual from the group. In systematics, a monothetic species is a monotypic species. In philosophy, a monothetic group is generally called a CLASS. A polythetic group is one formed from a *majority* or threshold value of shared properties, and maps in systematics to a polytypic group or taxon. In philosophy, Wittgensteinian classification by similarities are often called polythetic classes.

NAMES AND NOMENCLATURE

Nomenclature is a major facet of the species problem, but it is a problem of operational convenience. Although Linnaeus thought that once identified, binomials would be stable, it turns out that neither genus nor species names are in fact stable. In one group I examined—pinnipeds—as many as 75% of species names have been lost or amalgamated into other species, and varietal or subspecific names are equally unstable. In the Phocidae (earless or "true" seals) there are eight cases of genera being reassigned or created from existing genera out of the ten current genera (most of which are formed from *Phoca*, a genus assigned by Linnaeus himself). Therefore, there is a problem of metaphysics in terms of how the species category changes under evolutionary assumptions, and one of conventional and operational considerations forced by the nomenclatural instability due to splitting and lumping.

We should distinguish diagnosis from causation; the epistemological from the etiological aspects of species and other taxa. In short, the appropriate way to approach the problem is to look at species as the results of history rather than of investigation; the "history, not characters" approach. If species are the results of a physical—that is to say, biological and causal—historical sequence, and it turns out that no empirically-based discipline can fully and reliably recognize them or talk about them, so much the worse for the science. That would be a fact about biologists and their capacities rather than biological organisms. But there is no need to be an epistemic nihilist. So far, the science does very well, and the prospect of imminent catastrophic failure is remote. The point is, however, that the species problem cannot be resolved by convention, fiat, or practical decisions, for it is a problem about the organic world more than it is about the scientific community, even though there is an element of social construction in any discussion of a scientific concept (because, in order *to* discuss it, we must enter into the scientific community to some degree). How we resolve this depends a lot on what we individually conceive the relationship to be between scientists and their study objects. But we need not be scientific realists in order to think that a scientific term should have its meaning fixed by features of the objects to which it refers. All that is required is a strong empiricism, and experimentalism—the fitting of models to data, and not data to models. More abstract questions concerning the status of such concepts as "cause," "truth," "reference," and "meaning" can be bracketed off from any discussion of the causal role and referential status of a term like "species" and deferred to a more competent forum, since whatever is the case with "species," "gene," or any other disputed or accepted general scientific term is true of all scientific terms or concepts. If "species" is theory-dependent, then so is "organism." If either are empirically dependent, all scientific terms may be. If one can refer or fail to refer, all of them can.

However, terms do not exist in Plato's heaven—at least, not those derived from science. They exist in the practices of scientists in the scientific community, in scientific communication. Moreover, they do not remain constant; concepts evolve. They can have quite different meanings at one time or at another. Agassiz "refuted" Darwinism by effectively arguing out of a dictionary: species are by definition static entities, and hence they could not evolve; Darwin could not therefore account for species. To which Darwin reacted with a redefinition: species are temporary things. *Argumentum ad lexicon* can be countered by a different translation manual.[81] Are Darwin's temporary species "the same" as Agassiz's permanent ones, or is Darwin just changing the subject? This could be an example of taxonomic incommensurability,[82]

---

[81] I owe the translation manual point to Paul Griffiths.
[82] Sankey 1998.

but I prefer to think of theoretical terms evolving like species, at different rates to be sure, sometimes abruptly, sometimes smoothly. Still, since the terms are occasioned by the things they refer to, empirically or realistically, they are "the same" in some relevant sense.[83] As a child, I believed that journalists wrote journals. Recovering from the shock of finding out, as a teenager, what they really wrote, I nevertheless continued to use a term that still referred to the same entities as when I was a child. The properties I ascribed to them had changed because I had learned more. Learning more meant I could ascribe more accurately the properties journalists actually have, and would no longer think of the local newspaper hack as if he or she were an essayist on a par with Voltaire or Stephen Jay Gould, but the term denoted "the same" people as it had before. It isn't important that these properties were collected under the term by my language community. A whole community could still be mistaken about the properties of that profession (as the misguided trust in journalistic integrity and objectivity in some sub-communities evidences). Likewise, scientific terms can mistakenly carry connotations, such as static existence or timeless denotation, even if it subsequently is learned that the things referred to carry none of the critical properties connoted.

The study of the history and theoretical senses of a key term is dependent upon the reconstruction of the way a discipline has learned about the referents of the term. It may be that the term never denoted (e.g., "phlogiston"), and it may be that the consensus of the present meaning of the term is ambiguous, incomplete, or inconsistent within its uses or with the data. In these cases, the role of the philosopher of science is to attempt, with the methods of philosophical analysis, to clarify or prune the term, or recommend its abandonment. This latter recommendation is, for example, made by Ereshefsky[84] and others for "species" on the grounds that it has no unambiguous referents and that other terms cover what is needed, while the implicit absolute ranking of the term is positively misleading if not outright false.

But what is this method of philosophical analysis that critics bring to bear? Is it something that only philosophers can exercise? Is it something quite distinct from the critical methods available to scientists and other theoretical specialists? I think not. There is nothing magical or *sui generis* to philosophical analysis beyond it being the evolved practice of a tradition and community that has addressed abstract topics over many years. Consequently, to properly address the use of a scientific concept or term (I treat the two as roughly synonymous), knowledge of the science is a prerequisite, and the validity of the philosophical take on species stands or falls on that knowledge. This being said, there are philosophical concepts, arguments, and considerations that are *not* available to the critical scientist, *qua* scientist (although they may be available to a philosophically educated scientist, *qua* philosopher). Of particular relevance are the resources of metaphysics and formal logic (classes, sets, individuals, universals, and particulars), and of linguistic philosophy and epistemology (natural kinds, sortals, theory-dependent objects). Hence, Michael Ghiselin in his role as analytic philosopher (although he is a biologist) and David Hull

---

[83] I am therefore rejecting the Kuhn-Feyerabend notion of incommensurability, at least for some categorial terms, including *species*, in favor of a "baptismal" notion of reference. Cf. Hull 1988, 500–502.

[84] Ereshefsky 1991, 1992, 1999, 2000.

(a philosopher, although he studied biology and history of biology) have brought those resources to bear on species from a metaphysical and analytic perspective. Elliot Sober has likewise invoked epistemology to deal with issues of phylogenetic systematics and evolution.[85]

This philosophical influence on biology is nothing new. Apart from the ubiquitous Popper, mathematical philosophers have strongly influenced the development of systematics. Russell and Whitehead's *Principia Mathematica* inspired J. H. Woodger to recast biology in terms of symbolic logic.[86] This in turn greatly influenced Hennig's phylogenetic systematics, and the resulting terms of grouping and order of branching that are now common property to biologists of all philosophies.[87]

When Mill discussed logical classification, he revived the scholastic conception of the relation between *differentiae* and *relata*. Unlike the medieval philosophers, however, Mill was concerned with *causal*, not formal or logical, *differentiae* and *relata*. It is this tension, between wanting to carve nature at its joints and yet returning to carving language at *its* joints, that continually recurs throughout western philosophy. To pre-empt this recurrence, consider how species differentiate *themselves*, rather than how we might do that, in the hope that systematics can retrieve the natural self-classification of the organisms themselves, but not in the expectation that it always will.

FAMILY RESEMBLANCE

WITTGENSTEIN AND RESEMBLANCE

Are species family resemblance terms?[88] In the famous and often-discussed sections 66 and 67 of the *Philosophical Investigations*, Wittgenstein discusses what is common to all games, and argues that if one does not assume that the term *game* has something common by definition but instead "looks and sees,"

> ... you will not see something that is common to all, but similarities, relationships, and a whole series of them at that. ... we see a complicated network of similarities overlapping and criss-crossing: sometimes overall similarities, sometimes similarities of detail. ... I can think of no better expression to characterize these similarities than "family resemblances"; for the various resemblances between members of a family: build, features, colour of eyes, gait, temperament, etc. etc. overlap, and criss-cross in the same way.—And I shall say, 'games' form a family."

This is an account of names, and it is curious that Wittgenstein applied an analogy between naming and resemblance based on inheritance. For species terms are names, and their extension is just such a crisscrossing of inheritance lines and related individuals. As I understand the family resemblance predicate (FRP) notion, it suggests that the extension of such a predicate is a cluster of intersecting sets, but not

---

[85] Sober 1988, Sober and Steel 2002, Sober 2008.
[86] It should be noted that Woodger was a zoologist by training.
[87] See Varma 2013 for a review.
[88] Several authors have proposed that they are: e.g., Pigliucci 2003, Ereshefsky 2010, Kull 2016.

proper subsets, of cases. Their "identity" lies in the general similarity of the individual terms to each other in some metric. FRPs can be cases of (i) general similarity of sufficient but not necessary conditions; (ii) identity by descent but not unity of defining character; or (iii) practical identity; that is, it is useful to group terms together. What they are not are classes. All members of a class instantiate a necessary and sufficient set of characters. As Pitcher observes, there are predicates, like "brother" and "vixen" that are defined by essential conditions like being male, or being female[89] in addition to having external relations. How to characterize the distinction between FRPs and these other relational predicates is the subject of a great deal of literature, but for our purposes here it is enough to note that an FRP is one that relies on similarity but a general lack of essence. *Species* can be a family-resemblance term in three ways:

A. As a Taxon Concept

Notions of species are often distinguished into various categories, and there is nothing much common between them that is not true of any class notion. However, with fine irony, the categorizations themselves appear to be FRPs, as they also criss-cross but are not reducible to each other. These may not be inconsistent, but they are clearly different ways of defining the biological group. Other classifications that are called species include the morphologically similar group, the karyotypic group (i.e., any group that has the same genetic structure and a restricted set of alleles for various loci), and the undivided lineage.

B. As a Classification of Organisms

If a taxonomist has three species to classify in relation to each other, the defining characteristics of each species (i.e., those that are different across all three) are used in cladistic and morphological taxonomy to establish the similarity relations.[90] The result is mostly just the sort of relationship one finds in Wittgenstein's games. Species A and B may share some characters that C lacks, and A and C, and B and C, may share characters exclusively as well. A complete list of characters may give numerous set inclusions and intersections that are not consistent across the whole list of characters. (This is due to several evolutionary cases in the theoretical situations where an explanation is given in terms of phylogeny: convergence, reticulation, canalization, and other cases covered by the cladistic term *homoplasy*.)

C. As a Measure of Conspecificity

The literal case of family resemblance can be extended to cover all members of inclusion within a species. Individual organisms cover a distribution curve for any given trait in a sizable census, but location in the curves are not the uniform for all individuals—that is, an individual at the mean for one or most traits may be on either tail for any other trait. Use of an FRP to describe an individual in order to justify

---

[89] Pitcher 1964, 221n.

[90] In the common meaning of similarity; a cladist would say the relations are formed solely on the basis of monophyly, but see Vanderlaan et al. 2013 on the complexities of the notion of monophyly.

inclusion in a species classification is in effect the only alternative to a typological or essentialist approach, with the treatment of variance as degradation that it involves.

The use of a term of morphological similarity due to ancestry—"family resemblance"—by Wittgenstein indicates that the sort of classification principles appropriate to the natural biological world (and by extension to the sociolinguistic world) are very different from those appropriate in the mathematical sciences that served as the source for so much of modern western thought. There are some deep issues involved: lineages of information or structure appear to generate groupings that are complex, fractal, and stochastic, and the "linear" treatment of natural kinds that may (or then again may not) be useful and effective in the physical sciences, where history and individuality are not relevant to the state description of a system or process, are less useful and less effective in biology. The older views that consider species to be natural kinds of the physical/mathematical variety are severely Procrustean, and eventually artificial and therefore subjective. I think that Wittgenstein's FRP is an abiological attempt to generalize a class of "nonlinear" predicates, ones that are appropriate in biological work.

Do Family Resemblance Predicates Work for Biological Species?

The taxonomic movement known as "numerical taxonomy" at the time, of its formulation in the 1960s and later as "phenetics," attempted to use something like Wittgenstein's FRP. Indeed, the originators of phenetics, Sokal and Sneath, drew for inspiration on the work of Morton Beckner, who had explicitly used the Wittgesteinian FRP to discuss species.[91] However, phenetics ran up against a problem—their clustering through the use of computer analysis of characters in a morphological metric space led to instability and sensitivity of the results to the choice of characters chosen. Effectively, phenetics was an attempt to find the natural groupings of the populations themselves, in the hope that species would appear as clusters in a Cartesian metric that were isolated from other populations, and that this would be possible no matter which characters were used.

The problem appears to rely on two facts about actual species: first, there is no absolute measure of difference in, say, morphological metrics or genetic distance that is either common to all separate species or absolutely enough to prevent interbreeding in sexual organisms. Second, species do not all vary commensurately in all characters. Some characters are evolutionarily conserved for various reasons: they may be strongly buffered against change due to developmental linkages (pleiotropy), or they may be held constant by stabilizing selection because they are functionally important. So, the choice of characters to use is significant. Typically, taxonomists use "useless" ("nonfunctional") characters, which can change more or less at a constant rate over evolutionary time, because homoplasies (convergently evolving characters) lose phylogenetic information the way that a pathway that joins another pathway is no longer as informative about the prior journey of a walker as a single pathway would be—the walker may have traveled along two pathways now, and knowing the current location of the traveler is no help in determining which way they went.

---

[91] However, Chaney 1978 argues that the FRP and polythetic groups are different beasts, and the use of polythesis by numerical taxonomists is independent of the FRP.

So, the FRP, while true, is no real help in specifying which groups of organisms are distinct unless you already know which characters to use. When I asked one specialist who uses cladistic methods how he selected the right characters to use, his reply was that he should hope that he knew his organisms well enough by now to choose correctly. Again, we see the reciprocal illumination of which Hennig spoke—knowing something about one's group of organisms enables one to choose the characters that will uncover the relationships that allow one to test which characters one should use. The circularity is not vicious—each step acts to further test the next and to iteratively refine the overall picture.

Not surprisingly, many of the algorithms used to analyze the character sets by pheneticists have been incorporated into cladistic analysis, only now they are used to cluster basal taxa (species, as represented by a specimen, or by a "wild-type" set of characters which are known or suspected to be invariant across the species, and which are unique to it—autapomorphies, in other words). Cladistics is often accused of being typological for this reason. There is a sense in which this is harmlessly true, and a sense in which it is malign, but false. The type that is usually implicit in cladistic analyses is a modal type;[92] it is the mode that most, if not all, of the members of the species bear at the same stage in their lifecycle. This is gained empirically by field work, large-scale sampling, and biogeographical methods. Such typology is universal in alpha taxonomy and studies of organisms. If we had perfect knowledge of a species, then we could draw the distribution curves for each trait and specify the modal value for use in the data matrices used in taxonomic analysis. But the "essentialistic" sense that derives from logic, that a species is to be *defined* by the presence or absence of some character/s, is no more a part of cladism than it is of any other approach which is not "Aristotelian." Indeed, Aristotle's empirical work, where he recognizes "the more and the less" in species essences, is not, in the popular understanding of the term, Aristotelian, either.

THE QUA PROBLEM

Species are thought by some[93] to be named in a causal account similar to the Kripkean baptismal notion of general names. We first encounter a Tiger, and give it a name and a kind to match. When we meet other tigers, we learn what can vary and what is common to all. In this way, the first tiger acts as a kind of baptismal "type specimen" for subsequent uses of the name. Now this goes to a matter in the philosophy of language over meaning and (semantic) essentialism that to most if not all biologists is rather recondite, but, like essentialism itself, it has direct implications in biology. One of these is the issue of essentialism as a way of defining kinds, of course, but another derives from an extension of a problem raised first by Willard Van Orman Quine in his *Word and Object*, known as the Gavagai Problem.[94]

---

[92] In the statistical sense of a mode, not the philosophical sense of a modal operator. The discovery of modal distributions of homological characters that are not homoplasious in other ways is independent of phylogenetic analysis, and we might be able to do a cluster analysis of populations to distinguish species, but there can be no prior values that will settle the issue, because species can be polytypic, and the variation within them can be greater than the variation between them.

[93] For example, Hull 1976, Hull 1984; see Kitts and Kitts 1979.

[94] Quine 1960, 29.

Quine is wondering how words in one language or idiolect can be translated into another and imagines a field anthropologist trying to work out what the locals mean by their words purely empirically, in terms of the stimuli that elicit the verbal responses. A rabbit scurries by and the local says, "Gavagai!" What does "gavagai" mean? According to Quine it might mean "rabbit," "rabbit part," "extended rabbit," and so forth. Since the stimulus in each instance of the word so far encountered is to the anthropologist's eyes identical, how could we ever know?

The idea that species names are anchored in the biological world in terms of the type specimen, or the initial experience of naturalists and taxonomists, is commonly held, at least implicitly, and goes by the name the "baptismal" account. Why else do specialists apply the name to the holotype, and go to pains to assign a replacement specimen (lectotype) if the holotype is lost?[95] The problem with this is that the causal account underdetermines the extension of the term. Devitt and Sterelny call this the *Qua* Problem.[96] How, they ask, do we know what it is that we are referring to when we name a general kind term like Tiger? Is it the striped fur (if so, what about albinos), or the teeth (then what about toothless old and circus tigers?), and so forth. What is it that makes us refer to the species *Leo tigris qua* tiger?

The *Qua* Problem is known to taxonomists. There are numerous cases in which the extension of a named taxon, particularly species but also genera, turns out to be other than the biologist expected, sometimes requiring that the species name be subsumed to a prior designation, sometimes requiring that the taxon be divided into separate taxa. There are cases where taxon names have been given independently to male and female forms in birds, and where the females of a genus are so similar that they were put into one taxon while the gaudy males, each individually distinguishable by humans, have been placed into different taxa, and the mistake not realized until the two morphs were observed actually mating.[97]

As a response to essentialism, the baptismal account has something to recommend it; general kind terms do not have to have *definienda*, since the commonality is provided by the causal relations between the initial specimen and the other members of the kind. However, it does not manage to release us from the problems of *knowing* that other specimens are in the same kind.

In order to resolve the *Qua* Problem, we would have to range too far from the present subject, and so I shall merely note here that this problem occurs for any and all general terms, whether they are terms that cover only an extensional set, a causal lineage, or an abstract class. The more interesting problem in biology is how we know that a holotype "refers" to all other members of the species.[98] As always, there is the issue of what is, and the issue of what we can know about it. I take it that the idea of species as interlocking lineages (the Hennigian notion of tokogenetic, or

---

[95] Levine 2001.

[96] Devitt and Sterelny 1987.

[97] For example, the male superb warbler (*Malurus cyaneus*, now called the superb fairywren) was described in *The voyage of Governor Phillip to Botany Bay* [Phillip et al. 1789], and the female then given its own entry after the initial description of the male had been reported, based on later studies:

> Except from the size and shape, this bird would not be suspected at first sight to belong to the same species as the male: the epithet of *superb* applies very ill to the female.

[98] Levine 2001, LaPorte 2003, but see Colless 2006.

reticulating, lineages) answers to the reality of sexual species, and that species in asexuals are largely ecologically determined as quasispecies. So, the ontology of these species is due to a set of causal processes (as yet unspecified) which maintains the species' identity. What we can know, unsurprisingly, depends on what methods we have of determining the differences and similarities between organisms (such as genetic sequence distance, or Nei distance, or perhaps the degree of hybridization of DNA[99]); in short, what assays we have. Assuming that we happen to have or can develop an assay that touches on the actual causal mechanisms (whether we know that or not), then we can identify the *rest* of the species through a nearest-neighbor analysis until we reach some threshold, which may be abrupt or not.

ASEXUAL MICROBIAL SPECIES

Species and speciation are, respectively, the fundamental units of microbial diversity. Firm understanding of the scientific bases for species concepts and proposed mechanisms of speciation will achieve more than simple provision of an internally consistent language for taxonomy and systematics.[100]

In order to elaborate on these issues, let us consider what constitutes a microbial species, either a bacterial or other asexual single-celled species. Let us begin by noting that this is a long-standing problem in microbiology. In 1962, G. D. Floodgate noted the following points about the state of thinking about bacteria at the time. Arguably we are in much the same straits now as then.

Of all the words in the taxonomist's vocabulary, probably the most difficult to define is "species." The perplexities which bacteriologists encounter in trying to give the word an exact meaning are well known … . A species has been described as more real or as having "a greater degree of objectivity" than any other taxon … , but it has also been called a man-made fiction … . Again, according to one author … a species is a dynamic system, but according to another it is as outdated as phlogiston … . Further, a meaning given to "species" in one biological discipline may not be used at all in another. For example, the bacteriologist does not usually refer to interbreeding or exchange of genetical material when describing the meaning of species as other biologists sometimes do, though recently an attempt has been made to introduce species in this sense in bacteriology as well … . Yet again, some bacteriologists have considered a species to be a discrete segment of a phyletic line evolving independently of other segments … .[101]

To address this issue, I want to pose the Problem of Cohesion,[102] or, Why Are There Microbial Species and Not Just a Mess of Strains?

I shall first consider two leading microbial species concepts. As this is a very large field of literature, we shall restrict ourselves to two very general conceptions

---

[99] Berlocher 2000.

[100] Ogunseitan 2005, 19.

[101] Floodgate 1962.

[102] Initially, in the original paper, I called this the Problem of Homogeneity. Peter Godfrey-Smith suggested in review that it be renamed the Problem of Cohesion, to make the analogy with Templeton's Cohesion Species Concept clearer, and to link this to the HPC kinds account.

rather than discuss all possible permutations and combinations. Later I shall appeal to several current conceptions directly. Then we will consider three possible reasons why microbes are not homogeneous and undifferentiated. The first is due to stochastic processes of extinction and diversification, in what are called Branching Random Walks.[103] The second is due to the lateral genetic exchange of parts of bacterial genomes and plasmids, and is called the Core Genome Hypothesis.[104] The third is due to selection for adaptation to local fitness peaks maintaining a relatively stable cluster of genomes and phenotypes. I shall then appeal to the notion developed initially for viral species by Manfred Eigen—the "quasispecies," a collection of close neighboring genomes. In the first instance, "microbial" (that is, asexual single-celled) species form quasispecies.

Finally, I shall attempt to locate the different processes and species concepts in order to draw a general point about asexual organisms, sexual organisms, and the evolutionary groups they form in different ways. I then will argue that the basic notion of "species" is, in fact, the quasispecies, and that the definitions developed to serve metazoans and metaphytes are secondary rather than primary conceptions of what species are.

WHAT ARE WE TALKING ABOUT?

The literature on microbial species refers to a number of groupings of organisms: prokaryotes, protists, monerans, bacteria, and so on. There appears to be no single term to cover them all, so I will here adopt "microbe" as a convenient label, although it must not be thought that this implies microbes are a natural group. None of these groups appear to be "natural" (which in a cladistic sense means monophyletic). Almost all of them are defined as some complement of Life after other groups, eukaryotes, animals or metazoans, plants or metaphytes, and fungi, have been taken out. In other words, "microbe" is a paraphyletic or polyphyletic group.

Under the "three-domain" hypothesis of Woese, the three multicellular "kingdoms" (animals, fungi, and plants) are small branches on a much larger tree, part of the Eukaryota but not even the bulk of them.[105] Sex as we know it is a very restricted and minor part of the diversity of life. "Prokaryotes" refers broadly to a group of organisms that are basically not-eukaryotes.[106] Nowadays we refer instead to several groups: Bacteria, Archaea, and Eukaryota. This privative classification is under challenge, and many think that the three-domain deep phylogeny is an artifact of the genes used, and that prokaryotes are indeed a monophyletic group.[107] Whether or not they are, one

---

[103] Pie and Weitz 2005.
[104] Dykhuizen and Green 1991, Dykhuizen 1998, Wertz et al. 2003, Coleman et al. 2006.
[105] Woese 1998.
[106] Pace 2006; but for a dissenting view see Martin and Koonin 2006.
[107] See, for example, essays in Sapp 2005; particularly Doolittle 2005. Cavalier-Smith 2010 argues that the Eukaryote/Prokaryote division is fundamental (the six-kingdom, two-empire classification). Whitman 2009 accepts the three domains, but groups Archae and Bacteria as Prokaryota. Foster et al. 2009 make Eukaryota a sister group to Chrenarchaeota, and Euryarchaeota and Eubacteria sister groups, thus making Archaebacteria paraphyletic. More recent work indicates what McInerney et al. 2015 call a "ring of life" scheme, cf. Lake 2015. None of this affects the point made here: that metazoan eukaryotes with sexual reproduction are a tiny fraction of the tree of life. Thanks to Laurence Moran for this information.

thing the not-eukaryotes have in common is a lack of a nuclear membrane, allowing the transfer of genetic material between genomes. While the term "microbe" refers to single-celled organisms, whether eukaryotic or not, most of the discussion that follows can be regarded as covering the prokaryotic domain, and particularly the bacteria.

THE PROBLEM OF COHESION

In ... an asexual group, systematic classification would not be impossible, for groups of related forms would exist which had arisen by divergence from a common ancestor. Species, properly speaking, we could hardly expect to find, for each individual genotype would have an equal right to be regarded as specifically distinct, and no natural groups would exist bound together like species by a constant interchange of their germ-plasm.

The groups most nearly corresponding to species would be those adapted to fill so similar a place in nature that any one individual could replace another.

R. A. Fisher[108]

Many bacterial and archaeal species either do not exchange genes to reproduce or they can but do not need to. The question is sometimes raised whether these organisms form species at all, or if there is some replacement term or concept for groups of them.

It took a long time for some to even accept that there were asexual organisms. Darwin discussed hermaphroditic species, but they still had mating types or genders; it was just that a single individual had both kinds. It was long recognized that some plants could propagate vegetatively, but the notion that there were obligately asexual organisms was doubted, for instance, by Fisher as late as 1958, saying that "organisms in which sexual reproduction was entirely unknown" could not with certainty be ascribed to any known group.[109] In general, early members of the Modern Synthesis of genetics and Darwinian evolution had real problems with asexual taxa. Theodosius Dobzhansky, the famous joint architect of the synthesis and first author of the "biological" species concept, simply denied that asexuals formed species. Call them something else, he said.[110]

As pointed out by Babcock and Stebbins... , "The species, in the case of a sexual group, is an actuality as well as a human concept; in an agamic complex it ceases to be an actuality."[111]

And while many modern writers, such as Templeton[112] or Coyne and Orr[113] now, rather grudgingly accept the reality of asexual species, others, such as Meier and Willmann, continue to deny species rank to asexual taxa, instead calling them "agamotaxa."[114]

---

[108] Fisher 1958, 135.
[109] *Loc. cit.*
[110] Dobzhansky 1941, 320f.
[111] Dobzhansky 1951, 275.
[112] Templeton 1998.
[113] Coyne and Orr 2004.
[114] Meier and Willmann 2000. Other terms include *binoms* [Grant 1957], or *agamospecies* [Cain 1954].

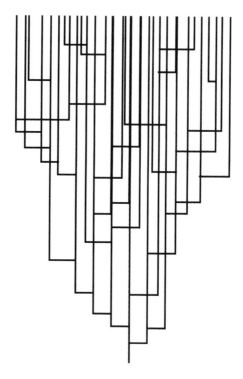

FIGURE 13.2   The "carpet model" of asexual strains. This is a two-dimensional representation of a high-dimensional genome space. See Gavrilets[116] and Skipper[117] for a discussion of the concept and representation of a genome spaces and fitness spaces. Although the distribution shown here is even, random strains are likely to form a Poisson distribution from the initial genome coordinate.

Even so, bacteriologists continued during the Synthesis and after to name and describe species, even though they could not really make use of the Biological Species Concept of Dobzhansky and Mayr. They relied, among other characters, on their organism's staining properties, the colony shape, the microscopic morphology of the cells, and of course the ecological conditions under which they lived. What was lacking, though, was a clear definition of what a species could be for these organisms.[115]

Part of the problem is that if a species were obligately clonal, then each mutation should make a new clonal lineage and we might naively expect to find not clusters but a carpet of strains more or less evenly distributed (Figure 13.2). How can we account for this? I will call this the *Problem of Cohesion*: why are asexual lineages ever found as groups at all? Why are they homogeneous over time, and stable enough to be called "species"?

[115] Istock et al. 1996, Moreno 1997, Cohan 2001, 2002.

[116] Gavrilets 2004.

[117] Skipper 2004.

THE PHYLOTYPE

One attempt to resolve the issue is an operational approach. The cluster of genomes of asexual organisms forms what is called a *"phylotype,"* a term coined by C. W. Cotterman in unpublished notes dated 1960.[118] *Phylotype* is a rank-neutral term, though, that is determined entirely by the arbitrary level of genetic identity chosen. For example, "species" in asexuals might be specified as being 98%+ similarity of genome, or it might be 99.9%+ (both are in the literature). A phylotype of, say 67% or 80% might be used for other purposes (such as identifying a disease-causing group of microbes):

> A genomic species is one described by DNA sequence homology. Here the most commonly adopted standard is a collection of strains the chromosomal DNA sequence homology of which is greater than 70% measured under optimal renaturation conditions with $\Delta T_m < 5\%$.[119]

The phylotype concept, while useful in other respects, reinvents, or preinvents, phenetics—1960 predates most of the work of Sokal and Sneath. A phenetic taxon was called the operational taxonomic unit (OTU) and it used an arbitrary measure of similarity and difference of morphological traits: an 80% "phenon line" was the arbitrary measure for species based on phenotypic similarities. Different phenon measures would give different taxa, and there was no principled reason other than convention for adopting one over another. Moreover, if "barcode" genes like 16sRNA are used for diagnosis of taxa,[120] then just as with phenetics, the taxa will vary according to the components of the analysis.

BRANCHING RANDOM WALKS

It might be thought that clustering in genome space doesn't need explanation. Clustering of phylotypes might in the first case simply be due to stochastic processes of elimination and divergence. To reject that assumption, I will appeal here to a study on null models in morphospace.[121] Given that the morphological traits here are heritable, and that they are in effect asexually transmitted (because the morphology here is typical of a species that diverges in a branching random walk), it is applicable to our case. This is an argument by simulation, and lacks as yet any empirical foundation, but as we have no clear prior assumptions about what the default expectations for asexually reproducing lineages may be, this may help to give us some. Further simulations might change these expectations. The authors understood their simulation to apply in the case of asexual lineages.[122]

Pie and Weitz simulated speciation in four cases (Figure 13.3). In the first case, they allowed speciation without extinction (A), which generated the expected carpet

---

[118] Denniston 1974.
[119] Denniston 1974; see also Johnson 1973.
[120] Hebert et al. 2003, Blaxter 2004, 2005, Wheeler 2005.
[121] Pie and Weitz 2005.
[122] Pie, *pers. comm.*

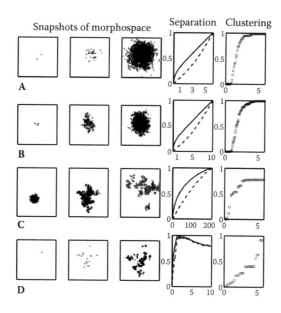

FIGURE 13.3 Pie and Weitz' simulation of speciation. Branching random walks under four assumptions. **A.** Speciation without extinction. **B.** Speciation and stochastic extinction. **C.** Speciation, extinction, and logistic diversification. **D.** A–C plus developmental entrenchment. The column "Separation" indicates the mean pairwise distance between taxa (solid line) and the mean pairwise distance squared (dotted line). The column "Clustering" indicates cluster size distribution over time. Note that under developmental entrenchment, mean separation decreases after initial increase.

model. In the second case (B), they allowed stochastic extinction, and the surprising result here is that the clustering, while tighter, still spreads out evenly. In the third case (C), however, they allowed logistic diversification, equivalent to "niche packing" according to carrying capacity, and now we see the scatter of clusters that we do find in genotypic space in asexual lineages. A fourth case (D) allowed developmental entrenchment, in which the longer the trait (in our case, gene sequence) has been in the lineage, the less likely it is to go extinct, and the clustering is even more diverse and tightly packed. Since A and B are stochastic, while the other two cases answer to selection for niche occupancy and internal cohesion respectively, there is little help to be garnered from simple stochasticity, and we must consider the other two cases separately.

However, as this is only a simulation, other realistic factors not included here, such as differential extinction due to habitat fracturing, may still form clusters in a stochastic manner. There can be, and must certainly often be, other reasons why the carpet is not smooth but patchy. Extinction can cause there to be patches in genome space. Some varieties die out. While this can be because of selection, often it will be due to plain old genetic drift, and contingency. A lineage of asexual genomes can stochastically drift due to random biases in mutational direction, for instance, while earlier forms can go extinct simply because of random drift or random termination of that clone. Moreover, if a genome evolves in a habitat that is sensitive to sudden changes in climate or even geological change that degrades it, then that genome and its neighbors will become extinct. Evolution doesn't explore the same coordinates in genome space all the time and everywhere.

Even so, Pie and Weitz's model leads us initially to expect that when patchiness occurs over large numbers of genome lineages, it will be due either to selection or to the phylogenetic integration of certain genes in the developmental process, which is how I am interpreting "developmental entrenchment" in their morphological model.

The Recombination Model

Another solution to the Problem of Cohesion for asexual organisms is what I will call the *Recombination Species Concept*. Proposed by Dykhuizen and Green, it revises the Biological Species Concept for lineages that occasionally share genes or gene fragments through lateral transfer:

> ...the phylogenies of different genes from individuals of the same species should be significantly different, whereas the phylogeny of genes from individuals of different species should not be significantly different. Thus we have an operational criterion for the defining of bacterial species.[123]
>
> Horizontal gene transfer across species should be rare enough that it will not be a problem.[124]

There are several mechanisms by which lateral transfer of DNA can occur among "prokaryotes." One is *DNA fragment reuptake*, in which DNA from a cell that has lysed (its membrane or wall has disintegrated, releasing the cell contents into the medium) is taken up by another cell, and rather than being digested it becomes active. There are variations on this. Entire small chromosomal rings, called *plasmids*, can be taken up this way. Or a cell can "bleb," forming vesicles, or compartments, of lipids, containing DNA (including plasmids), which then attach to the receiving cell, opening up to the interior.

A process that is directly analogous to sex in bacteria is *conjugation*. This is a case where part of the genetic component, usually plasmids, which are secondary small chromosomes, or the main nucleoid, can be inserted into another cell via processes called *pili*, which are part of the Type IV secretory system used for other purposes and which is homologous to flagella in Gram-negative bacteria. The typical mode of conjugation is that one mating type (often called the "male") is activated by pheromones from another mating type ("female") to attach the pilus to the recipient, and insert the genetic material. This is "almost-sex," because there is no genetic reassortment, but other processes will tend to shuffle genes into the nucleoid over many generations, as well as utilizing the taken-up plasmids. But again, while the mating types seem to act as cohesive mechanisms, conjugation can be profligate across large, even vast, phylogenetic distances (and hence genetic distances). It has been observed between bacteria and yeast, bacteria and plants, and there has even been a case in which it was observed between bacteria (*Escherichia coli*) and mammalian (hamster) cells, although there is no evidence this caused any evolutionary

---

[123] Dykhuizen and Green 1991, 7266.
[124] *Loc. cit.*, 7267.

outcome.[125] Some microbes can have multiple mating types at different parts of the developmental cycle. *Plasmodium falciparum*, the pathogen that causes malaria in humans, for instance, has seven, and many of the Plasmodia have more (a condition known as heterothallism).

The Recombination Model of microbial species is based on the claim that the greater the genetic distance between strains, the less likely it is that the transferred genes will be functional and useful in the receiving strain, and this is what serves to maintain the homogeneity of bacterial and other microbial species (the phylogenetic entrenchment of Pie and Weitz). One of the claims of Dykhuizen and Green is that in fact that the differences in genetic structure, and in some cases differences in restriction enzymes, that break DNA sequences, will make the DNA fragment non-coding, or insert it in a nonfunctional location in the target genome. This is sometimes referred to as the *Core Genome Hypothesis* or *Genomic Island Hypothesis*—what makes laterally transferred genes functional are the core "housekeeping genes" that do not get transferred because they are critical to the functioning and viability of the organism, and are closely entrenched in that organism's genome.[126] This means that they will likely not be useful in distantly related organisms with their own entrenchments.

These compatibility issues act in the same way sex does in sexual species to maintain the overall "location" in genome space of the population (Figure 13.4). Those strains that deviate too far from the mode will be unable to take up the functionally useful lateral genes, and so will be more susceptible to extinction through genetic load.

So, the Recombination Model is a mix of Maynard Smith's theory of sex and Mayr's notion of biological species. The problem is that it is not consistently true. Recombination via lateral transfer is rather more profligate than it first seemed,[127] while there are some "species" of bacteria, such as the Lyme disease-causing spirochete *Borrelia burgdorferi*, that do not share autapomorphic genes much, if at all (as Dykhuizen himself observes[128]). So, if in those latter cases clustering occurs, it is not due to lateral transfer, but some other processes. While it may be that recombination of lineages through partial lateral genetic transfer operates as a reason for *some* phylotypes occurring, it does not account for all, and LGT is therefore not a *sine qua non* of specieshood, or homogeneity, among all microbes.

THE PHYLO-PHENETIC SPECIES CONCEPT (POLYPHASIC SPECIES CONCEPT)

The two disparate concepts of phylotypes and core genomes have been rather awkwardly combined into an operational and a biological conception by Roselló-Mora and Amman to form the "phylo-phenetic" or "polyphasic" species concept, according to which a bacterial species is:

[125] Waters 2001.
[126] Wertz et al. 2003, Coleman et al. 2006.
[127] Beiko et al. 2005.
[128] Dykhuizen and Baranton 2001.

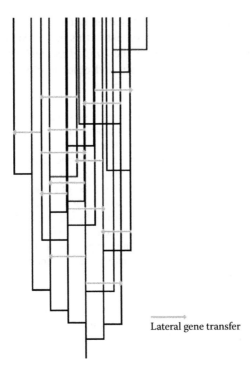

Lateral gene transfer

FIGURE 13.4 Lateral transfer. In the Recombination Model, functional lateral or horizontal transfer (thick horizontal lines) maintain the cohesion of the cluster of genomes.

... a monophyletic and genomically coherent cluster of individual organisms that show a high degree of overall similarity with respect to many independent characteristics, and is diagnosable by a discriminative phenotypic property.[129]

Such definitions are matters of convenience rather than useful theoretical accounts, however. Bacterial taxonomists, asked about their ideas about the bacterial species, are caught between Scylla and Charybdis: either they stick to a coherent species definition without it necessarily being a biological reality, or they visualize bacterial species as condensed nodes in a cloudy and confluent taxonomic space. The latter view implies that classification is a frame for the condensed nodes where some isolated internodal strains must also get a (provisional) place and name. Loosening the 70% rule often allows a compromise between the two views ...[130]

There is no phylogenetic standard for species, genus, or family delineation, nor is there a "gold standard" to identify bacterial species, and this is the major reason why a polyphasic approach is valuable.[131]

The polyphasic approach uses genotypic and phentotypic techniques to isolate bacterial species. Although an operational definition employed in taxonomic

---

[129] Rosselló-Mora and Amann 2001.
[130] Vandamme et al. 1996, 409.
[131] *Loc. cit.*, 430.

practice, the polyphasic account is not based on a theoretical account of microbial species, and can be passed over in the context of this paper. Another version of this, more theoretical and less operational, is the *Unified Bacterial Species Concept*, in which the ecological approach to bacteria and other microbes relies on a partial exchange of genes that are ecologically adaptive:

> ... a bacterial species is defined as a group of organisms that exploit a common ecology and, as a result, exhibit effective rates of recombination that are greater among members of the group than with other organisms; that is, although interspecific recombination may occur, the resulting recombinants are, on average, less successful because the hybrids do not effectively exploit either parental environment... . Over time, the divergence of nucleotide sequences reduces the likelihood of homologous DNA exchange between lineages, effectively imposing premating genetic isolation.[132]
>
> ... a microbial species is a concept represented by a group of strains, that contains freshly isolated strains, stock strains, maintained in vitro for varying periods of time, and their variants (strains not identical with their parents in all characteristics) which have in common a set or pattern of correlating stable properties that separates the group from other groups of strains.[133]

In this version, what maintains the homogeneity of clones is that they can exchange genes that have a crucial role in the adaptive capacity of the organisms. Therefore, it is a combination of the "biological" concept of species and the ecological concept.

THE QUASISPECIES MODEL

The third main approach to a *natural* conception of microbial species (by which I mean, as opposed to *operational*, *practical*, or *conventional* ones, collectively called "artificial" conceptions, such as the polyphasic account) is what I will call the Quasispecies Model, similar to the Phylotype model in intent. According to the concept developed by Manfred Eigen[134] for viral species, a *quasispecies* ("as-if" species) is a cluster of genomes in a genome space of the dimensionality of the number of loci. A quasispecies is in effect a cloud of genomes (Figure 13.5) with a "wild-type" coordinate (that is, the most central genome in the space) that may or may not actually have an extant or extinct instance.

A genetic cluster of the quasispecies kind is therefore not unlike Mallet's Genotypic Cluster Concept,[135] although his is more an operational definition than a substantive underlying account of species. However, we still need to account for microbial quasispecies existing in the first place. We have considered one possible mechanism—gene sharing by lateral transfer—and found it to be insufficient.

---

[132] Lawrence 2001, 481.

[133] Gordon 1978, Rosselló-Mora and Amann 2001, 53.

[134] Eigen 1993. Eigen and those who have followed him have presented a hypercycle model of self-organization for why there are quasispecies. However, while this is consonant with the view I present here, what is important at this point is the phenomenal nature of quasispecies' genomes—that they are clusters of genotypes in gene-space—not that they are caused in any general manner offered by quasispecies researchers.

[135] Mallet 1995.

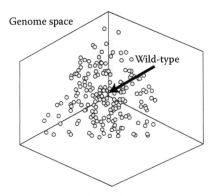

FIGURE 13.5 Quasispecies. A quasispecies is a cluster of genomes in a space determined by the number of loci in the genomes, and the alleles at those loci, about a "wild-type" mode. Developed first to cover molecular and then viral species, Eigen's notion assumes that there is an optimal wild-type genome, which is here represented as a circle in an abstract "sequence space," divergences from which form a cloud of sequences constrained by selection. This applies to asexual organisms not otherwise constrained (e.g., by developmental constraints).

Is there something else we might make use of? The proposal by Alan Templeton, devised for sexual organisms, defines a species as a genetic cluster, as Mallet's does, but accounts for it either by genetic exchange, as in the Recombination Model, or ecological interchangeability. This latter notion is what we might call the "Fitness Peak Conception" of quasispecies.

Each coordinate in genome space—that is, each genome—has a fitness value associated with it that is imposed by ecological factors. If the adaptive landscape is relatively smooth, which means that adjacent coordinates are correlated in their fitness values, we should expect in the absence of all other causes of clustering that the cloud of genomes will tend to center upon the most adaptive genome (Figure 13.6). Of course, this is an abstraction and a gross simplification—genomes are not independent of each other, or from fitness values. Organisms create their ecological conditions to a degree, and how fit a genome is depends also upon what other genomes exist in a population (at least, in sexual organisms), but we can leave these complications to one side for the moment.

So, one potential reason why quasispecies exist, why genomes cluster, is that they track local fitness peaks. Take a pathogen that is clonal. It needs to employ certain features of the host species in order to infect and exploit that host. Assuming these features remain constant—say they are recognition molecules on a cell surface—the quasispecies will cluster about those points in the genome space that are more effective than others with respect to the capacity to infect and exploit.

So now we have the two cohesion mechanisms proposed by Templeton—cohesion due to shared genes, and cohesion due to the need to exploit the environment better than competitors. Anything that can do well at the latter will tend to be better represented in the average population. Hence quasispecies. But fitness peaks typically do *not* remain constant or decoupled from the populational structure. Does this

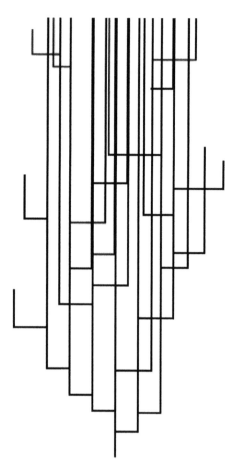

FIGURE 13.6 Tracking fitness peaks. Clustering could be explained in terms of tracking fitness peaks—deviations from the local adaptive peak would be eliminated by selection.

mean that quasispecies are not *real* species because they are ephemeral? Of course not, as all species are, over a suitably extended timescale, ephemeral. That is in the nature of evolution. What the fitness peak conception means is that a quasispecies will remain homogeneous so long as there is a more or less unitary fitness peak. If the peaks shift, there will be "speciation" or, stopping the pretense that quasispecies aren't real species, there will be real speciation.

When we attempt to apply to organisms that are not obligately sexual (that is, which don't *have* to have sex to reproduce) concepts that were specified to use with those that are sexual, we have problems. The Recombination Model is one such attempt. It is true that some microbial species exchange genes. Others do it more frequently and more completely. However, there appears to be a continuum of gene exchange all the way from almost never to almost every generation. So why should we expect that gene transfer will provide us with the sort of cohesion of lineages and quasispecies that it does in obligate sexual species?

In part, this is because we always start from what we know. As mentioned, the existence and ubiquity of asexual organisms has been resisted for a long time, and treated as exceptional rather than the rule, because biology began with large-scale plants and animals where the paradigm cases were encountered for the biospecies concept. As exceptional cases were encountered, these concepts were stretched and modified to serve the increasingly "deviant" cases, until now we realize that deviance is relative to the exemplary conceptions rather than a fact about the organisms. Instead, we should treat this continuum idea more seriously and as the basis from which the metazoan and metaphyte conceptions are drawn out. It is worth considering how the two factors of the Templeton conception—exchange and ecological niche—play differing roles according to the degree of sex a lineage has (see Figure 13.7).

So, with respect to the notion that sex is not an all-or-nothing affair, if the continuum ranges from 0% gene exchange (total asexuality) to 50% gene exchange (total sexuality), then it follows that at the asexual end, exchange can have only limited to no effect on maintaining homogeneity, while at the sexual end, it can have a very great role in maintaining homogeneity (Figure 13.8A). Likewise, at the asexual end, homogeneity not due to stochastic effects will be due largely to ecological selection (fitness peak tracking), while for sexuals this will not be so great a cause. For sexual organisms at the upper end, reproductive cohesion permits the taxon to track many ecological fitness peaks when the reproductive cohesion is sufficient to overcome the divergent selection of multiple fitness peaks (Figure 13.8B). However, in cases where so-called sympatric speciation is possible, as set out by Gavrilets, this will not prevent speciation.

This plurality of cluster causation is why gene exchange doesn't have a simple relationship to genomic homogeneity. It will depend on the rate and number of genes shared across lineages case by case, as well as the degree to which the quasispecies is maintained by ecological interchangeability. Following on from a discussion

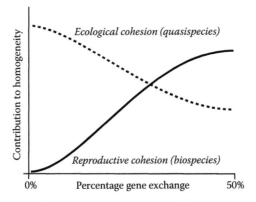

FIGURE 13.7 Contributions to species cohesion. The contribution of the cohesion of a microbial species of reproductive cohesion increases with the percentage of gene exchange between lineages, while the proportion of the contribution of niche tracking will decline.

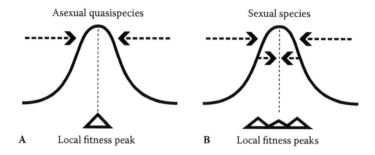

FIGURE 13.8   Asexual and sexual species and niche selection. Pure asexual species are maintained by niche selection (**A**), while pure sexuals are maintained also by reproductive cohesion (**B**), allowing multiple fitness peaks to be tracked to a degree.

about speciation and an earlier paper on species concepts,[136] where I employed the phylogeny of sex to argue that being a biospecies is an evolved trait, not a natural kind, we ought to expect that each species and group of species has its own unique evolutionary history and therefore properties, just as limbs and lungs and livers do. In effect, the modality of "being a species" is an evolved property. There will be a more general theoretical context of adaptive landscapes, gene dynamics, and so on, but given that each evolutionary group has encountered different conditions, this means that each modality will be shared in a fairly limited way.

Moreover, this implies that *biospecies* is not the most basal notion of species. In fact, reproductive isolation conceptions of species in general are based on derived modalities. The basal notion is *quasispecies*, as all species are at least quasispecies; some are in addition reproductive cohesion or isolation species. Hence some standard requirements for biospecies, such as reciprocal monophyly, do not need to apply to all species. But there is one rather interesting aspect to this approach that I find illuminating: a simple quasispecies is cohered by more or less one adaptive peak, while a biospecies famously can be and often is adapted as a generalist or have polytypic traits for differing adaptive peaks (Figure 13.8 above).

What maintains a sexual species in the polytypy case is therefore the combination of extrinsic selection and internal or intrinsic cohesion, due to selection for reproductive compatibility with potential mates. "Microbial" species, on the other hand, will tend to depend on adaptive peak cohesion inversely to the degree that they do exchange their genes, and directly to what functional value of those genes they share have ecologically.

"Microbial" species form a continuum of gene exchange from 0% to a possible 50% in the case of obligately sexual microbes, and as that percentage drops, the explanation for the cohesion of homogeneous quasispecies clusters is increasingly ecological niche adaptation, in the absence of stochastic explanations. As it approaches 50% exchange, the explanation must also include some role for the coadaptation of mating types and the lack of fitness of interspecific hybrids. All species are quasispecies first, and sexual species last, as it were.

---

[136] Wilkins 2003, 2007.

SPECIES DEFINITIONS AS SOCIOLOGICAL MARKERS

A final observation: It is often remarked upon that the specialty of the investigator has a marked effect upon the concept of species that they use. Microbiologists, for example, tend to apply what could be seen as a "phenetic" notion of species since gene exchange is not usually done by sex (in prokaryotes, at any rate), and what genes are exchanged or moved laterally need not be between closely related taxa. Ornithologists, on the other hand, have no issue with the reproductive isolation conception, as no birds do not reproduce sexually, and hybridism tends to be ignored.[137] Similar concordances between specialties and definitions occur throughout biology.

However, there is a further correlation between definitions and theoretical traditions. Often adherence to a particular definition or conception is a statement of adherence to a particular school or one leading figure over against other figures. For example, many adherents of the Reproductive Isolation Species Conception (RISC) (including those who are evolutionary systematists) use it as a tribal marker of those who are non-cladists, while many cladists use their preferred version of the phylospecies conceptions to identify against other cladists. The Hennig Convention in particular is one such: it tends to be adopted by German-speaking Hennigians rather than Anglophone systematists. De Queiroz's General/Universal conception is also often an academic marker in this way. I have noticed this happening for a while now. Since philosophers are very often beholden to a particular biological writer or tradition (especially Mayr, who overcame his adversaries by the dual strategies of prolixity and longevity), they tend to see the issues in terms specified by those particular authorities.

This is not to say that such definitional boundary-work within science is independent of real concerns and interests of the scientific work itself. I very much doubt that any scientist adopts a view *just* because their advisors or mentors propound it. However, the viciousness of the academic debate (as Wallace Stanley Sayre memorably noted, vicious because the stakes are so low[138]) over species means that the intensity with which proponents defend and attack serves just such a social identity-defining function within science.

Empirical work is still needed on how biologists actually do use species concepts, if indeed they *use* them at all in their work, but a recent survey of biologists showed some interesting results.[139] One is that the majority of specialists do not think there is a monistic conception of species (124 of 155), and that it is not desirable to have one (80 of 155). Even more interesting is that roughly the same number think species are monothetically defined (80) as polythetically defined (76), and that roughly the same number think species are real entities (70) as those who think they are not (71), while a minority thinks they are conventional (nominalistic) labels (28). The notion that

---

[137] However, hybridism in birds has been known for a long time [e.g., Elliot 1892, Suchetet 1897, Deane 1905, Ridgway 1909, Evans 1911] and even Mayr studied it [Mayr and Gilliard 1952].

[138] Although the stakes could not be higher for a professional academic. Their careers, funding, status and grants all depend on being of the "right" view. There is a tradeoff between marginality and originality on the one hand, and the likelihood of professional advancement on the other. "Deviant" views may pay off handsomely if successful, but more mainstream views have a higher chance of return on investment, if a lower payoff expectation.

[139] Pušić et al. 2017.

species are individuals is very much a minority opinion, possibly because it is a philosophical thesis rather than a substantive or operational one in biology. Also in the minority is the view that species are units of evolution (13 of 176). There are some limitations on this survey and its results: the samples are from the EU and the US and do not take into account biological practice outside these regions (the views of South American biologists would be particularly interesting). Also, there may be some self-selection bias in the results. However, this is the first such study and improves on the intuitive and anecdotal views ascribed by philosophers and historians.

BIBLIOGRAPHY

Ao, P. 2005. Laws in Darwinian evolutionary theory. *Physics of Life Reviews* 2 (2):117–156.
Armstrong, David M. 1978. *Universals and Scientific Realism*. Cambridge: Cambridge University Press.
Ayers, Michael R. 1981. Locke versus Aristotle on natural kinds. *The Journal of Philosophy* 78 (5):247–272.
Bailey, Liberty Hyde. 1896. The philosophy of species-making. *Botanical Gazette* 22 (6): 454–462.
Barker, Matthew J., and Robert A. Wilson. 2010. Cohesion, gene flow, and the nature of species. *The Journal of Philosophy* 107 (2):61–79.
Baum, David A. 2009. Species as ranked taxa. *Systematic Biology* 58 (1):74–86.
Beckner, Morton. 1959. *The Biological Way of Thought*. New York: Columbia University Press.
Beiko, Robert G. et al. 2005. Highways of gene sharing in prokaryotes. *Proceedings of the National Academy of Sciences, USA* 102 (40):14332–14337.
Beltrán, Margarita et al. 2002. Phylogenetic discordance at the species boundary: Comparative gene genealogies among rapidly radiating heliconius butterflies. *Molecular Biology and Evolution* 19 (12):2176–2190.
Berlocher, Stewart H. 2000. Radiation and divergence in the *Rhagoletis pomonella* species group: Inferences from allozymes. *Evolution; International Journal of Organic Evolution* 54 (2):543–557.
Birky, C. William, Jr. et al. 2010. Using population genetic theory and DNA sequences for species detection and identification in asexual organisms. *PLoS ONE* 5 (5):e10609.
Blaxter, Mark L. 2004. The promise of a DNA taxonomy. *Philosophical Transactions: Biological Sciences* 359 (1444):669–679.
Blaxter, Mark et al. 2005. Defining operational taxonomic units using DNA barcode data. *Philosophical Transactions: Biological Sciences* 360 (1462):1935–1943.
Boyd, Richard. 1991. Realism, anti-foundationalism and the enthusiasm for natural kinds. *Philosophical Studies*.
——. 1999a. Homeostasis, species, and higher taxa. In *Species, New Interdisciplinary Essays*, edited by Robert A. Wilson, 141–186. Cambridge, MA: Bradford/MIT Press.
——. 1999b. Kinds, complexity and multiple realization. *Philosophical Studies* 95 (1–2, Reduction and Emergence "The Thirty-Third Oberlin Colloquium in Philosophy"):67–98.
——. 2010. Homeostasis, higher taxa, and monophyly. *Philosophy of Science* 77 (5):686–701.
Brigandt, Ingo. 2009. Natural kinds in evolution and systematics: Metaphysical and epistemological considerations. *Acta Biotheoretica* 57 (1):77–97.
Cain, Arthur J. 1954. *Animal Species and Their Evolution*. London: Hutchinson University Library.
Caplan, Arthur L. 1980. Have species become déclassé? *PSA: Proceedings of the Biennial Meeting of the Philosophy of Science Association I: Contributed Papers*:71–82.

Caron, D. A. et al. 2004. The growing contributions of molecular biology and immunology to protistan ecology: Molecular signatures as ecological tools. *Journal of Eukaryotic Microbiology* 51 (1):38–48.

Cavalier-Smith, Thomas. 2010. Deep phylogeny, ancestral groups and the four ages of life. *Philosophical Transactions of the Royal Society B: Biological Sciences* 365 (1537):111–132.

Chaney, Richard Paul. 1978. Polythematic expansion: Remarks on Needham's polythetic classification. *Current Anthropology* 19 (1):139–143.

Cohan, Frederick M. 2001. Bacterial species and speciation. *Systematic Biology* 50(4):513–524.

—. 2002. What are bacterial species? *Annual Review of Microbiology* 56:457–487.

Coleman, Maureen L. et al. 2006. Genomic islands and the ecology and evolution of Prochlorococcus. *Science* 311 (5768):1768–1770.

Colless, Donald. 2006. Taxa, individuals, clusters and a few other things. *Biology and Philosophy* 21 (3):353–367.

Colyvan, Mark, and Lev R. Ginzburg. 2003. The Galilean turn in population ecology. *Biology and Philosophy* 18:401–414.

Coyne, Jerry A., and H. Allen Orr. 2004. *Speciation*. Sunderland, MA: Sinauer.

Dawkins, Richard. 1989. *The Selfish Gene*. New ed. Oxford UK/New York: Oxford University Press.

de Queiroz, Kevin. 2007. Species concepts and species delimitation. *Systematic Biology* 56 (6):879–886.

Deane, Ruthven. 1905. Hybridism between the Shoveller and Blue-winged Teal. *The Auk* 321–321.

Denniston, Carter. 1974. An extension of the probability approach to genetic relationships: One locus. *Theoretical Population Biology* 6 (1):58–75.

DeSalle, Rob et al. 2005. The unholy trinity: Taxonomy, species delimitation and DNA barcoding. *Philosophical Transactions: Biological Sciences* 360 (1462):1905–1916.

Devitt, Michael. 2008. Resurrecting biological essentialism. *Philosophy of Science* 75 (3):344–382.

—. 2011. Natural kinds, and biological realisms. In *Carving Nature at Its Joints: Natural Kinds in Metaphysics and Science*, edited by Joseph Keim Campbell et al., 155–174. Cambridge, MA/London: MIT Press.

Devitt, Michael, and Kim Sterelny. 1987. *Language and Reality*. Oxford: Blackwell/MIT Press.

Dobzhansky, Theodosius. 1935. A critique of the species concept in biology. *Philosophy of Science* 2:344–355.

—. 1941. *Genetics and the Origin of Species*. 2nd ed. New York: Columbia University Press.

—. 1951. *Genetics and the Origin of Species*. 3rd rev. ed. New York: Columbia University Press.

Doolittle, W. F. 2005. If the tree of life fell, would it make a sound? In *Microbial Phylogeny and Evolution: Concepts and Controversies*, edited by Jan Sapp, 119–133. Oxford UK, New York: Oxford University Press.

Dumsday, Travis. 2012. A new argument for intrinsic biological essentialism. *Philosophical Quarterly* 62 (248):486–504.

Dupré, John. 1993. *The Disorder of Things: Metaphysical Foundations of the Disunity of Science*. Cambridge, MA: Harvard University Press.

Dykhuizen, Daniel E. 1998. Santa Rosalia revisited: Why are there so many species of bacteria? *Antonie Van Leeuwenhoek* 73 (1):25–33.

Dykhuizen, Daniel E., and Guy Baranton. 2001. The implications of a low rate of horizontal transfer in Borrelia. *Trends in Microbiology* 9 (7):344–350.

Dykhuizen, Daniel E., and L. Green. 1991. Recombination in *Escherichia coli* and the definition of biological species. *Journal of Bacteriology* 173 (22):7257–7268.

Eigen, Manfred. 1993. Viral quasispecies. *Scientific American* July 1993 (32–39).

Elliot, D. G. 1892. Hybridism, and a description of a hybrid between *Anas boschas* and *Anas americana. The Auk* 9 (2):160–166.

Ellis, Brian David. 2001. *Scientific Essentialism, Cambridge Studies in Philosophy.* Cambridge: Cambridge University Press.

Ereshefsky, Marc. 1991. Species, higher taxa, and the units of evolution. *Philosophy of Science* 58 (84–101).

—. 1992. Eliminative pluralism. *Philosophy of Science* 59:671–690.

—. 1999. Species and the Linnean hierarchy. In *Species, New Interdisciplinary Essays,* edited by R. A. Wilson, 285–305. Cambridge, MA: Bradford/MIT Press.

—. 2000. *The Poverty of Linnaean Hierarchy: A Philosophical Study of Biological Taxonomy.* Cambridge, UK/New York: Cambridge University Press.

—. 2010. Darwin's solution to the species problem. *Synthese* 175 (3):405–425.

—. 2014. Species, historicity, and path dependency. *Philosophy of Science* 81 (5):714–726.

Ereshefsky, Marc, and Thomas A. C. Reydon. 2015. Scientific kinds. *Philosophical Studies* 172:969–986.

Evans, Wallace. 1911. Hybridizing game birds in captivity. *Journal of Heredity* (1):100–102.

Fisher, Ronald Aylmer. 1958. *The Genetical Theory of Natural Selection.* 2nd rev. ed. New York: Dover.

Fitzhugh, Kirk. 2009. Species as explanatory hypotheses: Refinements and implications. *Acta Biotheoretica* 57 (1):201–248.

Floodgate, G. D. 1962. Some remarks on the theoretical aspects of bacterial taxonomy. *Bacteriological Review* 26:277–291.

Foster, Peter G. et al. 2009. The primary divisions of life: A phylogenomic approach employing composition-heterogeneous methods. *Philosophical Transactions of the Royal Society B: Biological Sciences* 364 (1527):2197–2207.

Gavrilets, Sergey. 2004. *Fitness Landscapes and the Origin of Species, Monographs in Population Biology.* Vol. 41. Princeton, NJ/Oxford, UK: Princeton University Press.

Gelman, Susan A. 2004. Psychological essentialism in children. *Trends in Cognitive Sciences* 8 (9):404–409.

Ghiselin, Michael T. 1997. *Metaphysics and the Origin of Species.* Albany: State University of New York Press.

Godfrey-Smith, Peter. 2009. *Darwinian Populations and Natural Selection.* Oxford: Oxford University Press.

Goodsir, Joseph Taylor. 1868. *The Bases of Life: A Discourse in Reply to Professor Huxley's Lecture on "The Physical Bases of Life."* London: Williams & Norgate.

Goodwin, Brian C. 1994. *How the Leopard Changed Its Spots: The Evolution of Complexity.* New York: Charles Scribner's Sons.

Goodwin, Brian C. et al. 1983. *Development and Evolution.* Cambridge; New York: Cambridge University Press.

Gordon, Ruth E. 1978. A species definition. *International Journal of Systematic Bacteriology* 28 (4):605–607.

Gould, Stephen Jay. 2002. *The Structure of Evolutionary Theory.* Cambridge/MA: The Belknap Press of Harvard University Press.

Graff, Delia, and Timothy Williamson. 2002. *Vagueness.* Aldershot: Ashgate Dartmouth.

Grant, Verne. 1957. The plant species in theory and practice. In *The Species Problem,* edited by Ernst Mayr, 39–80. Washington, DC: American Association for the Advancement of Science.

Griffiths, Paul E. 1994. Cladistic classification and functional explanation. *Philosophy of Science* 61 (2):206–227.

—. 1999. Squaring the circle: Natural kinds with historical essences. In *Species, New Interdisciplinary Essays,* edited by Richard A. Wilson, 209–228. Cambridge, MA: Bradford/MIT Press.

Hacking, Ian. 2007a. Natural kinds: Rosy dawn, scholastic twilight. *Royal Institute of Philosophy Supplement* 82 (Supplement 61):203–239.

—. 2007b. Putnam's theory of natural kinds and their names is not the same as Kripke's. *Principia: An International Journal of Epistemology* 11 (1):1–24.

Hanage, William P. et al. 2005. Fuzzy species among recombinogenic bacteria. *BMC Biology* 3 (6). Available at https://bmcbiol.biomedcentral.com/articles/10.1186/1741-7007-3-6 (accessed 3/17).

Hebert, Paul D. N. et al. 2003. Biological identifications through DNA barcodes. *Proceedings of the Royal Society of London B* 270 (1512):313–321.

Hennig, W. 1975. "Cladistic analysis or cladistic classification?" A reply to Ernst Mayr. *Systematic Zoology* 24 (2):244–256.

Hull, David Lee. 1976. Are species really individuals? *Systematic Zoology* 25 (2):174–191.

—. 1980. Individuality and selection. *Annual Review of Ecology and Systematics* 11:311–332.

—. 1984. Can Kripke alone save essentialism? A reply to Kitts. *Systematic Zoology* 33 (1):110–112.

—. 1988. *Science as a Process: An Evolutionary Account of the Social and Conceptual Development of Science*. Chicago: University of Chicago Press.

—. 2002. A career in the glare of public acclaim. *BioScience* 52 (9):837–841.

Huxley, Julian. 1957a. *New Bottles for New Wine: Essays*. London: Chatto & Windus.

—. 1957b. The three types of evolutionary process. *Nature* 180 (4584):454–455.

Istock, Conrad A. et al. 1996. Bacterial species and evolution: Theoretical and practical perspectives. *Journal of Industrial Microbiology and Biotechnology* 17 (3–4):137–150.

John, David M., and Christine A. Maggs. 1997. Species problems in eukaryotic algae: A modern perspective. In *Species: The Units of Biodiversity*, edited by Michael F. Claridge et al., 83–107. London: Chapman & Hall.

Johnson, John L. 1973. Use of nucleic-acid homologies in the taxonomy of anaerobic bacteria. *International Journal of Systematic Bacteriology* 23:308–315.

Johnstone, James. 1914. *The Philosophy of Biology*. Cambridge: Cambridge University Press.

Khalidi, Muhammad Ali. 2013. *Natural Categories and Human Kinds: Classification in the Natural and Social Sciences*. Cambridge University Press.

Kitcher, Philip. 1984. Species. *Philosophy of Science* 51 (2):308–333.

—. 1987. Ghostly whispers: Mayr, Ghiselin, and the "Philosophers" on the ontological status of species. *Biology and Philosophy* 2 (2):184–192.

Kitts, David B., and David J. Kitts. 1979. Biological species as natural kinds. *Philosophy of Science* 46 (4):613–622.

Knowles, L. Lacey, and Bryan C. Carstens. 2007. Delimiting species without monophyletic gene trees. *Systematic Biology* 56 (6):887–895.

Kull, Kalevi. 2016. The biosemiotic concept of the species. *Biosemiotics* 9 (1):61–71.

Lake, James A. 2015. Eukaryotic origins. *Philosophical Transactions of the Royal Society B: Biological Sciences* 370 (1678): 20140321. Available at http://rstb.royalsocietypublishing .org/content/370/1678/20140321.

LaPorte, Joseph. 1997. Essential membership. *Philosophy of Science* 64 (1):96–112.

—. 2000. Rigidity and kind. *Philosophical Studies* 97 (3):293–316.

—. 2003. Does a type specimen necessarily or contingently belong to its species? *Biology and Philosophy* 18 (4):583–588.

—. 2004. *Natural Kinds and Conceptual Change*. Cambridge, UK: Cambridge University Press.

—. 2018. "Modern essentialism for species and its animadversions." In Routledge Handbook of Evolution and Philosophy, edited by Richard Joyce, 182–193. Abingdon, UK: Routledge.

Lawrence, Jeffrey G. 2001. Catalyzing bacterial speciation: Correlating lateral transfer with genetic headroom. *Systematic Biology* 50 (4):479–496.

Levine, Alex. 2001. Individualism, type specimens, and the scrutability of species membership. *Biology and Philosophy* 16 (3):325–338.

Mallet, James. 1995. The species definition for the modern synthesis. *Trends in Ecology and Evolution* 10 (7):294–299.

—. 2010. Why was Darwin's view of species rejected by twentieth century biologists? *Biology and Philosophy* 25 (4):497–527.

Martin, Cristopher H. et al. 2015. Complex histories of repeated gene flow in Cameroon crater lake cichlids cast doubt on one of the clearest examples of sympatric speciation. *Evolution* 69 (6):1406–1422.

Martin, William, and Eugene V. Koonin. 2006. A positive definition of prokaryotes. *Nature* 442 (7105):868–868.

Martínez, Manolo. 2015. Informationally-connected property clusters, and polymorphism. *Biology and Philosophy* 30 (1):99–117.

Mayr, Ernst. 1963. *Animal Species and Evolution*. Cambridge, MA: The Belknap Press of Harvard University Press.

Mayr, Ernst, and Walter J. Bock. 2002. Classifications and other ordering systems. *Journal of Zoological Systematics and Evolutionary Research* 40 (4):169–194.

Mayr, Ernst, and E. Thomas Gilliard. 1952. Altitudinal hybridization in New Guinea honeyeaters. *The Condor* 54 (6):325–337.

McInerney, James et al. 2015. The ring of life hypothesis for eukaryote origins is supported by multiple kinds of data. *Philosophical Transactions of the Royal Society B: Biological Sciences* 370 (1678):1–10.

Meier, Rudolf, and Rainer Willmann. 2000. The Hennigian species concept. In *Species Concepts and Phylogenetic Theory: A Debate*, edited by Quentin D. Wheeler and Rudolf Meier, 167–178. New York: Columbia University Press.

Mellor, David H. 1977. Natural kinds. *British Journal for the History of Science* 28 (4):299–312.

Moreno, Edgardo. 1997. In search of a bacterial species definition. *Revista de Biologia Tropical* 45 (2):753–771.

Murray, Bertram G., Jr. 2001. Are ecological and evolutionary theories scientific? *Biological Reviews of the Cambridge Philosophical Society* 76 (2):255–289.

Nagel, Ernst. 1961. *The Structure of Science: Problems in the Logic of Scientific Explanation*. London: Routledge and Kegan Paul.

Nelson, Gareth J. 1989. Cladistics and evolutionary models. *Cladistics* 5 (3):275–289.

Neto, Celso. 2016. Rethinking cohesion and species individuality. *Biological Theory* 11 (3):138–149.

Ogunseitan, Oladele. 2005. *Microbial Diversity: Form and Function in Prokaryotes*. Malden, MA: Blackwell.

Okasha, Samir. 2002. Darwinian metaphysics: Species and the question of essentialism. *Synthese* 131 (2):191–213.

Oken, Lorenz. 1847. *Elements of physiophilosophy*. Translated by Alfred Tulk. London: The Ray Society. Original edition, Lehrbuch der Naturphilosophie (Jena: Friedrich Frommann, 1831, 2nd edition).

Pace, Norman R. 2006. Time for a change. *Nature* 441 (7091):289–289.

Padian, Kevin. 1999. Charles Darwin's view of classification in theory and practice. *Systematic Biology* 48 (2):352–364.

Papillon, Fernand. 1875. *Nature and life. Facts and doctrines relating to the constitution of matter, the new dynamics, and the philosophy of nature*. Translated by August Rodney Macdonough. New York: Appleton.

Pedroso, Makmiller. 2013. Origin essentialism in biology. *The Philosophical Quarterly* 64 (254):60–81.

Phillip, Arthur et al. 1789. *The voyage of Governor Phillip to Botany Bay: with an account of the establishment of the colonies of Port Jackson & Norfolk Island; compiled from authentic papers which have been obtained from the several departments, to which are added the journals of Lieuts. Shortland, Watts, Ball, & Capt. Marshall; with an account of their new discoveries.* London: John Stockdale.

Pie, Marcio R., and Joshua S. Weitz. 2005. A null model of morphospace occupation. *American Naturalist* 166 (1):E1–E13.

Pigliucci, Massimo. 2003. Species as family resemblance concepts: The (dis-)solution of the species problem? *BioEssays* 25 (6):596–602.

Pitcher, George. 1964. *The Philosophy of Wittgenstein.* Englewood Cliffs, NJ: Prentice-Hall.

Purvis, Ole William. 1997. The species concept in lichens. In *Species: The Units of Biodiversity*, edited by Michael F. Claridge et al., 109–134. London: Chapman & Hall.

Pušić, Bruno et al. 2017. What do biologists make of the species problem? *Acta Biotheoretica* 65 (3):179–209. doi:10.1007/s10441-017-9311-x.

Queller, David. 2011. A gene's eye view of Darwinian populations. *Biology and Philosophy* 26 (6):905–913.

Quine, Willard Van Orman. 1960. *Word and Object, Studies in Communication.* Cambridge, MA: Technology Press of the Massachusetts Institute of Technology.

Raxworthy, Christopher J. et al. 2007. Applications of ecological niche modeling for species delimitation: A review and empirical evaluation using day geckos (Phelsuma) from Madagascar. *Systematic Biology* 56 (6):907–923.

Richards, Richard A. 2016. *Biological Classification: A Philosophical Introduction, Cambridge Introductions to Philosophy and Biology.* Cambridge, UK: Cambridge University Press.

Ridgway, Robert. 1909. Hybridism and generic characters in the Trochilidæ. *The Auk* 26 (4):440–442.

Rieppel, Olivier. 2008. Origins, taxa, names and meanings. *Cladistics* 24 (4):598–610.

—. 2009. Species as a process. *Acta Biotheoretica* 57 (1):33–49.

—. 2010. New essentialism in biology. *Philosophy of Science* 77 (5):662–673.

Rosenberg, Alexander. 1994. *Instrumental Biology, or, The Disunity of Science.* Chicago, IL: University of Chicago Press.

Rosselló-Mora, Ramon, and Rudolf Amann. 2001. The species concept for prokaryotes. *FEMS Microbiology Reviews* 25 (1):39–67.

Ruse, Michael. 1987. Biological species: Natural kinds, individuals, or what? *British Journal for the Philosophy of Science* 38 (2):225–242.

Sankey, Howard. 1998. Taxonomic incommensurability. *International Studies in the Philosophy of Science* 12 (1):7–16.

Sapp, Jan. 2005. The prokaryote-eukaryote dichotomy: Meanings and mythology. *Microbiology and Molecular Biology Reviews* 69 (2):292–305.

Scriven, Michael. 1959. Explanation and prediction in evolutionary theory. *Science* 130 (3374):477–482.

Sites, Jack W., and Jonathon C. Marshall. 2003. Delimiting species: A Renaissance issue in systematic biology. *Trends in Ecology and Evolution* 18 (9):462–470.

Skipper, Robert A. 2004. The heuristic role of Sewall Wright's 1932 adaptive landscape diagram. *Philosophy of Science* 71 (5):1176–1188.

Slater, Matthew H. 2016. *Are Species Real? An Essay on the Metaphysics of Species.* Basingstoke, New York: Palgrave Macmillan.

Smellie, William. 1790. *The philosophy of natural history.* Edinburgh: Printed for the heirs of Charles Elliot.

Sober, Elliott. 1980. Evolution, population thinking, and essentialism. *Philosophy of Science* 47 (3):350–383.

—. 1984. *The Nature of Selection: Evolutionary Theory in Philosophical Focus.* Cambridge, MA: MIT Press.

—. 1988. *Reconstructing the Past: Parsimony, Evolution, and Inference.* Cambridge, MA: MIT Press.

—. 2000. *Philosophy of Biology.* 2nd ed, Dimensions of Philosophy Series. Boulder, CO: Westview Press.

—. 2008. *Evidence and Evolution: The Logic behind the Science.* Cambridge/New York: Cambridge University Press.

Sober, Elliott, and Michael Steel. 2002. Testing the hypothesis of common ancestry. *Journal of Theoretical Biology* 218 (4):395–408.

Spencer, Herbert. 1884. *The principles of biology.* London: Williams and Norgate.

Staley, James T. 2006. The bacterial species dilemma and the genomic-phylogenetic species concept. *Philosophical Transactions of the Royal Society B: Biological Sciences* 361 (1475):1899–1909.

Stamos, David N. 1998. Buffon, Darwin, and the non-individuality of species—A reply to Jean Gayon. *Biology and Philosophy* 13 (3):443–470.

—. 2003. *The Species Problem: Biological Species, Ontology, and the Metaphysics of Biology.* Lanham, MD: Lexington Books.

Stewart, Michael A. 1980. Locke's mental atomism and the classification of ideas: II. *The Locke Newsletter* 11:25–62.

Strawson, Peter Frederick. 1964. *Individuals: An Essay in Descriptive Metaphysics.* London: Methuen.

Suchetet, A. 1897. *Des hybrides à l'état sauvage: Règne animal. (Classe des oiseaux).* Vol. 1. Lille: Le Bigot frères.

Templeton, Alan R. 1989. The meaning of species and speciation: A genetic perspective. In *Speciation and its Consequences*, edited by D. Otte and J. A. Endler, 3–27. Sunderland, MA: Sinauer.

—. 1998. Species and speciation: Geography, population structure, ecology, and gene trees. In *Endless Forms: Species and Speciation*, edited by Daniel J. Howard and Stewart H. Berlocher, 32–43. New York: Oxford University Press.

Templeton, Alan R. et al. 2000. Gene trees: A powerful tool for exploring the evolutionary biology of species and speciation. *Plant Species Biology* 15 (3):211–222.

Thiel, Udo. 2002. *Locke: Epistemology and Metaphysics, The International Library of Critical Essays in the History of Philosophy.* Aldershot, Hants/Brookfield, VT: Ashgate: Dartmouth.

Turner, George F. 2002. Parallel speciation, despeciation and respeciation: Implications for species definition. *Fish and Fisheries* 3 (3):225–229.

Van Valen, Leigh M. 1988. Species, sets, and the derivative nature of philosophy. *Biology and Philosophy* 3 (1):49–66.

Vandamme, Peter et al. 1996. Polyphasic taxonomy, a consensus approach to bacterial systematics. *Microbiology Review* 60 (2):407–438.

Vanderlaan, Tegan A. et al. 2013. Defining and redefining monophyly: Haeckel, Hennig, Ashlock, Nelson and the proliferation of definitions. *Australian Systematic Botany* 26 (5):347–355.

Varma, Charissa Sujata. 2013. Beyond set theory: The relationship between logic and taxonomy from the early 1930 to 1960. PhD dissertation, Institute for the History and Philosophy of Science and Technology, University of Toronto.

Waters, Virginia L. 2001. Conjugation between bacterial and mammalian cells. *Nature Genetics* 29 (4):375–376.

Wertz, J. E. et al. 2003. A molecular phylogeny of enteric bacteria and implications for a bacterial species concept. *Journal of Evolutionary Biology* 16 (6):1236–1248.

Wheeler, Quentin D. 2005. Losing the plot: DNA "barcodes" and taxonomy. *Cladistics* 21 (4):405–407.

Whewell, William. 1840. *The Philosophy of the Inductive Sciences: Founded upon Their History.* 2 vols. London: John W. Parker.

Whitman, William B. 2009. The modern concept of the procaryote. *Journal of Bacteriology* 191 (7):2000–2005.

Wilkerson, T. E. 1995. *Natural Kinds, Avebury Series in Philosophy.* Aldershot/Brookfield, VT: Avebury.

Wilkins, John S. 2002. Darwinism as metaphor and analogy: Language as a selection process. *Selection: Molecules, Genes, Memes* 3 (1):57–74.

—. 2003. How to be a chaste species pluralist-realist: The origins of species modes and the synapomorphic species concept. *Biology and Philosophy* 18 (5):621–638.

—. 2007. The dimensions, modes and definitions of species and speciation. *Biology and Philosophy* 22 (2):247–266.

Wilkins, John S., and Malte C. Ebach. 2013. *The Nature of Classification: Kinds and Relationships in the Natural Sciences,* edited by Steven French, New Directions in the Philosophy of Science. London: Palgrave Macmillan.

Will, Kipling W. et al. 2005. The perils of DNA barcoding and the need for integrative taxonomy. *Systematic Biology* 54 (5):844–851.

Wilson, Robert A. et al. 2007. When traditional essentialism fails: Biological natural kinds. *Philosophical Topics* 35 (1/2):189–215.

Wittgenstein, Ludwig. 1968. *Philosophical Investigations.* Translated by G.E.M. Anscombe. 3rd ed. Oxford: Basil Blackwell.

Woese, Carl Richard. 1998. A manifesto for microbial genomics. *Current Biology* 8 (22): R780–R783.

Woodward, Jim. 2001. Law and explanation in biology: Invariance is the kind of stability that matters. *Philosophy of Science* 68 (1):1–20.

Wu, Chung-I. 2001a. Genes and speciation. *Journal of Evolutionary Biology* 14 (6):889–891.

—. 2001b. The genic view of the process of speciation. *Journal of Evolutionary Biology* 14 (6):851–865.

Zalta, Edward N. 1988. *Abstract Objects: An Introduction to Axiomatic Metaphysics.* Dordrecht: D. Reidel.

14 Species Realism[1]

It is really laughable to see what different ideas are prominent in various naturalists' minds, when they speak of "species;" in some, resemblance is everything and descent of little weight—in some, resemblance seems to go for nothing, and Creation the reigning idea—in some, descent is the key,—in some, sterility an unfailing test, with others it is not worth a farthing. It all comes, I believe, from trying to define the undefinable.

Darwin to Hooker
24 December 1856[2]

No one definition has satisfied all naturalists; yet every naturalist knows vaguely what he means when he speaks of a species.

Darwin
Origin of Species[3]

Many scientists and philosophers have said something like "species are the units of evolution" or "biodiversity"; it is even in the titles of some well-known books on the subject.[4] What does this even mean? In this section I argue that species are real, *phenomenal*, objects rather than objects of any biological theory, let alone of evolutionary theory.

In addressing the impact of evolution upon the concept of species we must first ask what *are* the units of evolution? The ontology depends a lot on what theory is being employed. When talking about population genetics, the basic units, of course, are the *allele* and the *locus*.[5] When talking about development, then the unit is the *organism* or *lifecycle*, as it also is when you are talking about ecological interactions, although "species" is used here as the term for a class of ecologically exchangeable organisms; that is, organisms that play roughly the same role in the local ecosystem. Although organisms are pretty well all different (which is the point of population genetics), for the purpose of trophic webs (the food webs of ecology), conspecifics are treated as being interchangeable elemental units. Then there are the larger units of evolution: *populations*,[6] and the particular revision of that concept, the *"deme."*[7] A deme is basically the population that can interbreed—the term in the equations of population genetics is N_e, the number of effective, or reproductive, individuals. However, non-breeders also play a role in many species in contributing to the fitness of their kin, by helping raise them, or finding food, so the ontology here depends

---

[1] This section is revised from Wilkins 2018, with permission from the editor.
[2] Darwin 1888, vol. 2, 86.
[3] Darwin 1859, 38.
[4] Ereshefsky 1992, Claridge et al. 1997.
[5] Griffiths and Stotz 2007.
[6] Godfrey-Smith 2009.
[7] Winsor 2000.

solely upon what the research issues are. While "population" is itself somewhat fuzzy[8]—the more migration there is between two (sexual) populations the more they start to look like a single population—the term is a theoretical object.[9]

But species? No biological theory *requires* them. True, ecologists and conservation biologists *use* species, but what they are doing is using field guides as a surrogate for the ecological roles played in an ecosystem by individuals of the species who are more or less normal—the "wild type." Likewise, medical and biological researchers do the same thing with their model organisms. *Mus musculus*, or the common mouse, is used as a model because it is assumed that each individual member of that species shares the same properties (developmental cycles, phenotypes). But in practice they use "strains" that are specially bred to see the effects of gene knockouts, for example. The "objects" here are the genetic strains and the organisms.[10] Systematists "use" (the word) *species* because they describe and name them, but the explanations given of species *being* species are manifold. The notion of a "gene pool" or "metapopulation" is the foundation of one such explanation. But the theories used, the explanations, are not theories of species; they are theories of gene exchange, reproduction, fitness, adaptation, and so on. Species are being *explained*; they do no work in explaining.[11]

So, there are two ways we might go if species are not theoretical objects. One is we may deny that species exist, and a lot of people do this, and have done since the *Origin* and even before, Lamarck being the most prominent. I call them *species deniers*, because they deny that species exist, although the usual term is *species conventionalists*, or *nominalists* (both philosophically and historically misleading terms). A version of species denial is to replace the term *species* with some "neutral" term. *Deme* was one of these, but it got subverted by population geneticists for the meaning given above. Other examples include *Operational Taxonomic Units*, *Least Inclusive Taxonomic Units*, *Evolutionarily Significant Groups*, and so on. In each case, the term *species* came or is coming back into use.

Why is *species* so durable? The alternative is *not* taking the term and concept as a theoretical term. *Species* is a useful term because species are real *phenomena*.[12] That is, they are things observed that call for explanation, they are *explicanda*. Theories of biology explain why there are species, although the same theories do not apply in the same way for all species. Biology is not that neat. Some species are explained in textbook fashion through the acquisition of reproductive isolating mechanisms formed

---

[8] Gannett 2003.

[9] Millstein 2006.

[10] Ankeny 2000.

[11] One possible exception is the work species do in "species selection" theories [Stidd and Wade 1995, Grantham 1995, Rice 1995, Gould and Lloyd 1999, Jablonski 2008], but it is arguable whether these are actually theories as such, and equally arguable whether the properties that are "theoretical," which play a role in causal explanations, are those of the species, populations within the species, or of the individuals or kin groups. If species selection is taken to mean that species whose members have a particular property (like eurytopy) tend to speciate more often, then "species" in this sense is merely a mass noun [Grandy 2008].

[12] This is not to deny that *species* is also maintained by conventional and social practices. If entire volumes are dedicated to describing species, anyone who wishes to be taken seriously in that field has to refer to those described objects.

in geographical isolation. Some are not. There are species formed by hybridization,[13] by sexual selection, and of course asexual or mostly asexual species that are maintained, as I argue above, by adaptation to niches.

If there is no general theoretical account of species, why do we have this category? It might be because we tend to name things that look similar to us. This is what species deniers think—it is all about us and our cognitive dispositions, not the things themselves. However, there *are* some general features of species that license us calling them all species: *Species are salient phenomenal objects.* They are salient not because of our perceptual tendencies alone but because they do exist. They are a bit like mountains. Each particular mountain is caused by definite processes, but every mountain is not caused by the same processes. We *identify* mountains, because they are there. We *explain* them with theories of tectonics, vulcanism, erosion, or (if they are dunes) wind. Analogously, species are clusters of genomes, phenotypes, and organismic lineages. We explain them because they need explaining. A species is (roughly) where the lineages of genes, genomes, parent-child relationships, haplotypes, and ecological roles all tend to coincide—in bacterial systematics this is the *polyphasic* approach. Not all of these need to coincide in every case, but so long as most of them do, they are species, and we must give an account of them. And we can and do.

PHENOMENAL OBJECTS

Perhaps the most crucial practical aspect of the species concept debate lies in its relevance to conservation, but it is not the most theoretically interesting. Biology, like most sciences, has a need for units of measurement, and like most sciences those units need to be grounded in the real world. So, *species*—the "rank" of biology that is agreed upon by most sides as the most or only natural one in the Linnaean hierarchy—determines many measures of biology in fields from genetics to ecology. If, as a significant number of specialists think, the rank is a mere convention,[14] then those measures become arbitrary and meaningless.

Therefore, we need to consider what sort of "unit" a species might be. I can think of three alternatives. The first is that species are, in fact, simply *a matter of convention*, which is to say, something that makes things convenient for us in communication, just as John Locke said in the *Essay*.[15] Instead, say researchers such as polychaete specialist Pleijel and geneticist Hey, we need to replace the notion *species* with something like a "least inclusive taxonomic unit" (LITU; Pleijel) or "evolutionary group" (Hey). There are other replacement concepts in the offing. And as I have argued, the so-called "phylogenetic species concept" is not really a concept of species, at least in one of the versions under that name, so much as something very like a LITU that gets *called* a species.

---

[13] For example, see Knobloch 1959, Wagner 1983, Barrington et al. 1989, Arnold 1992, Jolly et al. 1997, Dowling and Secor 1997, Muir et al. 2000, Detwiler 2003, Bergman and Beehner 2004, Birkhead and Balen 2007, Cortes-Ortiz et al. 2007, Mallet 2007, Rieseberg and Willis 2007, Mallet 2008.

[14] Similar concerns exist with the use of families and genera as a measure of "kill curves" in extinction [Raup and Sepkoski 1986; see Sepkoski 1994].

[15] This was the view of John Maynard Smith as well [1958].

The second alternative is that *species* plays a theoretical role in biology, and this seems intuitively right: we sometimes talk about species as the units of evolution, so they are supposed to be required by evolutionary biology, and likewise in ecology, species are the unit that is crucial in defining the biodiversity of a region or ecosystem. But if species are theoretical objects, we ought to find them as a *consequence* of theory, not as a "unit" that we feed into theoretical or operational processes, and this is not the case. Population genetics and evolutionary theory have populations, haplotypes, alleles, trophic nodes, niches, and so on, but what they do not have are species. In every case where species are *used* in theory, they are primitives, or stand as surrogate terms for the other things mentioned. Theory does not define species.

This might be challenged by adherents of Mayr's biological species concept, or one of the derivative or related conceptions—a species is a protected gene pool, as Mayr said. This is certainly the view of Coyne and Orr in their *Speciation* book. However, the vast bulk of life would not be in species if that were the case, and anyway, species were well described and identified long before genetics was developed, up to two centuries before. So, they must at least be things that can be observed in the absence of theory. Of course, some species are more difficult to identify than others, requiring techniques that are recent, but that still does not make species theoretical objects.

The third alternative is that species are not theoretical objects at all; they are *objects that have phenomenal salience.*[16] That is, we do not *define* species, we *see* them. Consider mountains. Mountains are hard to define, and they have a multitude of geological causes, including uplift, subduction, vulcanism, differential erosion, and so forth. "Mountain" is not a theoretical object of geology—subduction zones, tectonic plates, and volcanoes are. A mountain is just something you *see*, although there are no necessary sets of properties (or heights) that mountains must have, and differentiating between them can be vague. A mountain calls for an explanation, and the explanation relies on theory, but equally so do mesas, land bridges, and caves.

So, the proposed answer to the question, What is a species? is that a species is something one sees when one realizes that two organisms are in some relevant manner the same. They are natural objects, not mere conveniences, but they are not derived from explanations, they call for them.

THEORY-DEPENDENCE AND DERIVATION

Traditionally, a theoretical object—that is, an object that was only theoretical— was something that theory required or employed but which was not empirically ascertainable—"electron" c. 1920, "gene" prior to 1952, and, until recently, "Higgs boson." But this is a positivistic sense of *theory*—a formal system in which objects are either observational or theoretical. Whether or not one is now a logical empiricist instead of a logical positivist, objects are much more nuanced than that. There is a school of thought that treats scientific ontology, the set of objects that one thinks

---

[16] Gal Kober (*pers. comm.*) suggested that species are a fourth alternative: units of *classification*. This is consistent with my third alternative unless one thinks, as Kober does, that classification is a theoretical operation.

exists in a domain, as basically the bounded variables of the best theory of that domain: that is, Quine's "to be is to be the value of a variable," and the subsequent development of that view.[17] In this case, species would be theoretical objects if they were such variables of a theory. But if they are not, we need to establish what sort of ontological status they, and other phenomenal non-theoretic objects, may have.

Consider planetary orbits. Observed and debated for a very long time before Newton proposed a general physics that accounted for them (and made predictions about them), Newton demoted these orbits from theoretically important objects (heavenly spheres) to special cases of larger and more universal physics. "Planetary orbit" is thus a special instantiation of astrophysical dynamics, which aperiodic comets, entire star systems, and even entire galaxies obey. Even if no orbits actually existed (and we can perhaps envisage this in some world) under this physics, the movements of objects would be still covered by Newtonian dynamics. Likewise, species. They obey, and when they occur are *post hoc* explained by the biology of populations, interbreeding, selection, drift, and so on, but they are not theoretical objects, any more than planetary orbits are in physics. Species occur, and are explicable in a multiplicity of ways, but they do not follow formally from any theory of biology.

This chapter's general characterization of species is that they are the nexus of the coalescence of genes, haplotypes, parent-child lineages, and so on, at or about the same level; they are *polyphasic*. In abstract terms, species are these coalescences that are distinct from other such coalescences, and each and every one has a general set of properties and modes of speciation, and a unique set of these that only they have (the synapomorphies, or shared characters, which are causally active in maintaining separation).[18] Because each species is a unique historical object, that makes its *species modality*, as I have called it, something that they tend to share only with those taxa to which they are closely related. As a result, the modality of a species is as much an evolved trait as having a vertebral column or a nuclear membrane. The causal process whereby a species has evolved and is maintained depends upon shared ancestral traits such as developmental machinery, genetic sequences, and ecological resources. While these may be very similar between related species, they cannot be expected to be the same in each case—not all Rhagoletids will speciate by host race transfer, for instance—and so each species will have a special set of causes. No theory will capture all and only these causes (not every sexual species will be caused by allopatric isolation, and not every asexual species will be caused by a single niche adaptation) except at a level of generality that is so vague as to preclude explanation in terms of mechanisms. As a take-home exercise to the reader, try to imagine under which conditions organisms like ours would not form species at all.

There have been several proposals for what makes an object "theoretical."[19] The most well known is that of Quine: something exists just to the extent that our best theory of a given domain requires them (that is, bounds them with a quantifier). On this view, species are simply not theoretical, and indeed do not exist, because if I am

---

[17] Quine 1948.
[18] This may seem like essentialism, but the point is merely that if they did not share these properties we would not even notice them as species.
[19] See Brittan 1986, Ladyman 1998, French and Ladyman 2003.

right that no theory of biology *requires* species, then they are never the value of a bound variable in any model of biology.

Another, similar, but not so restrictive view, is "Ramseyfication"; what kinds of objects a theory requires is based on a formalization of the theory—a "Ramsey sentence."[20] Objects exist so long as they are represented either by primitive terms (values of variables, or constants) of the theory or combinations or derivations of those. A primitive here might be empirical, so that species might be primitives of biology but are not themselves explained by it. This is not the case with species, though, because in every such case of which I am aware, one can replace "the species $X y$" used with something like "a local population of $X y$" or "organisms that behave in such a way, which is typical of $X y$" for functional accounts such as ecological ones. In other words, the species $X y$ is replaceable with objects that the theory actually employs. The Ramsey approach, also called the "Canberra Plan,"[21] treats these objects as non-objects. Sometimes this is played out under "Structural Realism" in which a theory structure is true, but the objects it poses which are "unobservable" may or may not be real, so long as the theory is empirically adequate in other ways.[22] This is irrelevant here.

What makes an object theoretical, and are there other roles objects and their representation play in science? For our view to work, it must be that there are objects that are described by the theory, which in the domain of that theory have a certain coherence or unity as objects. Mechanisms like tectonic drift are obviously theoretical in that sense. But mountains are more difficult. Mountains are real things, but the category as a whole lacks theoretical coherence. That is, a mountain has no theoretical place *qua* "mountain," but as a particular mountain, say, Mount St. Helens or the Matterhorn, it calls for explanation.[23] Like sand dunes, they are real things—if you have to map them, travel around them, or climb over them, they are as real as anything can be,[24] but the choice of demarcation between peaks can be conventional or even just something that perception hands to us on a plate. Nothing in theory demands that this particular mountain exists, or even that there are mountains. On a planet with an atmosphere and no tectonics, after a reasonable period, there may be none. Species are like that. They are real facts about the world, which we perceive rather than define. Of course, this makes identifying them relative in a way to the rules and capacities of perception. If we had poor vision, we might not "perceive" mountains until we had telescopic surveyor's sights. Once we have that technology, though (which, note, does not rely upon the theories of geology) we *do*

---

[20] Psillos 2000, 2006. Also called a "Carnap-Ramsey sentence" or a "Ramsey-Lewis sentence," see Koslow 2006.

[21] Jackson 1998, Braddon-Mitchell and Nola 2009.

[22] Psillos 1999, 2000, 2006; see also French 2011 with an application to biology.

[23] Neil Thomason informs me that H. P. Grice in a seminar criticized Quine's view on bounded objects by remarking "Quine thinks we can't count the mountains in the Rockies." However, he never published this so far as I can find, and he may have been referring to vague objects or Sorites rather than a theory–phenomena distinction about objects.

[24] Echoing Hacking's comment about electrons, that if you can spray them, they are real [Hacking 1983, 22]. To avoid unnecessary metaphysical concerns, I will say they are real *enough*—that is, as real as anything else in biology.

see mountains. Similarly, we may need to use all kinds of assay techniques to see species, but when we have them they are seen.[25]

An instructive example is the discovery by Murray Littlejohn and his advisor of the different species that had previously been called *Rana pipiens*, the "leopard frog" of the southern United States.[26] The leopard frog is widespread, and Littlejohn was using a new piece of equipment designed for speech therapy, the sonograph, to graph the mating calls of these frogs. He discovered that there were up to six distinct mating calls.[27] Since mating calls in amphibians are highly species-particular, Littlejohn proposed that *R. pipiens* was a species complex in which morphology and ecology were indistinguishable, but that mating was restricted within the mating call groups, the species. Subsequent work proved this to be the case. The differentiation was always there, but you needed the right assay technique. This is not species being "constructed" or any other bad "postmodern" nonsense. While the *concept* we have of those species is being constructed (and reconstructed as new evidence comes in), the concept refers denotes realities, either of classes of things that are theoretical, such as populations, haplotypes, genes, developmental sequences or cycles, and so on, or of things that are not required by the theory. When we construct a concept, we are *learning* about the things we describe. It's like finding that Everest has a hitherto-hidden peak that is even higher. Our concept of Everest changes, but the thing itself was already as it is.

WHAT ARE SPECIES?

As I argued above: *There is only one species concept.* That is to say, there is only one concept that we are all trying to define in many ways, according to both our preferred theories of how species come into being and maintain themselves over evolutionary time, and what happens to be the general case for the particular group of organisms we have in our minds when we attempt our definitions. The former case is what we might call *theoretical* conceptions of species, where a "conception" is a definition of the word and concept of species. The latter are the *prototypical* conceptions of species, called by taxonomists "good species."[28] If you work in, say, fishes, then your conception of species has to deal with the usual facts about fishes.[29] If you are a fern botanist, then those organisms set up your prototype.[30] And the debate over what species are has been driven by differing prototypes as much as by different theories of speciation.

---

[25] This means that the current scientific fascination with "species delimitation by genomics," including the co-called DNA barcoding approach, are misguided. They mistake the assay used for a definition of "species."

[26] Littlejohn and Oldham 1968.

[27] Although new species have since been identified [Platz 1993].

[28] Amitani 2015.

[29] Rosen 1978.

[30] Wagner 1983.

In Appendix B, I give 28 conceptions[31] of species in the modern (post-Synthesis) literature (and four replacement conceptions). I am going to focus now on the few basic ideas that underlie nearly all of these. The first concept is based on *reproductive isolation*. Since the Synthesis of genetics and Darwinian evolution, the ruling notion of species generation (speciation) was based on the criterion of sexual populations that are isolated from each other, so that they evolve in divergent ways, leading to populations that, when they meet, if they do, in the same range, they no longer tend to interbreed, and their gene pools are now distinct over evolutionary time scales. The conception of species that the Synthesis adopted as a result of this genetic-evolutionary view is the *biological species concept* (BSC). It is called this because it was contrasted to the practices of museum taxonomists, who identified species based on differences in the morphology of captured or collected specimens, which was held to be a sterile methodology where the data was more in the heads of the taxonomists than in the real world. Hence the BSC was biological, while the museum approach was conventional (due to the conveniences of the taxonomists). But the leading idea of the BSC is not that things live, or that they are in messy populations, although that is part of it, but rather that these populations are *reproductively isolated* from each other. So, call this conception the *Reproductive Isolation Species Conception* (RISC), or "isolationist" conception for short. There are several versions of it, but the basic idea—that something inhibits interbreeding when they meet—is common to them all.

Criticisms of the RISC began early. For a start, it was observed that there was a disconnection between the theoretical justification for the RISC and the ways in which taxonomists who adopted it did their taxonomy. To ensure that you *have* a RISC taxon, you really need to do breeding experiments. Many quite diverse morphs in, say, butterflies, that were identified as distinct species in the nineteenth century, turned out to be different genders of the same species. "Aha!" said the isolationists, "This is a failure of morphology." But when similar cases occurred and were found to be different genders *before* the Synthesis, these so-called "morphologists" had no problem seeing them as the same species on that ground. It was understood that form was only a guide to the underlying biological reality (in the nineteenth century, this was often referred to as the *physiological species*), not an end in itself. Worse, isolationists themselves use morphology to identify their species. Breeding experiments, even when technically possible, take enormous time and resources. So, while isolationists are *theoretically* basing their work on reproductive isolation, *practically* they are doing just what their supposedly mistaken predecessors did. This might lead us to think that the older workers were not so silly after all.

The second of our broad conceptions of species is *ecological isolation*, often called the *Ecological Species Conception*, which goes back in one form or another to Linnaeus. However, it gained currency in modern times when Turesson did his studies during the 1920s of plant morphologies in different ecological conditions. Turesson coined the term *ecotype* to describe these differing morphologies. He

---

[31] A *conception* is a variant definition of a *concept*. This in turn is distinct from the various formulations of conceptions of the concept. For example Lherminer and Solignac 2000 give several hundred formulations, but mostly these are of a much fewer number of conceptions.

distinguished between ecotypes and *ecospecies*, which were populations prevented by adaptation to a particular ecological niche from interbreeding. In the 1970s, Leigh Van Valen offered a new version based on the fact that American oaks will freely interbreed, but that the ecological types remain constant.[32] In these cases, the "species" is effectively maintained by the ecological niche. Similar cases are common in plants and single-celled organisms, though less so among multicellular animals. Bacteria and other single-celled organisms which do not often exchange genes may be entirely maintained by this. Lacking sex, they cannot be RISC species, so another term was coined for them: *agamospecies* (meaning, sexless species). However, in animals, asexual reproduction has evolved from sexual species many times (*parthenogens* "virgin origins"), while in plants, it is even more common (*apomicts* "apart from mixing").

The third kind of species conception is known variously as *Morphological*, *Typological*, or *Essentialist*, but these labels are misleading, as I have argued. Sometimes it is called the *Linnaean Conception*, because it is supposed to have been the default view before genetics and evolution were discovered, and hence the view of Linnaean taxonomy. This is a bit unfair. Linnaeus never clearly defined a species concept, and the standard view at the time was that of John Ray, in which a species was twofold: a *form*, which is *reproduced*. This conception was never isolated from the notion of normal reproduction by parents. Moreover, Linnaean and Rayesque species were not defined by essences either. The important thing was that it was the overall organization of the organisms that defined them as a species, so long as it was reproduced. Ray's definition was not based on Aristotle or any logical system, but upon observation.

A fourth general conception is based on the convenience of biological work, including mutual communication between specialists: *species conventionalism*—the view that, as Locke had said, species are made for communication and nothing else. For Darwin, species were real but temporary things, and he believed there was no special rank or level in biology that was unique to species. Contrary to common opinion since the turn of the twentieth century (and earlier, *vide* Agassiz), Darwin was not a conventionalist, but evolutionary thinking made it harder to be exact about species.

This leads us to our final conception; based on *evolutionary history*, it has two main versions: the *phylogenetic* species conceptions based on cladistics, and the so-called *evolutionary* species concepts, which are often a mixture of the RISC, the ecological species conception, and phylogenetic accounts of reconstructed history. The former are often more like the RISC, because they rely on there being separation of lineages over a great deal of time as defined by their sharing or not of evolved traits, and this implies genetic isolation. The latter do not rely on RISC, but only that after the fact the lineages remain distinct for whatever reason (thus including ecotypes and ecospecies).

These conceptions are process-based, and are equally as non-operational as the RISC, but cladistics at least has a large number of mathematical and formal

---

[32] Van Valen 1976. Very similar issues arise with Australian eucalypts [Delaporte et al. 2001].

techniques for drawing up their cladograms.[33] The problem is that, without some way of saying what the level of separation is for species, cladistics can divide lineages up to a very small level (such as haplotype groups), leading to "taxonomic inflation."[34] Phylogenetic species can run to as much as nine or ten times in number compared to the ordinary ("Linnaean") kind. The debate rages through the modern systematics community.

After all that, what is a species? Any *universal* monistic concept of species has to range over the entire evolutionary tree, but the modes of being a species will depend on what ways they have evolved to remain distinct from each other. Hence, none of the particular conceptions are sufficient or necessary to cover being a species in all organisms. However, even these conceptions only tell us what species sometimes are. They do not tell us why these different things should even be *called* "species." For example, RISC proponents will often say that asexual organisms (*agamospecies*) are not really species at all, because they lack the defining property of species which is, of course, reproductive isolation. So we should call them something else— agamospecies, quasispecies, pseudospecies, paraspecies, and so forth. This has the unwanted consequence that the bulk of life does not exist in species, but only those few clades that happened to evolve sex. I think we should say that *all* organisms come in "kinds," some of which are sexual kinds. Others come in genetic bundles or are clustered for ecological reasons, and many are a mixture.

So here is a "definition" of the word "species": *A species is any lineage of organisms that is distinct from other lineages because of differences in some overall shared biological property set.* It has to be a lineage, to distinguish biological species (but not just RISC species) from species of chemical compounds, minerals, and symptomatic diseases. However, while all species are lineages, not all lineages are species, not even the monophyletic ones. It has to be a causal definition, because formal approaches do no explanatory work (in short, the formalist definition merely restates that there are differences). It has to be based on biological properties, because non-biological properties like range or geography are not enough to include or exclude populations and organisms from a species.[35] And it has to be "shared" properties, because differences in unique properties are non-explanatory. All the various conceptions try to give the differences in shared biological properties some detail—and when we look at them that way, it becomes clear why none of them are sufficient or necessary for all species: the mechanisms that keep lineages distinct evolved uniquely in every case, and so generalizations only cover some, not all, of life.

---

[33] Cladistics has fought a number of skirmishes and even all-out wars over the choice of techniques, mostly statistical, to be used in phylogenetic analysis. These often also involve the choice of software packages. The methodology and epistemology of cladistics is not well discussed in the literature except from a partisan perspective, such as Sober 1988, Faith and Trueman 2001, or Felsenstein 2004.

[34] Isaac et al. 2004, Padial and De la Riva 2006, Gippoliti and Groves 2013, Zachos and Lovari 2013, Zachos 2015.

[35] This is not to assert an essentialistic view of these properties; "shared" here means something like "commonly found in these tokogenetic lineages." HPC accounts can still apply.

PATTERN RECOGNITION AND ABDUCTION

We tend to classify epistemic activities into two kinds: *induction* (about which we have many arguments as to its warrantability) and *deduction* (with many arguments about its applicability). In recent years, the notion of *abduction*, or Inference to the Best Explanation (IBE) has been added to the list. But there is something else that we do to learn about what exists in the world. In my book with Malte Ebach, I argue that this is *classification*, but classification is typically regarded as one of the other kinds of inference.[36] Instead, we believe, it is a new kind, similar to abduction, but distinct.

When we classify in a theory-lacking domain we are not yet inductively constructing theory, and we are not able to deduce from theory (since there is not any yet) the classes of objects in the domain we are investigating, nor are we able to abductively explain the classes as such; we merely recognize them. What is happening here is *pattern recognition*.[37] We are classifier systems. It is one of the distinguishing features of neural network (NN) systems, such as those between our ears, that they will classify patterns. They do so in an interesting fashion. Rather than being cued by theory or explanatory goals, NNs are cued by stereotypical "training sets." In effect, in order to see patterns, you need to have prior patterns to train your NN.

Where do these training sets come from? There are several sources. One is evolution itself: we are observer/classifier systems of a certain kind. This gives us a host of cue types to which we respond by training our stereotype classifier system. For example, we respond to movement of large objects, to differences in color and shade, and so on, in our optical system. Quine referred to this as our "quality spaces"— fields of *discriminata*, to which we (in Quine's view, behavioralistically) react.[38] They are adaptations to the exigencies of survival for organisms of the kind that we are. The problem is that so long as our survival and reproductive success is ensured, evolution cannot guarantee us access to the way things "really" are. At best, it gives us a good balance between false positives and false negatives. It is good enough, as it were, for government work.[39] But is it good enough for science and metaphysics?

One of the standard accounts of the success of science is that it increasingly approaches the truth. This is called the *Ultimate Argument for Scientific Realism* by van Fraassen and the *[No] Miracle[s] Argument* by Putnam—that unless science does converge on reality, science would be a miracle.[40] It is quite clear that the received dispositions evolution has bequeathed our cognitive capacities are not enough. While one might reject Plantinga's argument against all naturalism based on this insufficiency of our evolved cognitive powers,[41] there remains a problem. How do we come to identify aspects of the world reliably and properly?[42]

Science proceeds by refining its categories of what exists in the world based on two main sources. These are evidence and explanatory force. In the case of a domain

---

[36] Wilkins and Ebach 2013.
[37] Bishop 1995.
[38] Quine 1953.
[39] Godfrey-Smith 1991.
[40] Putnam 1975, Van Fraassen 1980.
[41] Plantinga 2002.
[42] Griffiths and Wilkins 2014.

of investigation for which there is as yet no explanation, all we have is evidence, but apart from our evolved dispositions to respond to certain stimuli, called our *Umwelten* by Uexkull,[43] how can we identify the salient aspects of evidence? There is an almost infinite amount of possible information we might use, and so we must glean the *right* sources of information.

The other source is economic necessity. Over time, farmers and hunters will tend to respond to the features of the things they are engaged in acquiring and using that are more or less important for success, because those features which are not salient will impose a cost of time and effort that tends to reduce success. This is a process very like natural selection, and has been the basis for what came to be known as *evolutionary epistemology*, in which a parallel process to biological evolution occurs in the domain of knowledge. Cognitive traditions become better at acquiring reliable knowledge because ideas and approaches that do not aid this goal are costly and are abandoned.

However, we have a superfluity of cognitive and conceptual resources. We can retain ideas and practices that are not really natural for social reasons, such as rituals and "explanations" that have no counterpart with the reality being dealt with. The fact that a particular culture is successful at farming by relying upon a ritual calendar (as in pre-colonial Bali) does not warrant belief in Hindu gods. The functional aspects of the rituals act to transmit the information even if nobody in the culture (or in Western agribusiness) fully understands why those rituals make farming successful.[44] When a classifier recognizes patterns in economic circumstances, what counts is not the conceptual superstructure, the theories and ideologies, but the categories of what matters—in this case of water, soil, and landscapes. How might this explain the success of science? Taxonomists are classifiers in a particular economic context: that of professional science. When a taxonomist encounters organisms in the wild, they are in the same situation as a hunter who hunts in that ecology. To succeed at taxonomy, as to succeed at hunting, the agent must know the *right* things about the target objects. A hunter who does not know what different species of bird look like and how they behave and where they live is in exactly the same economic conditions as a taxonomist who also lacks knowledge. Neither will end up with dinner on the plate (*qua* hunter or taxonomist). In the case of the taxonomist, the gap between failure and hunger is somewhat more distal than for the hunter (although hunters typically get most of their food from foraging rather than hunting anyway, courtesy of the non-hunters, mostly women, in their village), but ultimately economic success depends directly upon correct pattern recognition. Mayr was fond of telling the story of how when he visited Papua in the 1930s, he and the local hunters identified the same species of bird, with an exception where western ornithologists also disagreed, and he used this as justification for the reality of those (and all) species. He inferred that science was able to discover kinds of things that were real in the world. However, when E. O. Wilson tried the same experiment about ants, a subject he knows intimately, instead of the locals counting the same species he did (several dozen) he

---

[43] Uexküll 1926.
[44] Lansing 2007.

found they did not discriminate them.[45] Why did Mayr's informants know their birds while Wilson's did not know their ants? The answer, I believe, is that birds were of economic importance to locals and ornithologists, while ants were of economic importance only to Wilson and other myrmecologists.

By "economic" here I do not mean fiscally, but the acquisition of resources, success at which gives the person or persons investigating a living. What distinguishes scientific success is a unique socioeconomic system of professionalism, credit in society, and access to funds and resources such as labs, students, and equipment. The motivations of the individuals concerned are several, often (but not always) based on personal curiosity, but curiosity is not enough if you do not get the resources to do the work. So, we are very good at turning our perceptual pattern recognition systems to scientific work. What evolution provides, science refines. It happens that pattern recognition and the subsequent classificatory activities can deliver reliable knowledge of the world when it matters. But being as it is parasitic upon those evolved capacities, and being as scientists are social organisms, this is not without its failures. Social influences, particularly the inherited traditions of ritual and conception that history bequeaths, can skew and bias our categories about the world. This is where theory and experiment come in.

Science, by way of its historical antecedents, also seeks to explain things in ways that can be tested. Here the ordinary philosophical issues come into play—we inductively generalize based on the patterns we have recognized, and form hypotheses, and from those hypotheses we derive deductive consequences, which we can test in ways that are not circular, which do not rely upon our original observations. And so we can eliminate hypotheses that are not fit to the facts, more or less. This is what Popper and the evolutionary epistemologists built their views upon. What evolutionary epistemology never explained, nor did Popper, was how we came up with those hypotheses in the first place. Pattern recognition does.

For a half century or more we have had the view that observation is theory-laden.[46] As Ebach and I argue, observation need not be laden with theory pertaining to the *domain under investigation* (DUI). What evolution has bequeathed need not be in the slightest theoretical, nor even reliable (as the massive literature on illusions shows). We can naively observe things that we know little about, but we never start by knowing, or at least being disposed to know, nothing. Lipton proposed *Inference to the Best Explanation*[47] (IBE), a principle that one should choose to explain an observation based on the best causal account, the most likely one based on background knowledge. A commenter on an earlier version of this section suggested that pattern recognition is a form of IBE. I think it is not. For a start, pattern recognition–based classification does not require positing an explanation, but it does require one once you *have* classified, which is to say, like observing a species, any classification sets up an *explicandum*. To make an IBE, one needs to already know enough to make some explanations more likely than others; Lipton calls this assessing the

---

[45] Wilson 1992, 39.

[46] Or the countervailing view that theory-ladenness is required because there is no observational language that is theory-free; as positivists proposed. I do not support this view exactly.

[47] Lipton 1991.

"loveliness" of competing hypotheses. But while pattern recognition involves prior knowledge—of the domain and its general properties, mostly what to look for—it does not involve assessing the loveliness of hypotheses. Instead it involves assessing the salience of differing stimuli. In order to make an IBE, you have to recognize things. For example, to make an IBE about what made footprints in the snow, you have to recognize the pattern of footprints. This is something you have learned to do, not least by making footprints. Second, to make an IBE that a deer made these particular tracks, you need to recognize the difference between bipedal and quadrupedal tracks (gotten from years of observing them), and between claws and hoofs (likewise) and so on. With all that categorical apparatus in play, you "leap" to the hypothesis that of the likely animals in the area, it was a deer, not a cat or horse.

Classification differs, at least when the DUI is unexplored. You know about the wider domains in which the DUI is situated, so you are primed to see some sorts of things, but you get an idea of what is in the DUI by looking, a lot. Experience trains you to see patterns, and then and only then can you make IBEs. Hence the argument above. There are those who think taxa are explanations; for example, Fitzhugh thinks species are explanations.[48] An *explanation* of why a species is a species is something independent of *recognizing* the species. Others have argued that phylogenies are explanations or hypotheses, in a Popperian fashion. In the case of phylogenies, the *explanation* is the theory of common descent (or, in some cases, lateral transfer and introgression through hybridization), but the phylogenies themselves are *patterns in data*. If a systematist works out the phylogeny of a group, then there is an IBE of common ancestry, but common ancestry is not the same thing as working out the phylogeny.[49] The relations of different kinds of cognitive activities here are not simple. While it may help to classify them as distinct activities, in practice we shift and change from one to another, or do them simultaneously. Science is not done by recipe. However, it pays to be clear about the differences.

WHAT KIND OF PHENOMENA ARE SPECIES?

Assuming species are phenomena, we must ask what kind of phenomena they are. A particular morph of a species, such as a regional plumage in a bird, or a gender morph, is not a species; all specialists would agree with that. So, even if species are phenomena, not all biological phenomena are species. We need to know when a phenomenon of clustering traits is a species in order to have something to explain. How do we arrive at this?

---

[48] Fitzhugh 2005, 2009. I must disagree here. *That* some phenomenon is a species is a hypothesis, assuming a particular conception of what species are for that group, and it is defeasible, but *calling* something a species is not an explanation of that phenomenon *per se.*

[49] A comment about cladistics that was made very early on is that common ancestry and phylogenies are different. The pattern cladists, so-called, forcefully made that point to the stage that their critics accused them of creationism and "typological thinking." However, once we recast the issue as *explicandum* and *explanandum*, much of the heat goes out of this debate. Even process cladists now concede that common ancestry, while it may be the cause of phylogenetic patterns, is a hypothesis in each individual case that may be confounded by parallelism or lateral inheritance, for instance.

In a way, this is similar to the phylogenetic species problem—given that not all species are monophyletic, and not all terminal clades are species, the specieshood is a given, not an outcome, of phylospecies conceptions. In the case of phenomenal accounts (phenospecies), we need to be able to specify the traits that count and those that do not, like regional variants or dimorphic characters. However, as we have argued and seen, there are no universal criteria for identifying species in order to either determine phylospecies *or* phenospecies.

Therefore, we are owed an account of how we identify species in order to subject them to explanation. I attempted to do this in the previous section, but some words on the general issue are due. A specialist identifies species through a bootstrapping process of knowing what are species, generally, in the organisms that they have identified as forming a natural group. Knowing that the nearest relatives for a specimen of, say, coleoptera, form species by reproductive isolation, or by ecological adaptation, or some mixture of these, permits the specialist to identify that the novel specimen forms a species in much the same way. Sometimes the novel specimen is not like its nearest relatives, as in the whiptail lizards that are parthenogens (*Aspidoscelis uniparens*) among a related group of sexual lizards. This sets up an explanatory contrast that the specialist must account for (in that case, polyploidy), and revises the notion of *species*, but only for *that* natural group. It does not generalize to, for example, all other members of the family Teiidae, of which they are members.[50] Thus the *notion* of what counts as a species is relative to the local group in which the organisms are placed. The inductive generalization about specieshood is limited to that group.

ARE SPECIES FORMS OF LIFE?

The philosophical ideas and terms of Wittgenstein have played an interesting and underappreciated role in the species debate: we saw Beckner appeal to family resemblance predicates, and Pigliucci revive that, as explanations of specieshood. I would like to appeal to another Wittgensteinian notion: forms of life (*Lebensformen*).[51]

In the *Philosophical Investigations*, Wittgenstein was discussing our ability to understand foreign points of view (including other minds) and famously said, "If a lion could talk, we could not understand him." A little later he stated, "What has to be accepted, the given, is—so one could say—*forms of life*."[52] For Wittgenstein, understanding another language-game depends upon shared points of reference in the form of life.[53] Since we do not share a *Lebensform* with a lion, his language-game

---

[50] So how does the Teilldæist know *A. uniparens* is a species? Because of the *other* criteria used to distinguish the *sexual* species: behavior, ecological niche, morphology, genetics, and so forth. Recent claims about "integrative taxonomy" (using an evolutionary species conception) appeal to this also [Padial and De La Riva 2010, Padial et al. 2010].

[51] Floyd 2016 presents the history and prior connections of the term in Wittgenstein's development.

[52] Wittgenstein 1968, 223, 226. See Hunter 1968 for an analysis of the *Lebensform* concept as an "organic" (that is, developmental) notion. While Gier 1980 has argued that Hunter is incorrect in his biological exegesis of Wittgenstein, and notes that there is also a strong social and linguistic aspect to *Lebensform* with respect to humans in Wittgenstein's work, we are free to interpret the notion organically as well in the present context, as Wittgenstein's attitude to biology is not to be trusted.

[53] "... if language is to be a means of communication there must be agreement not only in definitions but also (queer as this may sound) in judgements" Wittgenstein 1968, §242, 288.

would be opaque to us. However, any biologist would know that we *do* share a *Lebensform* with lions, and indeed all mammals, and more distally, all vertebrates, and so forth. And so, if we could speak lion, we would understand what he was saying to that degree of evolutionary relatedness.[54]

The notion of a form of life has recently been applied to species by the neo-Aristotelian ethicist, Roland Sandler,[55] as part of a justification of Aristotelian virtue ethics based upon the natural flourishing of organisms (especially humans, of course). Leaving aside the ethical arguments, this is intriguing. What would it mean to say a species is a form of life? In the ordinary sense, that is a truism, but a *Lebensform* is much more than its appearance. A *Lebensform* is the sum total of the relations of the individual to its community, to its environment, and between its needs and parts, leading to typical development. In the general sense required for biology, the *Lebensform* of a species is the interrelations of members of the population with each other and with the ecological context and history of that population.

Any organism is a developmental system. That is to say, the outcomes of its development are not predetermined merely by its genes (genetic determinism is a kind of preformationism) but also by the environment in which it develops to maturity and further reproduction.[56] Susan Oyama wrote:

> What shapes species-typical characters is not formative powers but a developmental system, much of which is bequeathed to offspring by parent and/or arranged by the developing organism itself. The same is true of atypical ones, which may result from developmental systems that are novel in some respect; an aberrant climate or diet may "play" on the genome in a different way, a mutation may eventuate in altered stimulus preferences or metabolic processes, thus altering the effective environment, etc.[57]

So, the issue with what makes a life-form is not that there is an "essential" set of biological properties in a given organism, but that the organismal outcome will depend on the interplay between endogenous and exogenous properties: genes, somatic inheritance, the abiotic environment, food source availability at different developmental typical stages, parental investment, and so forth. In an approach named "niche construction," several biologists have argued that preceding generations construct aspects of the organismic environment that can be usefully seen as inheritance for the organism, such as trackways, nests, cultural behaviors, and so on for animal species.[58] Thus, in addition to the usual *gene + environment = phenotype* "interactionist consensus,"[59] there arises a complex system of feedback loops and

---

[54] Wittgenstein was notoriously underwhelmed about evolutionary theory: "Darwin's theory has no more to do with philosophy than any other hypothesis in natural science." Wittgenstein 1922, 4.1122.

[55] Sandler 2007. I do not concur with Sandler that natural good is the *Lebensform* of a species, as it implies a natural state model of the evolution of each species. I appreciate Jay Odenbaugh providing me with this reference, and see his response to this approach which gives further references: Odenbaugh 2017.

[56] Griffiths and Gray 1994, Griffiths and Knight 1998, Griffiths and Stotz 2013.

[57] Oyama 2000, 139.

[58] Webb et al. 2002, Day et al. 2003, Odling-Smee et al. 2003. See philosophical commentary on this view: Griffiths 2002, Okasha 2005, Barker 2008.

[59] Kitcher and Sterelny 1988, Sterelny and Griffiths 1999, §5.3.

effects that make an organism the organism it is. Oyama's comment indicates that what "makes" a species includes, among other things, the extra-genetic heredity[60] within a metapopulation. In short, a biological life-form is neither going to involve some uniquely shared set of properties nor will it be a social or cognitive construct[61]; it is part of a set of more or less stable processes that we observe and report.

Wittgenstein's version of *Lebensformen* is that they have self-enclosed criteria for typical behaviors and cognitive styles, which are self-constructed through the use of language games.[62] It has no particular biological implications, and Sandler's use of the term is based upon species having self-justifying natural goods. Here, however, we can appeal to some more generalized features of the concept—that species construct through their variable ways of making a living in their environment, the properties that we see are clustered together.

Parenthetically, *Lebensform* is also a term that has a non-Wittgensteinian history in biology, particularly in botany.[63] Eugenius Warming used it for the forms of vegetation.[64] Raunkiær, a Danish ecologist, proposed what has come to be known as the Life-form Spectrum, which was early adopted as a taxonomy of ecotypes.[65] It is still in use today. Helmreich and Roosth argue that the term has always held the implication of "a space of possibility in which life might take shape," and so it is appropriate to employ it in this context.[66]

Recognition of a *Lebensform* is still something that is done iteratively and recursively. And in line with the notion of a family resemblance predicate, it is not to be expected that there will be sharp or universal tests for delimiting species.

CONSEQUENCES

Taking species as phenomena makes sense of several facts about biological science.

It explains why we recognized species well in advance of there being anything remotely like a theoretical explanation of them, from the sixteenth century onward. Ray's formal definition in 1686 may have been novel, but his view was implicit in the work of natural historians going back to Aristotle and Theophrastus. Genetic and developmental accounts of species did not arise until around 1900.

It explains why when replacement terms are proposed for *species*, they tend to settle on the same sorts of phenomena, and eventually *species* makes a comeback.

---

[60] The usual term in this case is "epigenetic." However, as "epigenetic" is used in multiple ways, I would prefer to restrict it to gene methylation and closely similar molecular inheritance, to avoid confusion.

[61] That is, it will not be *just* a cognitive or social construct. Pretty well everything that is thought, in science or society in general, is *at least* a cognitive and social construct; something else that Wittgenstein noted in his later philosophy.

[62] *Lebensform* is one of a number of similar concepts such as Worldview, Paradigm, Belief System, and so forth, in which the criteria for justification, sense, or inclusion are determined by a comprehensive set of beliefs that are incommensurate with other such sets. Wittgenstein's comment about the lion talking is meant to show the complete non-translatability of lion-talk and people-talk; but, as I have argued, on biological grounds, lions and humans share a considerable homology and analogy of lifestyles and interests such that we can understand a fair bit about their signaling calls.

[63] Helmreich and Roosth 2010.

[64] Warming 1895.

[65] Smith 1913, Raunkiær 1934.

[66] Similar suggestions have also been made by Mulder 2016 and Hähnel 2017.

It also explains why it is that when autochthonous peoples employ organisms economically, say by hunting them or raising them, they recognize the same sorts we do for scientific reasons.[67] Things are phenomenally salient if you have to interact with them.

But most of all it explains something about science. I would like to briefly sketch what I think are the implications of accepting phenomenal objects in the ontology of a science. As I noted, in the traditional view of science, observation is theory-dependent and objects are theoretical. I am proposing that some objects are *not* theory-dependent in the DUI. In doing so I can explain why it is that so much of biology is what Rutherford sneeringly called "stamp collecting." Before you can begin to formulate theories, you have to gather together the objects under explanation and organize that information into a taxonomy, otherwise it is not even clear what the domain of the theory *is*. The traditional view of science of the twentieth century ignored classificatory activities as uninteresting; I am suggesting it is one of the most crucial and essential aspects of a science. This has been hidden to some extent by focusing solely on theory-dependence.

One might object that of *course* these objects are theoretical: to observe them is to identify a difference by measurement, and that implies an assay or methodological protocol. This is usually true, although species and mountains do not need much if any theoretical ancillary assumptions. But the point is that they do not need the *theory under investigation* in order to be phenomenal objects. That is, if they are theory-dependent, they are dependent on theories *outside the domain in question* (the DUI). Moreover, they are often tokens of a class of phenomenal objects that call for explanation in *those* theories as well (consider optical theories, or genomic clusters in genetic theories). This is not extreme empiricism, but it does emphasize the empirical aspects of classification over the theoretical.[68]

Since the dependence here is a general kind (such as for optics), the theory-dependence is benign. With respect to our theory *T*, there is no special dependence on which the observations are being made, so the phenomena are *T*-independent. This does not mean there is such a thing as *completely* naïve observation—nobody ever starts from total naivety or from a *tabula rasa*.[69] Even observers in the mountains of Papua New Guinea are informed by prior ideas and experience. But we can say they observe species, and do not thereby need to define them.

SUMMARY

1. A species is something that forms phenomenal, salient lineages of populations of organisms and genes.
2. A species can have a particular modality based on evolved biological properties.

---

[67] Atran 1985.

[68] A fuller discussion is available in Wilkins and Ebach 2013, chapter 6.

[69] One response by discussants has been that we have an "evolutionary theoretical" heritage; all I can say is that this sense of "theory" beggars the term to the extent that it no longer refers to anything interesting in science. If cognitive dispositions are theories, then we no longer even need use the term. See Wilkins and Griffiths 2013, Griffiths and Wilkins 2014.

3. The species *conception* applied in each case depends on whether that species meets the conditions for that conception.
4. Each species is a phenomenon that calls for a conception and an explanation.

So, we do not need to have a *monistic* or singular definition of *species*, because species are things to be explained, they are *explicanda*, not an a priori category or rank into which every biological organism must be fitted.[70]

FINAL THOUGHTS

We have rejected classical essentialism as an adequate account of species, and along with it the idea they are classical (i.e., Millian) natural kinds. However, species taxa remain kinds, and they remain natural (at least, prospectively, until further investigated). So, we must finish with a discussion of the kind of kinds species taxa are.

There is a distinction made by Zachos between T-species (the species of taxonomic description) and E-species (the things that evolve), and another related distinction between *synchronic* views of species at a moment in time (Mayr's "non-dimensional" conception) and *diachronic* views of species over time. The debate about the species concept is over whether T-species are E-species, and under what conditions. It is my argument that there is no necessary connection between the two for all T-species, but they may be E-species in some individual cases. This lack of identity, however, suggests that *species* is not a theoretical category at all, but instead is an empirical classification that is a prelude to further investigation. As such, it is an *empirical* category, partially constructed but from evidence rather than convention alone. This is not to be a thoroughgoing empiricist, however, as theoretical and prior knowledge constrains the ways in which species are identified and studied. Instead, I presume that evidence matters to scientists even when theory and experience are employed.

One point that is not often discussed in this context is whether biology overall even *has* an ontology. If species are not "units" of evolution or life, what are they? I have argued they are phenomena in need of explanation in each particular case. This implies that there is no *rank* of species. I believe few if any ontological categories touted as biological kinds are in fact real object kinds. Instead, they are mostly either theoretical kinds of interest to those who use those kinds in their practices, as in the ways models are used in science, or they are assay-relative kinds, as in the use of genetic sequences as "barcodes." These have their utility in science, of course, but they stand as ways of organizing the data and applying them, not as ontic categories *per se*. Some, indeed quite a few, may be real (types of molecules, for example), but the more "biological" categories are, the less "real" they tend to be. We arrange the data we acquire (through naïve or sophisticated techniques) in ways that make the patterns in the data tractable and useful. Species are just such patterns. They are patterns in observations that permit us to do further work. They *may* be objects

---

[70] Ingo Brigandt first made the claim that species are phenomena, not theoretical objects [Brigandt 2003], to my knowledge.

of various kinds, or they may turn out not to be, but in the end, the etymology of *species*—form or appearance—is a guide to the nature of the term in biology.

The ontological turn in philosophy of science over the past century has tended to overlook the role of empirical constraints on concept formation in the sciences. In particular, the abandonment of the distinction between observation-language (*O*-languages) and theory-language (*T*-languages) after the collapse of logical positivism led to a dismissal of "pure" observation. I am not here supposing that unconditional observation, let alone observation language, is possible, but to deny that biologists observe in the absence of relevant theory is equally unfeasible. Perhaps useful and knowledgeable terms like *species* are an essential element in science, even if they are neither unitary nor well-defined in theory.

BIBLIOGRAPHY

Amitani, Yuichi. 2015. Prototypical reasoning about species and the species problem. *Biological Theory* 10 (4):289–300.

Ankeny, Rachel A. 2000. Fashioning descriptive models in biology: Of worms and wiring diagrams. *Philosophy of Science* 67 (Supplement. Proceedings of the 1998 Biennial Meetings of the Philosophy of Science Association. Part II: Symposia Papers) S260–S272.

Arnold, Michael L. 1992. Natural hybridization as an evolutionary process. *Annual Review of Ecology and Systematics* 23:237–261.

Atran, Scott. 1985. The early history of the species concept: An anthropological reading. In *Histoire du Concept D'Espece dans les Sciences de la Vie*, edited by Scott Atran, 1–36. Paris: Fondation Singer-Polignac.

Barker, Gillian. 2008. Biological levers and extended adaptationism. *Biology and Philosophy* 23 (1):1–25.

Barrington, David S. et al. 1989. Hybridization, reticulation, and species concepts in the ferns. *American Fern Journal* 79 (2):55–64.

Bergman, Thore J., and Jacinta C. Beehner. 2004. Social system of a hybrid baboon group (*Papio anubis* × *P. hamadryas*). *International Journal of Primatology* 25 (6):1313–1330.

Birkhead, T. R., and S. Van Balen. 2007. Unidirectional hybridization in birds: An historical review of bullfinch (*Pyrrhula pyrrhula*) hybrids. *Archives of Natural History* 34 (1):20–29.

Bishop, Christopher M. 1995. *Neural Networks for Pattern Recognition*. Oxford, New York: Clarendon Press/Oxford University Press.

Braddon-Mitchell, David, and Robert Nola. 2009. *Conceptual Analysis and Philosophical Naturalism*. Cambridge, MA: MIT Press.

Brigandt, Ingo. 2003. Species pluralism does not imply species eliminativism. *Philosophy of Science* 70 (5):1305–1316.

Brittan, Gordon G. 1986. Towards a theory of theoretical objects. *PSA: Proceedings of the Biennial Meeting of the Philosophy of Science Association* 1986:384–393.

Claridge, Michael F. et al. 1997. *Species: The Units of Biodiversity*. London/New York: Chapman & Hall.

Cortes-Ortiz, Liliana et al. 2007. Hybridization in large-bodied New World primates. *Genetics* 176 (4):2421–2425.

Darwin, Charles Robert. 1859. *On the origin of species by means of natural selection, or The preservation of favoured races in the struggle for life*. London: John Murray.

Darwin, Francis, ed. 1888. *The life and letters of Charles Darwin: Including an autobiographical chapter*. 3 vols. London: John Murray.

Day, Rachel L. et al. 2003. Rethinking adaptation: The niche-construction perspective. *Perspectives in Biology and Medicine* 46 (1):80.

Delaporte, Kate et al. 2001. Interspecific hybridization within Eucalyptus (Myrtaceae): Subgenus Symphyomyrtus, sections Bisectae and Adnataria. *International Journal of Plant Sciences* (6):1317–1326.

Detwiler, Kate M. 2003. Hybridization between Red-tailed Monkeys (*Cercopithecus ascanius*) and Blue Monkeys (*C. mitis*) in East African forests. In *The Guenons: Diversity and Adaptation in African Monkeys* 79–97.

Dowling, Thomas E., and Carol L. Secor. 1997. The role of hybridization and introgression in the diversification of animals. *Annual Review of Ecology, Evolution, and Systematics* 28:593–619.

Ereshefsky, Marc. 1992. Eliminative pluralism. *Philosophy of Science* 59:671–690.

Faith, D. P., and J. W. H. Trueman. 2001. Towards an inclusive philosophy for phylogenetic inference. *Systematic Biology* 50 (3):331–350.

Felsenstein, Joseph. 2004. *Inferring Phylogenies*. Sunderland, MA: Sinauer.

Fitzhugh, Kirk. 2005. The inferential basis of species hypotheses: The solution to defining the term 'species'. *Marine Ecology* 26 (3–4):155–165.

—. 2009. Species as explanatory hypotheses: Refinements and implications. *Acta Biotheoretica* 57 (1):201–248.

Floyd, Juliet. 2016. Chains of life: Turing, Lebensform, and the emergence of Wittgenstein's later style. *Nordic Wittgenstein Review* 7–89.

French, Steven. 2011. Shifting to structures in physics and biology: A prophylactic for promiscuous realism. *Studies in History and Philosophy of Science Part C: Studies in History and Philosophy of Biological and Biomedical Sciences* 42 (2):164–173.

French, Steven, and James Ladyman. 2003. Remodelling structural realism: Quantum physics and the metaphysics of structure. *Synthese* 136–141 (1):31–56.

Gannett, Lisa. 2003. Making populations: Bounding genes in space and in time. *Philosophy of Science* 70 (5):989–1001.

Gier, Nicholas F. 1980. Wittgenstein and Forms of Life. *Philosophy of the Social Sciences* 10 (3):241–258.

Gippoliti, Spartaco, and Colin P. Groves. 2013. "Taxonomic inflation" in the historical context of mammalogy and conservation. *Hystrix, the Italian Journal of Mammalogy* 23 (2):8–11.

Godfrey-Smith, Peter. 1991. Signal, decision, action. *Journal of Philosophy* 88:709–722.

—. 2009. *Darwinian populations and natural selection*. Oxford: Oxford University Press.

Gould, Stephen Jay, and Elisabeth A. Lloyd. 1999. Individuality and adaptation across levels of selection: How shall we name and generalize the unit of Darwinism? *Proceedings of the National Academy of Sciences of the United States of America* 96 (21):11904–11909.

Grandy, Richard. 2008. Sortals. In *The Stanford Encyclopedia of Philosophy*, edited by Edward N. Zalta. Available at http://plato.stanford.edu/archives/sum2008/entries/sortals; accessed 22 November 2016.

Grantham, Todd. 1995. Hierarchical approaches to macroevolution: Recent work on species selection and the Effect Hypothesis. *Annual Review of Ecology and Systematics* 26:301–321.

Griffiths, Paul E., and Karola Stotz. 2007. Gene. In *Cambridge Companion to Philosophy of Biology*, edited by Michael Ruse and David Hull, 85–102. Cambridge: Cambridge University Press.

Griffiths, Paul E., and John S. Wilkins. 2014. When do evolutionary explanations of belief debunk belief? In *Darwin in the 21st Century: Nature, Humanity, and God*, edited by Philip Sloan. Notre Dame, IN: Notre Dame University Press.

Griffiths, Paul E. 2002. Beyond the Baldwin Effect: James Mark Baldwin's 'social heredity', epigenetic inheritance and niche-construction. In *Evolution and Learning: The Baldwin Effect Reconsidered, Life and Mind*, edited by Bruce Weber and David Depew, 193–215. Cambridge, MA: MIT Press.

Griffiths, Paul E., and Russell D. Gray. 1994. Developmental systems and evolutionary explanation. *Journal of Philosophy* 91 (6):277–304.

Griffiths, Paul E., and Russell D. Knight. 1998. What is the developmentalist challenge? *Philosophy of Science* 65 (2):253–258.

Griffiths, Paul, and Karola Stotz. 2013. *Genetics and Philosophy: An Introduction.* Cambridge: Cambridge University Press.

Hacking, Ian. 1983. *Representing and Intervening: Introductory Topics in the Philosophy of Natural Science.* Cambridge: Cambridge University Press.

Hähnel, Martin. 2017. Blurring nature at its boundaries. Vague phenomena in current stem cell debate. *Medicine, Health Care and Philosophy* Online First.

Helmreich, Stefan, and Sophia Roosth. 2010. Life Forms: A keyword entry. *Representations* 112 (1):27–53.

Hunter, John Fletcher MacGregor. 1968. "Forms of Life" in Wittgenstein's "Philosophical Investigations." *American Philosophical Quarterly* 5 (4):233–243.

Isaac, N. J. B. et al. 2004. Taxonomic inflation: Its influence on macroecology and conservation. *Trends in Ecology and Evolution* 19 (9):464–469.

Jablonski, David. 2008. Species selection: Theory and data. *Annual Review of Ecology, Evolution, and Systematics* 39 (1):501–524.

Jackson, Frank. 1998. *From metaphysics to ethics: A defence of conceptual analysis.* Oxford/New York: Clarendon Press.

Jolly, Clifford J. et al. 1997. Intergeneric hybrid baboons. *International Journal of Primatology* 18 (4):597–627.

Kitcher, Philip, and Kim Sterelny. 1988. The return of the gene. *Journal of Philosophy* 85 (7):339–361.

Knobloch, Irving W. 1959. A preliminary estimate of the importance of hybridization in speciation. *Bulletin of the Torrey Botanical Club* 86 (5):296–299.

Koslow, Arnold. 2006. The representational inadequacy of Ramsey Sentences. *Theoria* 72 (2):100–125.

Ladyman, James. 1998. What is structural realism? *Studies in History and Philosophy of Science Part A* 29 (3):409–424.

Lansing, J. Stephen. 2007. *Priests and Programmers: Technologies of Power in the Engineered Landscape of Bali.* Princeton, NJ: Princeton University Press.

Lherminer, Philippe, and Michel Solignac. 2000. L'espèce: DÉfinitions d'auters. *Sciences de la Vie* 153–165.

Lipton, Peter. 1991. *Inference to the Best Explanation.* London: Routledge.

Littlejohn, Murray J., and R. S. Oldham. 1968. *Rana pipiens* complex: Mating call structure and taxonomy. *Science* 162 (3857):1003–1005.

Mallet, James. 2007. Hybrid speciation. *Nature* 446 (7133):279–283.

—. 2008. Hybridization, ecological races and the nature of species: Empirical evidence for the ease of speciation. *Philosophical Transactions of the Royal Society B: Biological Sciences* 363 (1506):2971–2986.

Maynard Smith, John. 1958. *The Theory of Evolution.* Harmondsworth, UK: Penguin Books.

Millstein, Roberta L. 2006. Natural selection as a population-level causal process. *British Journal for the Philosophy of Science* 57 (4):627–653.

Muir, Graham et al. 2000. Species status of hybridizing oaks. *Nature* 405 (6790):1016.

Mulder, Jesse M. 2016. A vital challenge to materialism. *Philosophy* 91 (2):153–182.

Odenbaugh, Jay. 2017. Nothing in ethics makes sense except in the light of evolution? Natural goodness, normativity, and naturalism. *Synthese* 194 (4):1031–1055.

Odling-Smee, F. John et al. 2003. *Niche Construction: The Neglected Process in Evolution, Monographs in Population Biology 37.* Princeton, NJ: Princeton University Press.

Okasha, Samir. 2005. On niche construction and extended evolutionary theory. *Biology and Philosophy* 20 (1):1–10.

Oyama, Susan. 2000. *The Ontogeny of Information: Developmental Systems and Evolution.* 2nd ed, Science and Cultural Theory. Durham, NC: Duke University Press.

Padial, José, and Ignacio De la Riva. 2006. Taxonomic inflation and the stability of species lists: The perils of ostrich's behavior. *Systematic Biology* 55 (5):859–867.

Padial, José M., and Ignacio De La Riva. 2010. A response to recent proposals for integrative taxonomy. *Biological Journal of the Linnean Society* 101 (3):747–756.

Padial, José M. et al. 2010. The integrative future of taxonomy. *Frontiers in Zoology* 7 (1):16.

Plantinga, Alvin. 2002. The evolutionary argument against naturalism. In *Naturalism Defeated? Essays on Plantinga's Evolutionary Argument Against Naturalism*, edited by James K Beilby, 1–13. Ithaca, NY: Cornell University Press.

Platz, J. E. 1993. *Rana subaquavocalis*, a remarkable new species of leopard frog (*Rana pipiens*; complex) from southeastern Arizona that calls under water. *Journal of Herpetology* 27 (2):154–162.

Psillos, Stathis. 1999. *Scientific Realism: How Science Tracks Truth.* London/New York: Routledge.

—. 2000. Carnap, the Ramsey-Sentence and realistic empiricism. *Erkenntnis* 52 (2):253–279.

Psillos, Stathos. 2006. Ramsey's Ramsey-sentences. In *Cambridge and Vienna: Frank P Ramsey and the Vienna Circle*, edited by Maria Carla Galavotti, 67–90. New York: Springer International.

Putnam, Hilary. 1975. *Philosophical Papers.* Cambridge: Cambridge University Press.

Quine, Willard Van Orman. 1948. On what there is. *Review of Metaphysics* 2 (5):21–38.

—. 1953. *From a logical point of view: 9 logico-philosophical essays.* Cambridge, MA: Harvard University Press.

Raunkiær, Christen Christiansen. 1934. *The life forms of plants and statistical plant geography. Being the collected papers of C. Raunkiaer.* Translated by Humphrey G. Carter et al. Oxford University Press: Oxford.

Raup, D. M., and J. J. Sepkoski, Jr. 1986. Periodic extinction of families and genera. *Science* 231 (4740):833–836.

Rice, Sean H. 1995. A genetical theory of species selection. *Journal of Theoretical Biology* 177 (3):237–245.

Rieseberg, Loren H., and John H. Willis. 2007. Plant speciation. *Science* 317 (5840):910–914.

Rosen, Donn E. 1978. Vicariant patterns and historical explanation in biogeography. *Systematic Zoology* 27 (2):159–188.

Sandler, Ronald L. 2007. *Character and Environment: A Virtue-Oriented Approach to Environmental Ethics.* New York: Columbia University Press.

Sepkoski, J. J., Jr. 1994. Extinction and the fossil record. *Geotimes* 39 (3):15–17.

Smith, William G. 1913. Raunkiær's "Life-forms" and statistical methods. *Journal of Ecology* 1 (1):16–26.

Sober, Elliott. 1988. *Reconstructing the Past: Parsimony, Evolution, and Inference.* Cambridge, MA: MIT Press.

Sterelny, Kim, and Paul E. Griffiths. 1999. *Sex and Death: An Introduction to Philosophy of Biology.* Chicago/London: University of Chicago Press.

Stidd, Benton M., and David L. Wade. 1995. Is species selection dependent upon emergent characters? *Biology and Philosophy* 10 (1):55–76.

Uexküll, Jakob von. 1926. *Theoretical Biology, International Library of Psychology, Philosophy and Scientific Method.* London: Kegan Paul, Trench, Trubner.

Van Fraassen, Bas C. 1980. *The Scientific Image.* Oxford: Clarendon Press.

Van Valen, Leigh M. 1976. Ecological species, multispecies, and oaks. *Taxon* 25 (2/3):233–239.

Wagner, Warren H. 1983. Reticulistics: The recognition of hybrids and their role in cladistics and classification. In *Advances in Cladistics*, edited by Norman I. Platnick and Vicki A. Funk, 63–79. New York: Columbia Univ. Press.

Warming, Eugenius. 1895. *Plantesamfund - Grundtræk af den økologiske Plantegeografi.* Kjøbenhavn: P.G. Philipsens Forlag.

Webb, Campbell O. et al. 2002. Phylogenies and community ecology. *Annual Review of Ecology and Systematics* 33:475–505.

Wilkins, John S. 2018. The reality of species: Real phenomena not theoretical objects." In *Routledge Handbook of Evolution and Philosophy*, edited by Richard Joyce, 167–168. Abingdon, UK: Routledge.

Wilkins, John S., and Malte C. Ebach. 2013. *The Nature of Classification: Kinds and Relationships in the Natural Sciences*, edited by Steven French, New Directions in the Philosophy of Science. London: Palgrave Macmillan.

Wilkins, John S., and Paul E. Griffiths. 2013. Evolutionary debunking arguments in three domains: Fact, value, and religion. In *A New Science of Religion*, edited by James Maclaurin and Greg Dawes, 133–146. Chicago: University of Chicago Press.

Wilson, Edward O. 1992. *The Diversity of Life*. London: Penguin.

Winsor, Mary Pickard. 2000. Species, demes, and the Omega Taxonomy: Gilmour and The New Systematics. *Biology and Philosophy* 15 (3):349–388.

Wittgenstein, Ludwig. 1922. *Tractatus Logico-Philosophicus*. Translated by Charles Kay Ogden. London: Kegan Paul, Trench, Trubner & Co.

—. 1968. *Philosophical Investigations*. Translated by G.E.M. Anscombe. 3rd ed. Oxford: Basil Blackwell.

Zachos, Frank E. 2015. Taxonomic inflation, the Phylogenetic Species Concept and lineages in the Tree of Life—A cautionary comment on species splitting. *Journal of Zoological Systematics and Evolutionary Research* 53 (2):180–184.

Zachos, Frank E., and Sandro Lovari. 2013. Taxonomic inflation and the poverty of the Phylogenetic Species Concept—A reply to Gippoliti and Groves. *Hystrix, the Italian Journal of Mammalogy* 24 (2):142–144.

Appendix A: Post-Linnaean Ranks

According to the *Oxford English Dictionary*, *phylum* is a term first coined by Cuvier, in *Le Regne Animal* Cuvier, (1817), to cover his four *embranchements*, although it does not occur in that work.[1] The term was either later adopted and made popular by Ernst Haeckel (1866, XIX) or as Valentine (2004, 8), probably correctly, credits him, he coined the term. I cannot locate any prior synonymous usage in English or French, but the term was used in German in 1851 by G. W. Körber as a synonym for Type (*Typus*) (Körber 1851, 61), so it was probably around before then; perhaps Haeckel, an inveterate coiner of terms, independently came up with it.

Family is most probably derived from Adanson's term (Judd et al. 1999, 40). The Strickland Code of 1842 (Strickland, 1843, 119) mentions "families," noting they ought to be ended in *-idea*, and this implies families were in common use by that time. It also allows subfamilies. Mayr and co-authors (1953, 272) give the introduction of "family" to Latrielle in 1796, but do not give any information regarding *phylum*.

In botany, *phylum* was not used until recently, and instead the rank was *division*, introduced in Alphonse de Candolle's 1867 *Rules* submitted to the Paris meeting that year of the International Botanical Congress (Candolle 1867).

The present International Code of Zoological Nomenclature does not regulate higher taxon ranks above *superfamily*, and so *phylum* is in effect an informal rank (Winston 1999, 32). Recent attempts to revise the rank of *kingdom* and add *empire*, or *domain* (Baldauf et al. 2000, Margulis and Schwartz 1998, Ruggiero et al. 2015, Syvanen and Kado 1998, Williams and Embley 1996, Woese 1998) are thus legitimated by tradition even if not yet widely accepted.

See Nicolson 1991, Smith 1957, and Weatherby 1949 for the history of botanical nomenclature. See Linsley and Usinger 1959 and Melville 1995 for the zoological history.

Recently, the needs of databases, ironically one of Linnaeus' own concerns (Müller-Wille and Charmantier 2012), have led to a regularizing and listing of all used ranks. The Integrated Taxonomic Information System (ITIS) initiative of the United States government lists the following:[2]

---

[1] Cuvier called them "great divisions" (*grande divisions*) in that work. He also referred to them as plans, models and forms. He introduced the term *embranchement* (branch) in his 1812 memoir [Cuvier 1812]. He had, however, previously used the term in a common sense, without any rank implied, when he divided insects into two *embranchements* in his *Lectures* in 1805 [Cuvier et al. 1805]. Thanks to Polly Winsor for the first citation and David Williams for the second.

[2] Anonymous July 20, 2015.

Plantae/Chromista	Animalia	Fungi	Archaea Bacteria/ Protozoa	ITIS rank_id
Kingdom	Kingdom	Kingdom	Kingdom	10
Subkingdom	Subkingdom	Subkingdom	Subkingdom	20
Infrakingdom	Infrakingdom		Infrakingdom (*Protozoa only*)	25
Superdivision	Superphylum			27
Division	Phylum	Division	Phylum	30
Subdivision	Subphylum	Subdivision	Subphylum	40
Infradivision	Infraphylum		Infraphylym (*Protozoa only*)	45
Parvdivision (*Chromista only*)			Parvphylum (*Protozoa only*)	47
Superclass	Superclass		Superclass	50
Class	Class	Class	Class	60
Subclass	Subclass	Subclass	Subclass	70
Infraclass	Infraclass		Infraclass	80
Superorder	Superorder	Superorder	Superorder	90
Order	Order	Order	Order	100
Suborder	Suborder	Suborder	Suborder	110
	Infraorder		Infraorder	120
	Section			124
	Subsection			126
	Superfamily		Superfamily	130
Family	Family	Family	Family	140
Subfamily	Subfamily	Subfamily	Subfamily	150
Tribe	Tribe	Tribe	Tribe	160
Subtribe	Subtribe	Subtribe	Subtribe	170
Genus	Genus	Genus	Genus	180
Subgenus	Subgenus	Subgenus	Subgenus	190
Section		Section		200
Subsection		Subsection		210
Species	Species	Species	Species	220
Subspecies	Subspecies	Subspecies	Subspecies	230
Variety	Variety	Variety	Variety (*Protozoa only*)	240
	Form			245
Subvariety	Race	Subvariety		250
	Stirp			255
Form	Morph	Form		260
	Aberration			265
Subform		Subform		270
	Unspecified			300

BIBLIOGRAPHY

Anonymous. July 20, 2015. ITIS, the Integrated Taxonomic Information System. Available at https://www.itis.gov (accessed 27/1/17).

Baldauf, Sandra L. et al. 2000. A kingdom-level phylogeny of eukaryotes based on combined protein data. *Science* 290 (5493):972–977.

Candolle, Alphonse de. 1867. *Lois de la nomenclature botanique adoptées par le Congrès international de botanique tenu à Paris en août 1867.* Paris: Genève et Bale.

Cuvier, Georges. 1812. Sur un nouveau rapprochement à établir entre les classes qui composent le règne animal. *Annales du Muséum d'Histoire naturelle* 19:73–84.

Cuvier, Georges et al. 1805. *Leçons d'anatomie comparée.* Paris: Crochard Fantin Baudouin.

Cuvier, Georges, and Pierre A. Latreille. 1817. Le règne animal distribué d'après son organisation, pour servir de base à l'histoire naturelle des animaux et d'introduction à l'anatomie comparée. 4 vols. Paris: Deterville.

Haeckel, Ernst Heinrich Philipp August. 1866. *Generelle Morphologie der Organismen. Allgemeine Grundzüge der organischen Formen-Wissenschaft, mechanisch begründet durch die von C. Darwin reformirte Descendenz-Theorie, etc.* 2 vols. Berlin: Georg Reimer.

Judd, Walter S. et al. 1999. *Plant Systematics: A Phylogenetic Approach.* Sunderland, MA: Sinauer Associates.

Körber, G. W. 1851. Ideen zur Geschichte der organischen Schöpfung. In *Öffentlich Prüfung aller Klassen des Elisabet-Gymnasiums*, edited by Karl Rudolph Fickert. Breslau: Grass, Bath und Comp.

Linsley, E. Gorton, and Robert L. Usinger. 1959. Linnaeus and the development of the International Code of Zoological Nomenclature. *Systematic Zoology* 8 (1):39–47.

Margulis, Lynn, and Karlene V. Schwartz. 1998. *Five Kingdoms: An Illustrated Guide to the Phyla of Life on Earth.* 3rd ed. San Francisco, CA: W.H. Freeman.

Mayr, Ernst et al. 1953. *Methods and Principles of Systematic Zoology.* New York: McGraw-Hill.

Melville, Richard V. 1995. *Towards Stability in the Names of Animals: A History of the International Commission on Zoological Nomenclature 1895–1995.* London: ICZN.

Müller-Wille, Staffan, and Isabelle Charmantier. 2012. Natural history and information overload: The case of Linnaeus. *Studies in History and Philosophy of Science Part C: Studies in History and Philosophy of Biological and Biomedical Sciences* 43 (1):4–15.

Nicolson, Dan H. 1991. A history of botanical nomenclature. *Annals of the Missouri Botanical Garden* 78 (1):33–56.

Ruggiero, Michael A. et al. 2015. A higher level classification of all living organisms. *PLoS ONE* 10 (4):e0119248.

Smith, A. C. 1957. Fifty years of botanical nomenclature. *Brittonia* 9 (1):2–8.

Strickland, Hugh E. et al. 1843. "Report of a committee appointed "to consider of the rules by which the nomenclature of zoology may be established on a uniform and permanent basis"." In *Report of the British Association for the Advancement of Science for 1842*, 105–121. London: John Murray.

Syvanen, Michael, and Clarence I. Kado, eds. 1998. *Horizontal Gene Transfer.* Boston, MA: Kluwer Academic Publishers.

Valentine, James W. 2004. *On the Origin of Phyla.* Chicago/London: University of Chicago Press.

Weatherby, Charles Alfred. 1949. Botanical nomenclature since 1867. *American Journal of Botany* 36 (1):5–7.

Williams, David M, and T Martin Embley. 1996. Microbial diversity: Domains and kingdoms. *Annual Review of Ecology and Systematics* 27:569–595.

Winston, Judith E. 1999. *Describing Species: Practical Taxonomic Procedures for Biologists.* New York: Columbia University Press.

Woese, Carl Richard. 1998. A manifesto for microbial genomics. *Current Biology* 8 (22): R780–R783.

Appendix B: A Summary List of Species Definitions

Groups and categories are distinctly different and there is no real connection between them. The tradition in codes of nomenclature artificially forces the use of taxonomic categories in a hierarchy for groups. A biological classification is a contrived system of categories used for the storage and retrieval of information about biological diversity, taxa, or groups. The concept 'category' is a class and has no separate existence from its use in organizing objects or thoughts; categories have no reality. Unlike groups, categories have no attributes; things or objects are not members of categories, but are parts of groups; and organisms are not members of any taxonomic category.

R. L. Mayden[1]

Warren Herb Wagner's Syncretic and Eclectic Species Definition

A convenient taxonomic category that defines a unit of organismic diversity with one or more ancestors in [a] given time frame and composed of individual organisms that resemble one another in all or most of their structural and functional characters, that reproduce true by any means, sexual or asexual, and constitute a distinct phylogenetic line that differs consistently and persistently from populations of other species in character state combinations including geographical, ecological, physiological, morphological, anatomical, cytological, chemical, and genetic, the character states of number and kind ordinarily used for species discrimination in the same and related genera and if partially or wholly sympatric and co-existent with related species in the same habitats, unable to cross or, if able to cross, able to maintain the species distinctness.

Circular sheet, dated 19 November 1998[2]

There are numerous species "concepts" at the research and practical levels in the scientific literature. Mayden's (1997) list of 22 distinct species concepts along with synonyms is a useful starting point for a review. I have added authors where I can locate them in addition to Mayden's references, and I have tried to give the concepts names, such as *biospecies* for "biological species," and so on, following George 1956, except where nothing natural suggests itself. I have used Mayden's abbreviations, except as noted, and added new ones for novel conceptions. A closely similar annotated list is found in Zachos 2016. In some cases I have adopted his terminology.

There have also been several additional concepts since Mayden's review, so I have added the views of Pleijel 1999 and Wu 2001a, b, and several newer revisions

---

[1] Mayden 1997, 386.
[2] *Pers. comm.*

presented in Wheeler and Meier 2000 as well as Hausdorf 2011. I also add some "partial" species concepts—the *compilospecies* concept and the *nothospecies* concept.

Conceptions with asterisks are what I consider to be "foundational" conceptions, from which the others are constructed in whole or part.

REPRODUCTIVE ISOLATION CONCEPTIONS (RISC)

Basing species on the ability of organisms to reproduce successfully is as old as natural history; as we have seen, Darwin considered (and rejected) it, as did Linnaeus and many others. It is not even the sole biological conception. As Van Valen noted, "The term 'biological species concept' is a propagandist ploy."[3] Many have noted that conceptions based on living things are always biological, and that this is better called something like a reproductive species concept; hence my use of RISC.

PHYLOSPECIES (PSC)

In addition to Hennig's conception (1950, 1966), I distinguish between two *phylospecies* concepts that go by various names, mostly the names of the authors presenting at the time, as in (Wheeler and Meier 2000). To remedy this terminological inflation, I have christened them the *autapomorphic species conception* and the *phylogenetic taxon species concept*.[4] Asterisks identify the "primary" conceptions (Mayden 1997), from which the others are formed. Mayden has criticized several of these as being insufficient for that purpose, but I have tagged those which seem to *function* as primary conceptions in scientific practice, if not in philosophical terms.

Definition: *The smallest unit appropriate for phylogenetic analysis, the smallest biological entities that are diagnosable and monophyletic, unit product of natural selection and descent. A geographically constrained group with one or more unique apomorphies (autapomorphies).*

There are two versions of this and they are not identical. One derives from Rosen and is what I call the *autapomorphic species* conception [conception 2]. It is primarily a concept of diagnosis and tends to be favored by the tradition known as pattern cladism. The other is what I call the *phylogenetic taxon species* conception [conception 23], and tends to be favored by process cladists.

Principal Authors: Cracraft 1983, Eldredge and Cracraft 1980, Nelson and Platnick 1981, Rosen 1979.

Synonyms: Autapomorphic phylospecies, monophyletic phylospecies, minimal monophyletic units, monophyletic species, lineages.

Related Concepts: Similar to: internodal species cladospecies, composite species, least inclusive taxonomic units.

---

[3] Van Valen 1988, 54.
[4] In the first edition, I called this the "synapomorphic" species conception.

INDIVIDUAL CONCEPTIONS

1. AGAMOSPECIES [ASC]

Definition: Asexual lineages, uniparental organisms (parthenogens and apomicts), that cluster together in terms of their genome. May be secondarily uniparental from biparental ancestors.

Quasispecies are asexual viruses or organisms that cluster about a "wild-type" due to selection.

Principal Authors: Cain 1954; Eigen 1993 for quasispecies.

Synonyms: Microspecies, paraspecies, pseudospecies, semispecies, quasispecies, genomospecies (Euzéby 2006), for prokaryotes, bacterial species (Cohan 2002).

2. AUTAPOMORPHIC SPECIES [APSC]

A phylospecies conception.

Definition: A geographically constrained group of individuals with some unique apomorphous characters, the unit of evolutionary significance (Rosen 1979); simply the smallest detected samples of self-perpetuating organisms that have unique sets of characters (Nelson and Platnick 1981); the smallest aggregation of (sexual) populations or (asexual) lineages diagnosable by a unique combination of character traits (Wheeler and Platnick 2000).

Principal Authors: Eldredge and Cracraft 1980, Nelson and Platnick 1981, Rosen 1979, Wheeler and Platnick 2000.

Synonyms: Diagnosable version, diagnosable and monophyly version (Zachos 2016).

3. BIOSPECIES [BSC]*

Definition: Inclusive Mendelian population of sexually reproducing organisms (Dobzhansky 1935, 1937, 1970) interbreeding natural population isolated from other such groups (Mayr 1940, 1942, 1963, 1991). Depends upon endogenous reproductive isolating mechanisms (RIMs).

Principal Authors: Defined by John Ray; Buffon, Dobzhansky 1935, Mayr 1942.

Synonyms: Syngen, speciationist species concept.

4. CLADOSPECIES [CISC]

A phylospecies conception.

Definition: Set of organisms between speciation events or between speciation event and extinction (Ridley 1989), a segment of a phylogenetic lineage between nodes. Upon speciation the ancestral species is extinguished and two new species are named.

Principal Authors: Hennig 1950, 1966, Kornet and McAllister 1993.

Synonyms: Internodal species concept, Hennigian species concept, Hennigian convention.

5. Cohesion Species [CSC]

Definition: Evolutionary lineages bounded by cohesion mechanisms that cause reproductive communities of demographically and/or genetically exchangeable individuals. The most inclusive population of individuals having the potential for phenotypic cohesion through intrinsic cohesion mechanisms.
Principal Author: Templeton 1989.

6. Compilospecies [CoSC]

Definition: A species pair where one species "plunders" (Latin: *compīlo*) the genetic resources of another via introgressive interbreeding.
Principal Authors: Aguilar et al. 1999, Harlan and De Wet 1963.

7. Composite Species [CpSC]

Definition: All organisms belonging to an internodon and its descendants until any subsequent internodon. An internodon is defined as a set of organisms whose parent–child relations are not split (have the INT relation).
Principal Authors: Kornet and McAllister 1993.
Synonyms: Phylospecies (in part; Hennigian species), internodal species (in part), cladospecies (in part).

8. Differential Fitness Species [DFSC]

Definition: Groups of individuals that are reciprocally characterized by features that would have negative fitness effects in other groups and that cannot be regularly exchanged between groups upon contact.
Principal Author: Hausdorf 2011.
Related Concepts: Genealogical concordance species, genic species, genotypic cluster species, reproductive competition species.

9. Ecospecies* [EcSC]

Definition: A lineage (or closely related set of lineages) which occupies an adaptive zone minimally different from that of any other lineage in its range and which evolves separately from all lineages outside its range.
Principal Authors: Simpson 1961, Turesson 1922, Van Valen 1976.
Synonyms: Ecotypes, ecological mosaics.

10. Evolutionary Species [ESC]*

Definition: A lineage (an ancestral–descendent sequence of populations) evolving separately from others and with its own unitary evolutionary role and tendencies.

Principal Authors: Simpson 1961, Wiley 1978, 1981.
Synonyms: Unit of evolution, evolutionary group.
Related Concepts: Evolutionary significant unit.

11. EVOLUTIONARY SIGNIFICANT UNIT [ESU]

Definition: A population (or group of populations) that (1) is substantially reproductively isolated from other conspecific population units, and (2) represents an important component in the evolutionary legacy of the species.
Principal Author: Waples 1991.
Related Concepts: Biospecies (in part) and evolutionary species (in part).

12. GENEALOGICAL CONCORDANCE SPECIES [GCC]

Definition: Population subdivisions concordantly identified by multiple independent genetic traits constitute the population units worthy of recognition as phylogenetic taxa.
Principal Authors: Avise and Ball Jr 1990.
Related Concepts: Biospecies (in part), cladospecies (in part), phylospecies (in part).

13. GENERAL LINEAGE CONCEPT [GLC]

Definition: A series of entities forming a single line of ancestry and descent.
Principal Author: de Queiroz 1998, 1999.
Synonyms: Unified species conception; universal species conception.

14. GENIC SPECIES [GeSC]

Definition: A species formed by the fixation of all isolating genetic traits ("speciation genes") in the common genome of the entire population.
Related Concepts: Genealogical concordance species, genetic species (in part), biospecies (in part), autapomorphic species (in part).
Principal Author: Wu 2001a, b.

15. GENETIC SPECIES [GSC]*

Definition: A group of organisms that may inherit characters from each other, a common gene pool, a reproductive community that forms a genetic unit.
Principal Authors: Dobzhansky 1950, Mayr 1969, Simpson 1943.
Synonyms: Gentes (singular: gens), Mendelian population.
Related Concepts: Biospecies, phenospecies, morphospecies.

16. Genotypic Cluster [GCD]

Definition: Clusters of monotypic or polytypic biological entities, identified using morphology or genetics, forming groups that have few or no intermediates when in contact.
Principal Author: Mallet 1995.
Synonyms: Polythetic species.

17. Hennigian Species [HSC]

A phylospecies conception.
Definition: A tokogenetic community that arises when a stem species is dissolved into two new species and ends when it goes extinct or speciates.
Principal Authors: Hennig 1950, 1966, Meier and Willmann 2000.
Synonyms: Biospecies (in part), cladospecies (in part), phylospecies (in part), internodal species.
Related Concepts: Agamospecies, biospecies, genetic species, Hennigian species, morphospecies, non-dimensional species, phenospecies, autapomorphic phylospecies, successional species, taxonomic species.

18. Internodal Species [ISC]

A phylospecies conception.
Definition: Organisms are conspecific in virtue of their common membership of a part of a genealogical network between two permanent splitting events or a splitting event and extinction.
Principal Author: Kornet 1993.
Synonyms: Cladospecies and Hennigian species (in part), phylospecies.

19. Morphospecies [MSC]*

Definition: Species are the smallest groups that are consistently and persistently distinct, and distinguishable by ordinary means. Contrary to the received view, this was never anything more than a diagnostic account of species.
Principal Authors: Aristotle and Linnaeus, and too many others to name, but including Owen, Agassiz, and recently, Cronquist 1978.
Synonyms: Classical species, Linnaean species, Linneons.

20. Non-Dimensional Species [NDSC]

Definition: Species delimitation in a non-dimensional system (a system without the dimensions of space and time).
Principal Author: Mayr 1963.
Synonyms: Folk taxonomic kinds (*speciemes*) (Atran 1990).
Related Concepts: Biospecies, genetic species, morphospecies, paleospecies, successional species, taxonomic species.

21. NOTHOSPECIES [NSC]

Definition: Species formed from the hybridization of two distinct parental species, often by polyploidy.
Principal Author: Wagner 1983.
Synonyms: Hybrid species, reticulate species.

22. PHENOSPECIES [PHSC]

Definition: A cluster of characters that statistically covary; a family resemblance concept in which possession of most characters is required for inclusion in a species, but not all. A class of organisms that share most of a set of characters.
Principal Authors: Beckner 1959, Sokal and Sneath 1963.
Synonyms: Phena (singular: phenon) (Smith 1994), operational taxonomic unit.

23. PHYLOGENETIC TAXON SPECIES [PTSC][5]

A phylospecies conception.
Definition: A species is the smallest diagnosable cluster of individual organisms within which there is a parental pattern of ancestry and descent (Cracraft 1983, Eldredge and Cracraft 1980); the least inclusive taxon recognized in a classification, into which organisms are grouped because of evidence of monophyly (usually, but not restricted to, the presence of synapomorphies), that is ranked as a species because it is the smallest important lineage deemed worthy of formal recognition, where "important" refers to the action of those processes that are dominant in producing and maintaining lineages in a particular case (Mishler and Brandon 1987, Nixon and Wheeler 1990).
Principal Authors: Cracraft 1983, Eldredge and Cracraft 1980, Mishler and Brandon 1987, Nixon and Wheeler 1990.
Synonyms: Monophyletic phylospecies, minimal monophyletic units, monophyletic species.

24. RECOGNITION SPECIES [RSC]

Definition: A species is that most inclusive population of individual, biparental organisms which share a common fertilization system.
Principal Author: Paterson 1985.
Synonyms: Specific mate recognition system (SMRS).

---

[5] Here I part ways with Mayden's list, dropping his subdivisions of the PSC.

25. REPRODUCTIVE COMPETITION SPECIES [RCC]

Definition: The most extensive units in the natural economy such that repro-
ductive competition occurs among their parts.
Principal Author: Ghiselin 1974, 1977.
Synonyms: Hypermodern species concept, Biospecies (in part).

26. SYNAPOMORPHIC SPECIES [SYSC]

A phylospecies conception.
Definition: The least inclusive taxon recognized in a classification, into which
organisms are grouped because of evidence of monophyly (usually, but not
restricted to, the presence of synapomorphies), that is ranked as a species
because it is the smallest "important" lineage deemed worthy of formal
recognition, where "important" refers to the action of those processes that
are dominant in producing and maintaining lineages in a particular case
(Mishler et al.). A species is a lineage separated from other lineages by
causal differences in synapomorphies (Wilkins).
Principal Authors: Mishler and Brandon 1987, Mishler and Donoghue 1982,
Wilkins 2003.

27. SUCCESSIONAL SPECIES [SSC]

Definition: Arbitrary anagenetic stages in morphological forms, mainly in the
paleontological record.
Principal Authors: George 1956, Simpson 1961.
Synonyms: Paleospecies, evolutionary species (in part), chronospecies.

28. TAXONOMIC SPECIES [TSC]*

Definition: Specimens considered by a taxonomist to be members of a kind
on the evidence or on the assumption they are as alike as their offspring of
hereditary relatives within a few generations. *Also:* Whatever a competent
taxonomist chooses to call a species.
Principal Authors: Blackwelder 1967, but see also Regan 1926, Strickland
et al. 1843.
Synonyms: Cynical species concept (Kitcher 1984).
Related Concepts: Agamospecies, genealogical concordance species, morphospecies, phenospecies, phylospecies.

REPLACEMENT CONCEPTIONS

Several so-called species conceptions are in fact replacement concepts that displace
species. Some (e.g., LITUs) are intended to replace *species* entirely; others are parallel concepts that are neutral regarding *species*.

1. OPERATIONAL TAXONOMIC UNIT [OTU]

Definition: Taxonomic units of different categorical ranks to be grouped into more inclusive aggregates during classification (Sneath and Sokal). In microbiology, a phylotype is an environmental DNA sequence or group of sequences sharing more than an arbitrarily chosen level of similarity of a particular gene marker (Moreira and López-García 2011).

Principal Authors: Sneath and Sokal 1973, Sokal and Sneath 1963.

Related Concepts: Microbial species, bacterial species, phenospecies, taxonomic species, phylotype.[6]

2. LEAST INCLUSIVE TAXONOMIC UNIT [LITU]

A phylospecies conception.

Definition: A taxonomic group that is diagnosable in terms of its autapomorphies, but has no fixed rank or binomial.

Principal Authors: Pleijel 1999, Pleijel and Rouse 2000, 2003.

3. METAPOPULATION

Definition: A population of populations that go extinct locally and recolonize (Levins 1970). A set of local populations of a single species that are linked by dispersal (Hanski and Gilpin 1991).

Principal Authors: Hanski and Gilpin 1991, Levins 1970.

Related Concepts: Ecospecies, Mendelian populations.

4. SMALLEST NAMED AND REGISTERED CLADE [SNaRC]

A phylospecies concept.

Definition: The smallest named and registered clade [in the Phylocode database].

Principal Authors: Mishler and Wilkins (2018).

Related Concept: Least inclusive taxonomic units.

BIBLIOGRAPHY

Aguilar, Javier Fuertes et al. 1999. Molecular evidence for the compilospecies model of reticulate evolution in *Armeria* (Plumbaginaceae). *Systematic Biology* 48 (4):735–754.

Atran, Scott. 1990. *The Cognitive Foundations of Natural History*. New York: Cambridge University Press.

Avise, John C., and R. Martin Ball Jr. 1990. Principles of genealogical concordance in species concepts and biological taxonomy. In *Oxford Surveys in Evolutionary Biology*, edited by Douglas J. Futuyma and Janis Antonovics, 45–67. Oxford: Oxford University Press.

---

[6] C. W. Cotterman in unpublished notes dated 1960, Denniston 1974.

Beckner, Morton. 1959. *The Biological Way of Thought*. New York: Columbia University Press.

Blackwelder, Richard Eliot. 1967. *Taxonomy: A Text and Reference Book*. New York: John Wiley & Sons.

Cain, Arthur J. 1954. *Animal Species and Their Evolution*. London: Hutchinson University Library.

Cohan, Frederick M. 2002. What are bacterial species? *Annual Review of Microbiology* 56:457–487.

Cracraft, Joel. 1983. Species concepts and speciation analysis. In *Current Ornithology*, edited by R. F. Johnston, 159–187. New York: Plenum Press.

Cronquist, A. 1978. Once again, what is a species? In *BioSystematics in Agriculture*, edited by L. V. Knutson, 3–20. Montclair, NJ: Alleheld Osmun.

de Queiroz, Kevin. 1998. The general lineage concept of species, species criteria, and the process of speciation. In *Endless Forms: Species and Speciation*, edited by Daniel J. Howard and Stewart H. Berlocher, 57–75. New York: Oxford University Press.

—. 1999. The general lineage concept of species and the defining properties of the species category. In *Species, New Interdisciplinary Essays*, edited by R. A. Wilson, 49–88. Cambridge, MA: Bradford/MIT Press.

Denniston, Carter. 1974. An extension of the probability approach to genetic relationships: One locus. *Theoretical Population Biology* 6 (1):58–75.

Dobzhansky, Theodosius. 1935. A critique of the species concept in biology. *Philosophy of Science* 2:344–355.

—. 1937. *Genetics and the Origin of Species*. New York: Columbia University Press.

—. 1950. Mendelian populations and their evolution. *American Naturalist* 74:312–321.

—. 1970. *Genetics of the Evolutionary Process*. New York: Columbia University Press.

Eigen, Manfred. 1993. Viral quasispecies. *Scientific American* July 1993 (32–39).

Eldredge, Niles, and Joel Cracraft. 1980. *Phylogenetic Patterns and the Evolutionary Process: Method and Theory in Comparative Biology*. New York: Columbia University Press.

Euzéby, Jean P. 2006. List of prokaryotic names with standing in nomenclature. Available at http://www.bacterio.cict.fr (accessed 17/2/06).

George, T. Neville. 1956. Biospecies, chronospecies and morphospecies. In *The species Concept in Paleontology*, edited by Peter C. Sylvester-Bradley, 123–137. London: Systematics Association.

Ghiselin, Michael T. 1974. *The Economy of Nature and the Evolution of Sex*. Berkeley: University of California Press.

—. 1977. On paradigms and the hypermodern species concept. *Systematic Biology* 26 (4):437–438.

Hanski, Ilkka, and Michael Gilpin. 1991. Metapopulation dynamics: Brief history and conceptual domain. In *Metapopulation Dynamics: Empirical and Theoretical Investigations*, edited by Michael Gilpin and Ilkka Hanski, 3–16. London: Academic Press.

Harlan, Jack R., and J. M. J. De Wet. 1963. The compilospecies concept. *Evolution* 17 (4):497–501.

Hausdorf, Bernhard. 2011. Progress toward a general species concept. *Evolution and Development* 65 (4):923–931.

Hennig, Willi. 1950. *Grundzeuge einer Theorie der Phylogenetischen Systematik*. Berlin: Aufbau Verlag.

—. 1966. *Phylogenetic Systematics*. Translated by D. Dwight Davis and Rainer Zangerl. Urbana: University of Illinois Press.

Kitcher, Philip. 1984. Species. *Philosophy of Science* 51 (2):308–333.

Kornet, Dina Joanna. 1993. Permanent splits as speciation events: A formal reconstruction of the internodal species concept. *Journal of Theoretical Biology* 164 (4):407–435.

Kornet, Dina Joanna, and James W. McAllister. 1993. The composite species concept. In *Reconstructing Species: Demarcations in Genealogical Networks*, 61–89. Rijksherbarium, Leiden: Unpublished phD dissertation, Institute for Theoretical Biology.

Levins, Richard. 1970. Extinction. In *Lectures on Mathematics in the Life Sciences*, edited by M. Gerstenhaber, 77–107. Providence, RI: American Mathematical Society.

Mallet, James. 1995. The species definition for the modern synthesis. *Trends in Ecology and Evolution* 10 (7):294–299.

Mayden, Richard L. 1997. A hierarchy of species concepts: The denoument in the saga of the species problem. In *Species: The Units of Diversity*, edited by Michael F. Claridge et al., 381–423. London: Chapman & Hall.

Mayr, Ernst. 1940. Speciation phenomena in birds. *American Naturalist* 74 (752):249–278.

—. 1942. *Systematics and the Origin of Species from the Viewpoint of a Zoologist*. New York: Columbia University Press.

—. 1963. *Animal Species and Evolution*. Cambridge, MA: The Belknap Press of Harvard University Press.

—. 1969. The biological meaning of species. *Biological Journal of the Linnean Society* 1 (3):311–320.

—. 1991. *One Long Argument: Charles Darwin and the Genesis of Modern Evolutionary Thought*. Cambridge, MA: Harvard University Press.

Meier, Rudolf, and Rainer Willmann. 2000. The Hennigian species concept. In *Species Concepts and Phylogenetic Theory: A Debate*, edited by Quentin D. Wheeler and Rudolf Meier, 167–178. New York: Columbia University Press.

Mishler, Brent D., and Robert N. Brandon. 1987. Individuality, pluralism, and the Phylogenetic Species Concept. *Biology and Philosophy* 2 (4):397–414.

Mishler, Brent D., and Michael J. Donoghue. 1982. Species concepts: A case for pluralism. *Systematic Zoology* 31 (4):491–503.

Mishler, Brent D., and John S. Wilkins. 2018 [in press]. The hunting of the SNaRC: A snarky solution to the species problem. *Philosophy, Theory, and Practice in Biology*.

Moreira, David, and Purificación López-García. 2011. Phylotype. In *Encyclopedia of Astrobiology*, edited by Muriel Gargaud et al., 1254–1254. Berlin, Heidelberg: Springer Berlin Heidelberg.

Nelson, Gareth J., and Norman I. Platnick. 1981. *Systematics and Biogeography: Cladistics and Vicariance*. New York: Columbia University Press.

Nixon, Kevin C., and Quentin D. Wheeler. 1990. An amplification of the phylogenetic species concept. *Cladistics* 6 (3):211–223.

Paterson, Hugh E. H. 1985. The recognition concept of species. In *Species and Speciation*, edited by Elisabeth S. Vrba, 21–29. Pretoria: Transvaal Museum.

Pleijel, Frederik. 1999. Phylogenetic taxonomy, a farewell to species, and a revision of *Heteropodarke (Hesionidae, Polychaeta, Annelida)*. *Systematic Biology* 48 (4):755–789.

Pleijel, Frederik, and Greg W. Rouse. 2000. Least-inclusive taxonomic unit: A new taxonomic concept for biology. *Proceedings of the Royal Society of London—Series B: Biological Sciences* 267 (1443):627–630.

—. 2003. Ceci n'est pas une pipe: Names, clades and phylogenetic nomenclature. *Journal of Zoological Systematics and Evolutionary Research* 41 (3):162–174.

Regan, C. Tate. 1926. Organic evolution. *Report of the British Association for the Advancement of Science* 1925:75–86.

Ridley, Mark. 1989. The cladistic solution to the species problem. *Biology and Philosophy* 4 (1):1–16.

Rosen, Donn E. 1979. Fishes from the uplands and intermontane basins of Guatemala: Revisionary studies and comparative biogeography. *Bulletin of the American Museum of Natural History* 162 (5):267–376.

Simpson, George Gaylord. 1943. Criteria for genera, species and subspecies in zoology and paleontology. *Annals New York Academy of Science* 44:145–178.

—. 1961. *Principles of Animal Taxonomy*. New York: Columbia University Press.

Smith, Andrew B. 1994. *Systematics and the Fossil Record: Documenting Evolutionary Patterns*. Oxford/Cambridge, MA: Blackwell Science.

Sneath, Peter Henry Andrews, and Robert R. Sokal. 1973. *Numerical Taxonomy: The Principles and Practice of Numerical Classification, A Series of Books in Biology*. San Francisco, CA: W. H. Freeman.

Sokal, Robert R., and Peter Henry Andrews Sneath. 1963. *Principles of Numerical Taxonomy, A Series of Books in Biology*. San Francisco: W. H. Freeman.

Strickland, Hugh. E. et al. 1843. Report of a committee appointed "to consider of the rules by which the nomenclature of zoology may be established on a uniform and permanent basis." In *Report of the British Association for the Advancement of Science for 1842*, 105–121. London: John Murray.

Templeton, Alan R. 1989. The meaning of species and speciation: A genetic perspective. In *Speciation and its Consequences*, edited by D. Otte and J. A. Endler, 3–27. Sunderland, MA: Sinauer.

Turesson, Göte. 1922. The genotypical response of the plant species to the habitat. *Hereditas* 3 (3):211–350.

Van Valen, Leigh M. 1976. Ecological species, multispecies, and oaks. *Taxon* 25 (2/3):233–239.

—. 1988. Species, sets, and the derivative nature of philosophy. *Biology and Philosophy* 3 (1):49–66.

Wagner, Warren H. 1983. Reticulistics: The recognition of hybrids and their role in cladistics and classification. In *Advances in Cladistics*, edited by Norman I. Platnick and Vicki A. Funk, 63–79. New York: Columbia Univ. Press.

Waples, Robin S. 1991. Pacific salmon, *Oncorhynchus* spp., and the definition of 'species' under the Endangered Species Act. *Marine Fisheries Review* 53:11–22.

Wheeler, Quentin D., and Rudolf Meier, eds. 2000. *Species Concepts and Phylogenetic Theory: A Debate*. New York: Columbia University Press.

Wheeler, Quentin D., and Norman I. Platnick. 2000. The phylogenetic species concept (*sensu* Wheeler and Platnick). In *Species Concepts and Phylogenetic Theory: A Debate*, edited by Quentin D. Wheeler and Rudolf Meier, 55–69. New York: Columbia University Press.

Wiley, Edward Orlando. 1978. The evolutionary species concept reconsidered. *Systematic Zoology* 27:17–26.

—. 1981. *Phylogenetics: The Theory and Practice of Phylogenetic Systematics*. New York: Wiley.

Wilkins, John S. 2003. How to be a chaste species pluralist-realist: The origins of species modes and the synapomorphic species concept. *Biology and Philosophy* 18 (5):621–638.

Wu, Chung-I. 2001a. Genes and speciation. *Journal of Evolutionary Biology* 14 (6):889–891.

—. 2001b. The genic view of the process of speciation. *Journal of Evolutionary Biology* 14 (6):851–865.

Zachos, Frank E. 2016. *Species Concepts in Biology: Historical Development, Theoretical Foundations and Practical Relevance*. Switzerland: Springer.

Index

Series List

SPECIES: THE EVOLUTION OF THE IDEA
John S. Wilkins

COMPARATIVE BIOGEOGRAPHY: DISCOVERING AND
CLASSIFYING BIO-GEOGRAPHICAL PATTERNS OF A
DYNAMIC EARTH
Lynee R. Parenti and Malte C. Ebach

BEYOND CLADISTICS
Edited by David M. Williams and Sandra Knapp

MOLECULAR PANBIOGEOGRAPHY ON THE TROPICS
Michael Heads

THE EVOLUTION OF PHYLOGENETIC SYSTEMATICS
Edited by Andrew Hamilton

EVOLUTION BY NATURAL SELECTION: CONFIDENCE,
EVIDENCE AND THE GAP
Michaelis Michael

PHYLOGENETIC SYSTEMATICS: HAECKEL TO HENNIG
Olivier Rieppel

Milton Keynes UK
Ingram Content Group UK Ltd.
UKHW021833071024
449327UK00021B/1490

9 780367 657369